GROUPS

Theory and Experience

Houghton Mifflin Company Boston

Dallas Geneva, Illinois Hopewell, New Jersey Palo Alto London

GROUPS

Theory and Experience

Second Edition

Rodney W. Napier / Temple University

Matti K. Gershenfeld / Temple University

To our children

Printed in the U.S.A.

Library of Congress Catalog Card Number: 80-82844

ISBN: 0-395-29703-6

Chapter opening photo credits:

Chapter One: Harry Wilks/Stock, Boston
Chapter Two: Peter Southwick/Stock, Boston
Chapter Three: Jean-Claude Lejeune/Stock, Boston
Chapter Four: Pamela Schuyler/Stock, Boston
Chapter Five: Elizabeth Hamlin/Stock, Boston
Chapter Six: Owen Franken/Stock, Boston
Chapter Seven: Chris Maynard/Stock, Boston
Chapter Eight: Ellis Herwig/Stock, Boston
Chapter Nine: D. Patterson/Stock, Boston

Contents

Six The Incredible Meeting Trap 311

Seven Group Problem Solving and Decision Making 351

Eight The Evolution of Working Groups: Understanding and Prediction 441

Preface

The first edition of GROUPS: THEORY AND EXPERIENCE has proven popular in a broad spectrum of fields and courses, including counselor education, group and social psychology, business, nursing, education, communications, and social work. The wide readership of this text supports the notion that the study of groups cannot be narrowly defined in terms of a particular area of academic learning and that information about groups can be applied in more than one setting. The primary source of interest in this book lies with those organizations and groups that are interested in human relations training, planned change, leadership development, and decision-making processes.

This text is designed to give readers an understanding of group processes and to improve their skills as group members or leaders. Our purpose is to provide a straightforward integration of group theory, research, and applied methods, both for students preparing to work with groups and for professionals presently working with groups. Throughout, our approach is based on the belief that successful group work can and should be less dependent on leader charisma or control than on the application of learned skills and tested methods. Following is a brief outline of the major content areas covered in the text.

We begin by exploring the observable and predictable communication patterns that tend to develop in every group. An awareness of these patterns is crucial for understanding the group and raising the level of effective interaction among group members. In addition, understanding one's own perception of the group and its members can dramatically improve communication in the group. Thus, perception and communication are the topics of the first chapter.

A second critical aspect for the success of a working group is an understanding of what makes individuals feel as though they truly belong to a group. One's feeling of belonging is directly related to the amount of cohesion found in a group and the ability of individuals to work effectively together. The concept of membership is explored fully in Chapter Two.

Chapter Three, "Norms, Group Pressures, and Deviancy," is closely related to Chapter Four, "Goals." Because norms can be both constructive and destructive in terms of a group reaching its goals, understanding what norms and goals are, how they develop, and how they can be changed is essential.

Leadership is central to the success of virtually any group. Our premises are that *any* group member can perform leadership functions, that appropriate leadership is determined by the needs of a group, and, because those needs change, so will leadership behavior and strategies. To be successful as a leader demands flexibility of role, a willingness to share authority when appropriate, and an interest in making the most of the resources of other group members, even when this calls for reducing one's own leadership role. Such is the dynamic role of leadership discussed in Chapter Five.

New to this edition is a chapter on meetings. Why meetings fail, what hinders progress and promotes conflict in meetings, and what can be done to make them more worthwhile and constructive is the thrust of the discussion. From this overview of meetings in general we move to an exploration of a critical aspect of any working group, that of problem solving and decision making. Techniques for improving the problem-solving capabilities of a group and various methods of reducing group conflict and reaching consensus are outlined.

Chapter Eight, "The Evolution of Working Groups," integrates much of the material presented in the earlier chapters by describing the developmental characteristics of working groups. Armed with an understanding of the stages of development, group needs, and the critical events in the life of a group, group members can increase the possibility that they will respond in appropriate and constructive ways and thereby contribute to the group's effectiveness.

Finally, we look at the use of humor in small groups, a new area of study in the field. As people who have worked with hundreds of groups, we know that humor plays many roles in the life of a group. From our observation, humor deliberately introduced into a group that is defensive and even hostile can reduce tension, provide a more constructive and positive climate, and thus help move a group toward its goals. Research in this area is only beginning and there is much to be learned about humor in group situations, but we feel it is important to share our thoughts and experience on this topic with readers.

In revising the first edition, our original intent was to conduct a thorough review of the literature and make appropriate changes. So much progress has been made in the study of groups that what resulted are two completely new chapters and a seventy percent revision of the others. We have attempted to be faithful to the reported literature and sensitive to the new theoretical constructs that have been developed. Perhaps more importantly, however, we have made a sincere effort to derive meaningful applications from the theoretical and abstract and to provide readers with an interesting, useful, and stimulating learning experience. End-of-chapter exercises are included to reinforce and make meaningful the concepts discussed in the text. To ensure greater participation on the part of readers, we have also included a number of "Reader Activities" within the body of each chapter. More exercises and concrete applications are provided in the instructor's manual. It is our hope that by integrating theory, experience, and practice, readers' cognitive grasp of the material is guaranteed and that their skills as group members and leaders will be increased.

We would like to thank the following people for their helpful suggestions and criticisms of the manuscript: Douglas Mickelson, University of Wisconsin; Joe Meier, Northeastern University; and Gerald Kushel, Long Island University. We also extend our thanks to the staff at Houghton Mifflin, who made a significant contribution to the quality of our product. Marty Klein and his team of research assistants made an extensive review of the literature of group process from the past decade—a review that was both manageable and meaningful for us. Finally, Paula Leder, Helen Ball, and Marvin Gershenfeld gave time and infinite care in helping us put it all together. To all of them our appreciation and thanks.

R.N.
M.G.

GROUPS

Theory and Experience

One

Perception and Communication

It looked as if they were pacing each other, each in her own lane scurrying down the hall, glancing to the right at intervals, searching for the right room number. The taller, older one was an almost exact replica of a modish, suburban type in September's *Vogue:* her styled, frosted hair; the color-coordinated, vibrant make-up; the tinted rimless glasses; her tweed suit with suede elbow patches, and a contrasting blouse with stockings the same color. Her obviously expensive, brass-trimmed leather briefcase punctuated the image. Her appearance had been carefully planned, down to the matching gold earrings, bracelets, and long neckchain dangling charms. She looked assured, sophisticated, and like a woman who would know how to order in a deluxe French restaurant.

The shorter, younger woman had also planned to make the right impression. She had washed her hair that morning and had set it full; she wore only a trace of blusher and natural lip shine. She was wearing her new menswear shirt, jeans that fit well (not too new), green stockings, and her comfortable clogs. Her punctuation was a canvas bookbag with *THINK* printed on it in lettering that began spaced but ended cramped together.

They continued to scurry and search. The younger one, annoyed at the seeming randomness of the numbers, said to the other, "This is a crazy place. Do you know where 302 is?" The older, surprised at being spoken to, hesitated and eventually said, "No, no, I don't. I'm looking for it too." As they now walked almost together, each began to think of the other—and what it would mean. The younger thought, "It's the professor! What will she think of me for saying this is a crazy place? Why did I say that? I wonder if she'll hold it against me all term? Why can't I learn to keep my mouth shut?"

The older woman was thinking, "I wonder if I came off rude. Why couldn't I talk, why did it take me so long to answer a simple question? I wonder if she thinks I'm dumb or insecure?"

At last, there was 302. Worn, but still legibly on the wall. They both prepared to enter. The older woman patted her hair, took a deep breath, fixed her smile, and walked in—arranging herself in a seat at the far end of the table. She remembered a blur of faces and a sense that everyone looked thirteen. She wondered if everyone could hear her heart pounding, and if they knew how dumb and scared and old she felt, in college at forty-two. The younger woman seemingly steadied her stomach and burst in, carefully scanning the room and its occupants before deciding where to sit. With a yell of recognition, two high-school classmates (not friends at home, but now a route to acceptance, friendship, and love and someone to sit with that terrible first day of college classes) beckoned. She raced toward them as they hugged, and laughed, and gushed about what each had done during the summer.

A few minutes later the professor arrived, and took his place at the head of the table. The older woman was in shock. He looked eighteen, at the most twenty-one. He looked like one of her son's friends. He couldn't be an expert, he was too young. How could the university entrust someone like him to teach sociology? What did he know of the world? How much had he lived? Then another startling thought came: young people are still working out their feelings about their mothers and rebelling against them. What if he decides I remind him of his mother and he doesn't like me? What will I do?

The younger woman also noted the professor. She was relieved that he wasn't the fancy lady she had come down the hall with. He looked pretty good, nice looking, crinkly eyes, wearing jeans, not the stuffy type. She was already feeling relaxed.

He began by saying that the interaction in the course was important, and that he wanted members of the class to be involved in conducting the seminar. One way to begin was for all the members of the class to introduce themselves with their name and something they would like to say about themselves "so that we all get to know each other."

How will each of them introduce themselves? How will each decide to present her best image?

We all have been there, standing on the threshold of a new group, carefully screening our own behavior and trying to communicate what we believe will be most acceptable. And, in turn, we select from the narrow world of experience what we believe is truth, spotlessly perceived information about the group facing us. We take the data projected at us, and after a process of filtering, sifting, and refining, we respond to the distortion we have created. The ideas and information become alloys of our own making. So, one new member of a group may see eight potential friends while another sees eight sources of potential rejection. One may observe dress, tone of voice, age, sex, posture, while another may focus immediately upon evidence of influence and power, indicated perhaps in the direction of word flow or the movement of eyes toward the source of approval. Whatever the processes and needs of the individual, the view that eventually enters the mind's eye will be, to some degree, a distorted vision of what actually is taking place, reduced by some and expanded by others.

Especially in new situations, when we lack the data of direct experience, and we are still too anxious to see beyond our global assumptions about the group situation, we are most likely to use ourselves as the primary reference point in assessing the group and the task and thereby increase our own distortions (Berelson & Steiner, 1964; Codol, 1974).

SELECTIVE PERCEPTION AND THE INDIVIDUAL

That we see what we need to see is not merely a psychologist's whim, it is reality. An ink blot reveals the wide range of responses among different individuals that can be lured from an ill-defined or nebulous stimulus. Each perception and its interpretation of virtually any event is based on a combination of historical experiences, present needs, and the inherent properties of the scene being perceived. Because what we see is always a combination of what we see and what is happening within us at that moment, it is unlikely that two people will ever perceive the same thing in exactly the same way (Harrison, 1976; Wrench, 1964).

It is necessary to begin with the assumption that we distort, and proceed to build on these distortions. Even with the most objective task, it is nearly impossible to keep our subjective views from altering that which really exists.

READER ACTIVITY

Before reading any further, look at the triangle figure below. Will you count the number of triangles in this diagram?

Write your number here ______ before reading on.

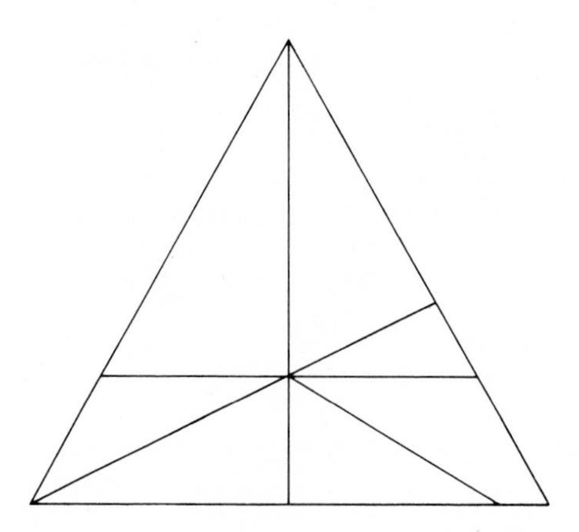

If, for example, the above diagram is presented to a group of fifty people, and they are simply asked to count the number of triangles (as you just were), it is almost certain there will be anywhere from ten to eighteen different responses, ranging from only one person giving a number to, at most, ten people agreeing to one number.

How is this possible among a representative group of normal, well-adjusted, and intelligent people? The task is clear and easy enough (a fifth-grader could handle it); yet there is rarely more than 20 percent agreement on one response. Following are only a few of the possible reasons for the differences in the perceived realities of this group.

1. One individual vigorously defends the fact that there is only one triangle in the diagram. Somewhere in the far reaches of her memory, she sees a triangle as having three equal sides.
2. Another person somehow discovers forty-three by counting every possible angle as a triangle.
3. A number of people count six triangles moving from left to right around the diagram but not counting the figure in the lower right hand corner.
4. Others find seven since they assume the figure in the lower right hand corner is a triangle.
5. Some people find seven, eight, nine by combining a few of the lines to form other triangles. Others push on and discover thirteen, fourteen, fifteen, and twenty triangles.

Later it is discovered that some who stopped short of discovering all of the possible combinations did not like puzzles or did poorly in geometry and still carried that fear with them. Others saw the whole thing as a game, perhaps a trick, and so felt why bother to try? Still others felt that they were being tested and thought it wiser to find fewer triangles and be correct (like SATs or Graduate Record Exams with their penalties for guessing) than to have many and be wrong. Then, of course, there are the competitors, some of whom managed to see triangles that were not even there. Some people assume that they have missed some (nobody's perfect) and add from five to ten just to be on the "safe" side (thinking that, like a jellybean contest, the number closest wins). Still others notice who puts up their hands at which number, and if that person is perceived as smart, they give up their own process and vote in accord with the smart one; to be aligned with talent is to experience a fleeting moment of being a winner. So even in a simple, straightforward task, among a responsible group of participants, one is able to ferret out unreasonable suspicion, fear of inadequacy, competitiveness, distortion of the instructions, and perhaps ten or twenty other variables that make such an enormous variety of responses predictable. (What was your own process in counting?) Because there are so many factors involved in one's choice on such a simple problem, imagine what happens when we add the additional variables that occur whenever the choice involves another human being.

In one way, our senses are overwhelmed by a thousand cues from the world we are attempting to understand. What we eventually perceive is the result of a complex sorting process that arranges stimuli in a manner most easily digestible, a process that facilitates our own sense of security. Colors, sizes, shapes, textures, smells, sounds, rhythms, and gestures, as well as the essence of time, place, and history impinge on us and are woven into a pattern of responding behavior. Responding becomes even more complex when we add what we think others think or feel about our behavior. It is little wonder that what we communicate and what we in turn receive back is so often dressed in grays and obscured by shadows we ourselves cast. Often, past distortions are infused into the present and compounded in such a way that the real and imagined, the past and the present, intertwine and form the present reality (Nidorf and Crockett, 1964; Nidorf, 1968).

Life Positions

One of the more interesting recent developments is the formulation in Transactional Analysis (Berne, 1976; Woollams and Brown, 1978) of the concept that, based on early experiences, people decide on a *life position*. Once they decide upon one, people are influenced by their life positions in how they think, feel, act, perceive, and relate to others.

The theory of Transactional Analysis essentially goes like this. We were all once children and, in the course of life's experiences, developed a concept of self-worth by the time we were six years old. At the same time we were formulating a sense of our own worth, we were also formulating the worth of others, especially those around us. We did this by crystallizing our experiences and making decisions about what kind of a life we will have (sad, happy); what parts we will play (a strong hero, a loner); and how we are going to act out the parts of our life script (adventurously, scared, slowly, with permission). These are our days of decision—time when we commit ourselves to acting in certain ways which become part of our character. These decisions, made very early in life about ourselves and others, may be quite unrealistic, although they seem logical and make sense to us at the time we make them.

For example, if as children we are ridiculed and regularly called stupid, we can decide that we are stupid and other people know it all. We will begin to think of ourselves that way and act that way. We base our life script on the position "I'm not OK, but you (other people) are OK." When we go to school, we may fail, feeling that we don't understand what's happening and have no brains. As we grow older, we will further fulfill our own prophecies by acting out our psychological position of constantly asking advice, doing what "they're" doing, and fearing being different. We often make mistakes, for which we are reprimanded; then we feel stupid, again maintaining our status quo. In this position, we may say to a classmate, "Everything comes so easily to you, you don't even have to study. Look at me, I don't understand what the teacher is saying, it's all Greek." Or we may become depressed, saying "I'm so stupid! How can you put up with someone like me?" Or we may phone around before a Saturday night party to ask what everyone is wearing, because we once again assume that everyone knows what to wear but that we don't. Everyone (almost) is perceived as smarter.

Life positions are related to three basic questions: Who am I? What am I doing here? Who are all those other people? The life positions can be generalized to four basic forms: "I'm not O.K., you're not O.K."; "I'm not O.K., you're O.K."; "I'm O.K., you're not O.K."; "I'm O.K., you're O.K."

TA theory holds that we not only perceive selectively in a given situation, but that by the time we are six we have already decided how we will perceive, what we will screen in, and what we will never see or hear. Three of the four positions involve the selective distortion of information and cues in order to maintain our early decided positions. Even as we think we are listening intently and being objective, we are screening out information.

Unconscious Factors

Often we never realize these most powerful influences on our perception. A group of people represents various degrees of acceptance and rejection, likes and dislikes, pleasant memories and distasteful ones, and it is from this complex assortment of stimuli that we conjure up a picture of our reality and build what appear to be appropriate responses—again, to maintain our own position and integrity within the group. Our needs may be so consistently present that nearly every perception is flavored in a similar fashion and our responding behaviors take on a consistency to match. Those who have limited tolerance for ambiguity create a structure for what they believe is, whether it is or is not (Livesley and Bromley, 1973). Those who are especially sensitive to cues of others as to whether they are liked are less likely to get across the information they want to express in an efficient manner because they are paying more attention to the interpersonal information (Winthrop, 1971). Those who make immediate decisions about liking or not liking someone and believe that they can size a person up and be right, set out to do just that. They make a decision on limited information and stay with it. They are "right," and are consequently resistant to change with new information (Reid and Ware, 1972; Erlich, 1969; Johnson and Ewens, 1971).

In many ways, this process can be illustrated by the experience of members of Congress. Their picture of the "facts" behind any one bill is far more complex than the views held by average citizens. Even the simplest issue may be confused by innumerable subtle pressures that encroach upon their perceptual fields. Lobbyists, personal biases, favors that must be repaid—these are all drawn into the picture, confounding what may have appeared to be a relatively simple and straightforward issue. When the issue is one involving emotion, with forceful, powerful proponents and opponents, members can be influenced by who does the asking, how they are dressed, and whom they represent. The Equal Rights Amendment, for example, had been in a committee of Congress since 1923. Each year until 1972, almost fifty years, there were women who asked that the bill be reported out. Think of how those women were viewed, and what factors influenced members of Congress to leave the bill in committee. Think of how, in the early 1970s, some members of Congress perceived the bill as being associated with stereotypical "radical, foul-mouthed, man-hating, bra-burning feminists," rather than with white, middle-class, educated women who actually introduced the bill shortly after women received the vote. And consider the strategy of the anti-ERA women, who in attempting to influence state legislatures to vote against the amendment, leave as their "calling card" with each legislator they visit, loaves of freshly baked, delicious cranberry bread. It may sound ridiculous, but the women report the cranberry bread has been a very successful device. Imagine what the equation is in the legislators' minds. And imagine how they vote as a consequence. Although a representative may laugh at the strategy of women marching on the legislature with frilly aprons and cranberry bread, we should not negate the influence of such behaviors. We perceive in highly idiosyncratic ways, and the effect of these strategies is certainly an illustration of the unconscious elements that can put pressure on people attempting to make decisions.

The Halo Effect

There may be those skeptics reading this who wonder whether these illustrations are exaggerations, and who prefer to believe that Congress and legislators render impartial judgments based on facts and information. Certainly people can render impartial judgments, uninfluenced by their personal knowledge of the people involved. You might even ask yourself the question, "Am I confident of my ability to render an impartial judgment uninfluenced by my personal knowledge of the people involved?" You probably will reply "yes"; most people do. Nisbett and Wilson, from the Institute of Social Research at the University of Michigan, were staunch believers that people could, until they conducted an experiment testing the psychological phenomenon known as "the halo effect" (Nisbett and Wilson, 1978). As a result, Nisbett is convinced that "one's objectivity is not to be trusted."

The experiment is an interesting one. The halo effect is simply defined as "the power of an overall feeling about an individual to influence evaluations of the person's individual attributes." For example, if you are usually annoyed when someone is consistently late, but find it charming in a woman friend whom you like, you have experienced the halo effect.

To test the extent of the halo effect and people's awareness of its influence, college students were shown one of two videotaped interviews with a college professor. In the first interview, the professor appeared to be quite likeable, expressing warm attitudes about his students and teaching. In the other, he conveyed the unlikeable attitude of distrust toward his students and rigidity in his teaching. Half the students saw the warm interview and half the cold interview. In addition, some of each half saw the interviews without the audio portion.

The students were then asked to rate how much they liked the teacher, his physical appearance, mannerisms, and distinct French accent. The majority who saw the warm interview rated the professor and his physical appearance and attributes positively, but split 50–50 on his accent. Those who saw the cold interview were unfavorable toward everything about the professor. Among those who saw the tapes without sound, however, there were only trivial differences in evaluations between the warm and cold professor. In other words, the professor's physical appearance and mannerisms did not differ between the two tapes. Rather, the differences perceived resulted solely from the overall attitude induced by what the professor said during the interview. "These results demonstrate that global assessment of a person can powerfully alter evaluations of particular attributes," the two researchers say.

Now, for the special part. To determine whether the subjects were aware of the cognitive processes underlying their evaluations, the researchers asked some of the students (as part of the design) whether their liking or disliking of the professor had influenced their evaluations of his personal characteristics. At the same time, they asked others the reverse question: had their ratings of individual characteristics influenced their overall liking?

Regardless of whether they had seen the warm or cold professor, the subjects who had been asked the first question said their evaluations of individual attributes had not been influenced by their liking of the man. When the question was reversed however, the subjects who saw the cold professor believed their negative evaluations of the individual traits had been responsible for their not liking the man. In the face of additional questioning, most of the students held firmly to these beliefs.

The students were not aware of how they had arrived at their evaluations. This finding parallels results from similar experiments conducted by the researchers. In all experiments, the subjects' explanations differed from the explanations that research showed had affected their judgment. Nisbett and Wilson observe, "People tend to rely on their prior assumptions about the causes of behavior instead of direct introspection." Therefore, they conclude, the validity of self-reporting is questionable, because people do not know why they do what they do. Nisbett and Wilson believe these findings have far-reaching implications for the establishment of rules concerning conflict of interest and nepotism, among others.

Even when we think we are making objective judgments, we aren't. Imagine how often we admit to not knowing how we arrived at a judgment. Combined, there is much to question about "objective perception" and "objective judgment."

SELECTIVE PERCEPTION AND GROUP BEHAVIOR

A need for recognition or a need to control will probably not alter greatly from one situation to the next. As we respond, we become predictable within a group. Similarly, groups become predictable in their perceptions and their responding behaviors. Consider an example.

Like a person, an advertising company must sell much more than is visible to the naked eye. It must sell confidence, and an aura of competence that may or may not be justified. Communication is the key to the entire enterprise. The Standish Agency is no exception. We walk through the door and immediately lose sight of our shoe tops in four-inch carpeting. We are surrounded by rich wood paneling and beautiful women, the latter looking busy indeed. An atmosphere of coolly directed business among the immediacy of copy deadlines pervades the office. We are attacked from every sense: visually, tactilely, emotionally. Taken into the unbelievably plush offices of the directors, we are surprised by their youthful dress, their easy manner, and their good humor amid all the work that is obviously being done. We are totally submerged in a feeling of efficiency, youth, up-to-dateness, and, above all, success. Just being there raises us above the ordinary, and we almost believe we belong in this atmosphere.

But what kind of communication lies behind the fantasy world of the front office? Actually, there are two worlds: the businesslike, efficient, Madison Avenue world of the directors and account executives who sell and maintain the advertising product; and the freewheeling, loose-hanging, liberal world of the creative team upon whom, eventually, all success depends. The two groups, each numbering about twelve, need each other and theoretically work closely together in creating meaningful concepts and designing means of expressing them. Whereas the creative group tend to be liberals, the executives are conservative; whereas the executives tend to *act* and *look* young, the creative staff tend to *be* young; whereas the executives reflect an unabashed desire to make profits and succeed, the creative people live in a world where the clever, artistic, and creative job is itself enough reward. They constantly find themselves justifying their own existence in such an "artificial" and "materialistic" environment and extol their own pure motivations in a system that owns neither their souls nor skills.

When members of the executive group sit down with members of the creative team (artist or writer), it is almost impossible for them to communicate except along clearly defined job lines. Usually such meetings are avoided until the last possible moment, when the pressure of time becomes an added factor in the tension that permeates the session. The executives desire a product that will sell, while the writers want a product that clearly represents their own abilities and creative thoughts. The executives expect a product in the shortest possible time, while the creative people resist (overtly or covertly) any attempt to push what must be by its very nature a spontaneous and unstudied process. The executives have talked with the client, but often the creative people have not, since they might prove an embarrassment with their long hair, outlandish clothes, and lack of understanding of the business aspects of advertising. Conversely, the writer or artist regularly interjects the feeling that it takes a "creative person" to take part effectively in the creative process. Moreover, the executives have more prestige, make more money, and receive the visible appreciation of a job well done. Although the creative group is relatively well paid for their services, they constantly feel the tensions inevitable in a "master-slave" relationship.

It is within this environment that ideas have to be communicated. The carpeting, the paneled walls, the beautiful secretaries mean little if effective communication fails to exist between these two groups. In a way, the problems here cut across the human life line of every group: status, influence, control, philosophy, dress, acceptance and personal worth, goals—each one in itself a potential source of tension. Together they succeed in stifling the potential of each to help the other. A dozen times a day stereotypes are reinforced, jokes and strategies generated at the expense of the other. Individual and group separateness is insured as each views the other through the jaundiced eye of personal needs. Each group and each individual must justify its existence.

The Influence of Stereotypes

In a group, individual stereotypes apparently feed on themselves, and we rapidly turn for support to those we believe share our own views (Kelley, 1951; Slater, 1955). The screening is quick, lightning fast, as we size up the opposition, gently test the climate, and feed the vibrations of other members into our own sorting system. We seek structure and support for our own insecurities in any situation we fail to understand or control. Especially in a group in which our roles are not determined clearly in advance, it is natural to seek confirmation that we are not alone, that there are potential allies among the strange faces (Festinger, 1950; Loomis, 1959). It takes but a few minutes to scan the superficial cues and sort out those with whom we can feel either safe or threatened, those with energy, anger, insecurity, power, softness, frayed nerves, or humor. We make our predictions and then spend a good part of our energy proving that we are correct (Erlich, 1969: Johnson & Ewens, 1971; Reid and Ware, 1972). The tapping fingers, nervous smile, loud talk, tightly folded arms, cultivated friendliness, reading in the face of potential conversation, unabashed sharing of one's self to a stranger—these and a thousand pieces of instant information are sifted, labeled, and shelved for later use in our effort to confirm our own identities and understand others in the group. They are used as evidence and, in the long run, can be destructive as well as helpful in the development of the group and our relationships within it.

Unless we are ready to test the untested assumptions upon which we base our perceptions, we can expect that there will be many breakdowns in communications. The problem, of course, is that if we attempted to test many of our assumptions about people, we would find it very difficult to classify them, stereotype them, or pigeonhole them. We would feel so much less secure much of the time because we would constantly have to adjust to other realities.

Life would be so much more complex. The fact is that people are a thousand things, but, first and foremost, they are what we want them to be in relation to our own needs. A person is fat, hostile, irrational,

Jewish, lethargic, smart, black, paranoid, handsome, homosexual, or militant. We take a very specific word with a very narrow definition and frame another complex human being.

READER ACTIVITY

Think back on some of your own experiences with others. How have you been "labeled" in a way that you did not like?

Label ______________________________

Why didn't you like it?

____

____

____

What do you wish had happened instead?

____

____

____

How did that label influence you?

____

____

____

READER ACTIVITY

This is another exercise to look at labeling.

What is your first name?

What image does that name conjure up for you?

(adjectives) ______________________________

How do you feel about those adjectives? ______________________

____

____

In the past, many whites thought of blacks as a separate and inferior species. The word *Negro* called up images and elicited behavior among whites that caused tremendous psychological and physical suffering. Judge Leon Higginbotham, of the U.S. Court of Appeals, Third Circuit, writes of his experience during World War II:

> In 1944, I was a 16-year-old freshman at Purdue University—one of twelve black civilian students. If we wanted to live in West Lafayette, Indiana, where the university was located, solely because of our color the twelve of us at Purdue were forced to live in a crowded private house rather than, as did most of our white classmates, in the university campus dormitories. We slept barracks-style in an unheated attic.
>
> One night, as the temperature was close to zero, I felt that I could suffer the personal indignities and denigration no longer. The United States was more than two years into the Second World War, a war our government had promised would "make the world safe for democracy." Surely there was room enough in that world, I told myself that night, for twelve black students in a northern university in the United States to be given a small corner of the on-campus heated dormitories for their quarters. Perhaps all that was needed was for one of us to speak up, to make sure the administration knew exactly how a small group of its students had been treated by those charged with assigning student housing.
>
> The next morning, I went to the office of Edward Charles Elliot, president of Purdue University, and asked to see him. I was given an appointment.
>
> At the scheduled time I arrived at President Elliot's office, neatly (but not elegantly) dressed, shoes polished, fingernails clean, hair cut short. Why was it, I asked him, that blacks—and blacks alone—had been subjected to this special ignominy? Though there were larger issues I might have raised with the president of an American university (this was but ten years before *Brown* v. *Board of Education)* I had not come that morning to move mountains, only to get myself and eleven friends out of the cold. Forcefully, but none the less deferentially, I put forth my modest request: that the black students of Purdue be allowed to stay in some section of the state-owned dormitories; segregated, if necessary, but at least not humiliated.

> Perhaps if President Elliot had talked sympathetically that morning, explaining his own impotence to change things but his willingness to take up the problem with those who could, I might not have felt as I did. Perhaps if he had communicated with some word or gesture, or even a sigh, that I had caused him to review his own commitment to things as they were, I might have felt I had won a small victory. But President Elliot, with directness and with no apparent qualms, answered, "Higginbotham the law doesn't require us to let colored students in the dorm, and you either accept things as they are or leave the University immediately."
>
> As I walked back to the house that afternoon, I reflected on the ambiguity of the day's events. I had heard, on that morning, an eloquent lecture on the history of the Declaration of Independence, and of the genius of the founding fathers. That afternoon I had been told that under the law the black civilian students at Purdue University could be treated differently from their 6,000 white classmates. Yet I knew that by nightfall hundreds of black soldiers would be injured, maimed, and some even killed on far-flung battlefields to make the world safe for democracy. Almost like a mystical experience, a thousand thoughts raced through my mind as I walked across campus. I knew then I had been touched in a way I had never been touched before, and that one day I would have to return to the most disturbing element in this incident—how a legal system that proclaims "equal justice for all" could simultaneously deny even a semblance of dignity to a 16-year-old boy who had committed no wrong.[1]

Being black meant being out of the system. That perception was intended to be the norm from the president to the newest freshman; the system was viewed as beyond change and to be accepted, and, despite its incongruity with the Declaration of Independence and a war, it was steadfastly upheld.

The word *black* is not the only label that elicits untested assumptions. There are a plethora of untested assumptions in the words *Arab* and *Jew*, *labor* and *management*, *male* and *female*. For example, what assumptions is a female making about a male when she says, "I hate that. I can tell by the look on your face—you're undressing me." In groups, women especially are often confused about how they will be perceived and how they should act. If they act sexy, will that lead to influence, or will influence be denied? Should they

[1]From *In the Matter of Color. Race and the American Legal Process: The Colonial Period* by A. Leon Higginbotham, Jr. Copyright © 1978 by Oxford University Press, Inc. Reprinted by permission.

accept a comment on their appearance as friendly, or should they hear it as a put down (women are only noticed for their bodies)? Should they be aloof rather than warm, because warmth will probably be perceived as sexiness.

We are in a day and age vastly different from the past. Attitudes about maleness and femaleness, about age, about modesty and appropriateness have been uprooted. Although men have traditionally been the aggressors in relationships with women, now women are acting as the aggressors. People are confused about what was once the most standardized and secure set of relationships. It is difficult at this time to point to rights and wrongs; do this, but not that. Rules and stereotypes once helped us keep our sanity; now, we don't know what to do.

Gestalt Theory

Just as we select and organize physical stimuli in a manner that is easiest and most convenient for us, so too we organize the complexities of human behavior in similar ways. A number of simple concepts developed by the Gestalt psychologists in relation to physical stimuli can be used to help understand what occurs when people get together in a group (Kohler, 1947). For example, we tend to create figure-ground relationships. In any one perceptual field (all the stimuli we are able to see at one time), certain figures are drawn forward into positions of dominance as others recede to form the background of the scene. In many cases, which objects are reduced to the background and which are drawn forward are dependent on the immediate needs of the viewer. For example, in the following

classic picture of the two profiles-goblet configuration, people will background. Similarly in a group, certain individuals will form a natural backdrop while others, for a variety of reasons, remain a clear part of the foreground.

Following this line of thought, there is a tendency to place objects in a natural order, thus making it easier to establish relationships out of the immediate confusion in a scene. The mind struggles to achieve order by grasping similarities that appear to be present or by perceiving certain continuities in the stimuli presented. This process of organization may take place when looking at dots on a piece of paper or individuals in a group.

READER ACTIVITY

This exercise is a simple example illustrating that we do organize dots. Below are nine dots, arranged three dots per row in three rows.

•   •   •

•   •   •

•   •   •

Connect the nine dots with four (4) lines so that the end of one line is the beginning of the next. It can be done.

Turn the page for the answer. Note that the solution would have been simple if the dots were to be connected with five lines. We structure the problem of connecting the dots within the imaginary boundaries of a square that is made up of the arrangement of the dots. And within that self-imposed box, we look for ways to solve the problem. Upon seeing the solution, individuals say, "But we didn't realize that we could go beyond the dots," as if the instructor had created the perceived box.

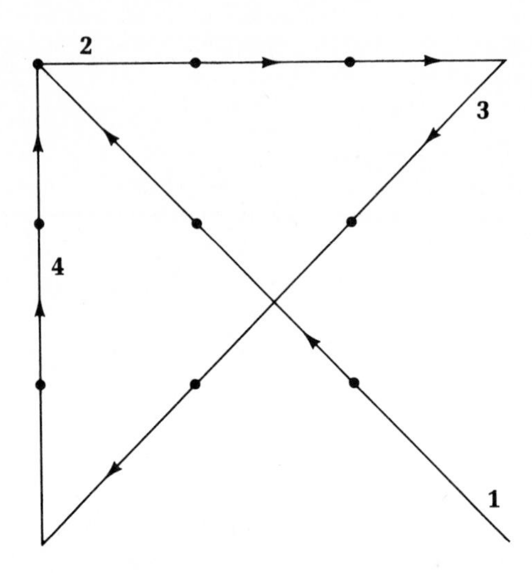

Using arrangement, size, sex, clothes, tone of voice, posture, and many other cues, we proceed to subtly organize the group into a variety of composite groupings. This ordering is merely a convenience, a way of handling the enormous amount of data that suddenly confronts us at any one moment in time.

Another concept discussed originally by Gestaltists concerns the tendency to take incomplete data and organize it into a meaningful whole. By that process, an incomplete circle will more often than not be seen by people as a full circle, rather than a curved line. Apparently, we have a need to bring closure to objects within our perceptual field.

In looking at the participants in a group, we will take the data they put forth such as voice, verbal gestures, and dress, then add our own stereotypes, and in this way develop a complete picture of the person. We bring closure to the incomplete object (in this case a person) and, in a sense, fill in the missing pieces so that we can more easily be content with the previously unknown commodity. By putting all the clues together into a meaningful package, we are better able to have a relationship that is consistent and comfortable for us. It provides a means of gaining a measure of safety for us in what is, perhaps, an incomplete, strange, and uncomfortable situation.

According to Gestalt theory, people have a tendency to take the various stimuli and focus on one set of stimuli that appears to be "good" in terms of similarity, continuity, closure, and symmetry. According to this concept, for example, we will immediately be attracted to those in a group who tend to fit our perception of a "good" group member, those who are least threatening to us and tend to create the least dissonance in terms of our own values and goals within the group.

If we were not able to impose this kind of order on the group and certain of its members, the situation might prove to be unbearably tense and difficult. So, if we are quiet and shy, we may seek order and some relief in the group by discovering those who are the least abrasive or dominating and those who show the greatest restraint. In this way, we can bring harmony to a dissonant situation; we can seek allies and support in fact or in fantasy.

SELECTIVE PERCEPTION AND COMMUNICATION

Our propensity to organize a group in a manner that is most comforting to us can prove to be a distinct liability to effective communication. It often generates inflexibility, restricted routes of information, and a need to verify and then justify our initial perception. As a result, we often begin with two strikes against us in our efforts to achieve understanding and insight into group processes. A first step, of course, is to confront the significance of our own feelings and uncover some of the untested assumptions and stereotypes that are helping to influence our behaviors. The fact that we hear what we wish to hear and draw on assumptions to support the view we wish to follow is vividly portrayed in the following situation.

Recently, parents and community members have become more involved with their school boards. Some participate because of their concern for the education of their children and for quality services. Others are concerned about the costs of education and the financial burdens of maintaining buildings, staffs, and services for declining enrollments. Recently independent candidates—not backed by any political party—have conducted campaigns for school board directorships and have won seats. Two such new members were elected to a small city school board. One was male, forty-eight, over six feet tall, a parent with two children in the high school, and an executive with a national business corporation. The other was a female, thirty-one, a parent with one preschooler and one child in elementary school, who was just over five feet tall, and a full-time homemaker.

School board members invited the male member to an informal dinner before his first school board meeting. At dinner, the president of the board made room for him at his left. Members passed the time jovially and informally explained each of the agenda items.

The following is the report of the other new member. "Throughout the evening, as was similar in the past when I attended meetings as a parent, I felt a distinct prejudice because of my size, my gender, and my age. The average age of the board was over fifty; the average size was at least six feet or taller; and I am the only female member of the school board. This discrimination exhibited itself when I found that others met unofficially for dinner before the school board meetings, neither informing nor inviting me. At the meeting, a seat at the far end had my name and materials. My opportunity to join a prime committee was difficult because louder voices seemed to grab attention and to be recognized. This prejudice and resentment of my election over the past president of the school board was further demonstrated by three members who completely avoided talking to me (not even hello, nor an acknowledgement of "congratulations" following the installation ceremony). Although a few members were genuinely polite and friendly, others displayed a "canned" sense of politeness and conveyed that they would only say the minimum necessary. My experience on the school board was very different from that of Tom (other new member). I felt unwanted, began to question my own abilities, and seriously wondered whether I could ever placate the others to at least accept me on the board. I felt that at that first meeting, I was not 'one of them,' and my excitement at being elected plummeted to feeling like a bad girl."

In this example, board members were acting according to their illusions of what competence looked like, and their fantasies of what attributes were necessary for an ideal board member. Two of the greatest sources of misinterpretation in the communications process are illusions and fantasies, arrangements of the past representing the present; and behavior that is based on these illusions. These behaviors, which elicit both structure and security, also spawn new problems in communication.

Communications theory postulates:

All behavior is communication. "One cannot *not* communicate."

All communication (therefore, all behavior) carries an explicit proffered definition of the relationship between the speaker and the listener.

Each successive act of communication not only responds to explicit task-oriented content of the message originally sent, but also offers a redefinition (similar/continuous or different) of the relationship between the speakers or group members.

As they adjust their expectations and relationship definitions to fit with one another, they work out (implicitly—without stating it) an arrangement that serves as a basis for further interaction.

The firmer one's membership in the system governed by a particular set of arrangements, the more blind one is to the distinctive features of the relationship definitions and redefinitions worked out. (Members are so used to the arrangement that they can't see the forest for the trees.)

So, the relationship among members is perennially the perceptual "ground," and the content/task is the "figure" for interaction within a group.

Real change or development arises only when the arrangement governing the relationship changes or develops further, and not when new content is being exchanged under the terms of the same old arrangement (Watzlawick, Beavin, and Jackson, 1967; Watzlawick, 1977).

The relationship definition played out in communication patterns is especially hard to understand for a number of reasons:

a. Members of the system who live by the rules of the agreed quid pro quo (or arrangement) are so used to the agreed order of interaction that they literally are oblivious to it.

b. Social norms/conventions require that we stay focused on the content/task aspects of communication rather than relationship aspects.

c. How one punctuates the sequence of interaction—where one starts and stops one's analysis—will play a large role in the understanding acquired. (An interesting example: the experimental rat "has the experimenter trained" to issue food pellets whenever commanded by the rat pressing the bar.)

d. Causality is circular (interaction of one to the other is a function of how the other acts to him), rather than linear (Watzlawick, 1977).

Language = Words; Communication = People

It has often been said that only people and not words have any meaning in our attempts to communicate (Fabun, 1965). Unless we are able to probe behind the easily flowing façade of words that screen us from one another, we will remain confused and often out of touch. So often it is the gesture, tone, inflection, posture, or eye contact that hold the key to the real message, while the clear, seemingly unambiguous words merely provide false starts and dead ends to the unwary listener. Even when we think we understand the meaning behind a word, there are usually three or four possible variations in meaning that could fit nicely into the sentence.

Usually, we draw upon the intent we believe the speaker has in mind. It's hard to imagine that a simple word such as *hard* has more than twenty-six possible definitions (*American Heritage Dictionary,* 1969). Among other things, that one word can describe the solidity of coal, the difficulty of a test, the type of binding on a book, the ability of a person to maintain his or her position despite pressure, the penetrating power of x ray, or the parsimony of an elderly person. There are so many subtle innuendos that flavor language and require a personal definition before they can be translated into the context of a particular statement. Often, what we end up with is nothing more than a makeshift assemblage of words spiced with half-known definitions and a variety of feelings. What eventually transpires depends upon the web created out of past experience, definitions, language skills, expectations, speed and clarity of the words spoken, and the general psychological climate that exists. How often are we led off the trail because we do not have enough information and drift further away rather than closer to the actual meaning. On the other hand, when history and experience are on our side, when we are familiar with the nonverbal communication that accompanies the words, then the group in which we are participating may respond in near unison to a message that to the casual listener may mean just the opposite.

There are other words that, when we hear them, evoke a special meaning for us depending upon our context, our experience, our culture. One such word is *management.* To much of American labor, the word *management* is adversarial; to the union member it means, "I have to get what I can get. Management is only concerned with their profits, and couldn't care less about me." In Japan, however, management is thought of as a caring father. Workers assume that there will be work for a lifetime. They feel that in working for a company, they can trust management and expect it to take care of them for their lifetime. Two very different perceptions of management.

Reward-Seeking Behavior and Communication Patterns

People desire to be liked and accepted. In some ways, that desire is our Achilles' heel, since it leaves us vulnerable to the subtle influence and control of those from whom we seek approval. Often, the pressures pushing us to adjustment and compliance, and eventual favor in the eyes of another are not even discernible by us or the other person (see Chapter Three relating to group norms). In our efforts to be accepted, we become sensitive to the minute behavioral clues that suggest the degree of approval on the part of the other persons; as we know them better or are in a group longer, we base how we act on those clues (Greenspoon, 1955; Sorensen and McCroskey, 1977; Verplanck, 1955). Indeed, we are as keen as any bloodhound in ferreting out and following these clues to acceptance. Rosenthal et al., in designing a test to discriminate between those more and those less sensitive to nonverbal messages—body images and tone of voice—found that five seconds was too long to show a film clip. At five seconds, everyone could discriminate. In order to differentiate sensitivity to nonverbal behavior, the film had to be shortened to two seconds (Rosenthal et al., 1974). Note how quickly we pick up clues; five seconds is too long. A nod of the head, the slightest murmur, a smile, a frown, or a seemingly innocuous "mm-mm" can put us on the track.

Although most of the research in this area of communication has been within the context of the one-to-one relationship, there is little doubt that the same holds true within any group in which people are concerned with their image and acceptability (Asch, 1956; Goffman, 1959, 1961, 1963; Slater, 1955; Strodtback et al., 1957). In fact, if the need is to be accepted by seven or eight people instead of only one, it is quite possible that we will work overtime to discover the sources of reward in the group as well as the favored behavior. This, in turn, alters communication patterns and overt behavior within the group, as we see in the example below.

A first meeting was being held with six women college faculty and a consultant to discuss relationship problems among departments and members of the faculty. Each of the women introduced themselves by their first names. They ranged in age from twenty-eight to fifty. Each was well-educated, articulate, conservatively dressed, and committed to the success and growth of the college. The consultant asked that each describe one incident that would illustrate the present problem of the college. All eyes gazed in the direction of the twenty-eight-year-old woman, the youngest of the group. She began by presenting an incident, and the others elaborated on it. No new incidents were described. The consultant then asked what was the one problem that they thought had to be resolved. This time the consultant called on the person to her left. There was a long pause, and finally she said, "I need time to think about it." A similar reply came from the next person, and the next. Again, the twenty-eight-year-old stated a problem of concern to her, and the others added information on that problem. The consultant was aware that something was happening, but what? How could the twenty-eight-year-old have such influence, and, without admonishing a person or saying a word, control the group so effectively? There was an answer. The consultant looked at the youngest, most powerful member and said, "Are you the newly appointed president of the college?" She was surprised and flustered; the others were incredulous that the consultant "somehow" knew.

It was clear that members of the committee were more influenced by the impression they wanted to make on the new president, than the goal of resolving the college problems. They were concerned with being liked by her, aware that, with her youth, she might be president for decades to come, fearful that she had a reputation for making major staff changes, and mindful that the trustees had described her as an "unanimous brilliant choice." What they would say, how they would act, would be much more a function of cues from the president than the explicit questions asked by the consultant.

FACTORS THAT INHIBIT COMMUNICATION IN A GROUP

Previous Experience of Group Members

Two of the greatest factors inhibiting communication in groups are the previous experience of members with groups, and disillusionment that success in a group is even possible. We remember conflicts in our families as we grew up, in which it seemed our parents represented their values and we were supposed to "be seen and not heard," and we recall our days as adolescents, when we had constant arguments with our parents and made exasperated, futile attempts to be understood. In school, we saw groups as a hassle. We had to deal with our desires for acceptance from our peers and the endless squabbling to plan a party or an outing or the silliness of school student council meetings and their intense discussions about whether to sell pretzels or tee shirts as fund raisers. Small group experiences on class projects were often unfair as one person did most of the work and the others shared in the grade; or involved such compromising that the product was an embarrassment that no one really wanted. In school, it was so easy to be made a scapegoat and being part of a "crowd" increased the opportunities to be labeled. We haven't known success in groups, we have never worked through the problems of being in a group, and consequently we don't expect a group to be successful.

READER ACTIVITY

Below is a list of ten statements about groups. Next to each item write *T* if you believe the statement is generally true, based on your experience; or *F* if you believe the statement is generally false from your experience.

_____ Less is accomplished in a group than by an individual.
_____ Groups waste time.
_____ I hate to be in a group.
_____ I could accomplish more if I could do it myself.
_____ Groups are inefficient—what with their bickering, arguing, compromising, and settling for a fifth-rate solution.
_____ Groups encourage conflict and arguing.
_____ Groups are a way to manipulate others.
_____ I feel manipulated in a group.
_____ Groups are basically the imposition of the will of the leader on others.
_____ Groups encourage cliques and subgroups, ins and outs.

How many did you answer true? _____ false? _____

Many people reply true to most of the above questions. Assume for the moment that you believed most of these statements were true. What would be the effect of these judgments on your behavior and on your communication, as you come into a new group? Much like the example of labor's perception of management in America, and the self-fulfilling prophecy that ensues, our negative experience with groups becomes another self-fulfilling prophecy at a group level, that groups are a waste of time and won't accomplish a thing. Most people have never experienced a successful group. (See Chapter Seven on problem solving to understand what it means to work through group problems and be successful.) Given the general feeling that being in a group will be difficult, and the pessimism about the success of the group and personal outcomes, members approach a group guardedly, often "acting."

False Assumptions

There are a number of false assumptions inhibiting communication in a group. These assumptions are so commonplace, so pervasive, that they must be brought to consciousness to be examined and, with understanding, rejected. These assumptions are not readily cast aside; they need to be brought regularly back into the glare of consciousness, and the decision made again and again to recognize these assumptions as false.

One faulty assumption is that we know what others mean, and another faulty assumption is that they know what we mean.

The assumptions are based on the premise that if people live or work together and share the same time and space, they are sharing the same events and experiences. And so, based on that proximity and like experiences, they are having the same experience. Therefore, it is logically assumed that there can be a direct connection between minds; each knows what the other saw, felt, and thinks. It is a major false assumption inhibiting communication. Let us look at the process as it really occurs, even given that two people have had certain common experiences:

Person A has an idea he or she wants to convey.

Person A then encodes (translates) the idea into particular words expressed in tone; pace of words; affect; facial expression; eye movements; and in body, arm, leg, and hand movements.

The words may be compatible with the intended message, or may not be formulated as intended.

The affect or body movements may be congruent (matching with the words) or may be incongruent (for example, saying "It is difficult for me to understand your position," while sounding angry and smiling).

Person B, receiving the message, hears it through his or her filters. Is he or she tired? Is he or she upset with person A? Is it a subject of little interest? Is person A older or younger, male or a female, a friend or an enemy, a supervisor or a colleague? All of these will affect the message he or she hears.

As a result of the complex pattern that comes through his or her filters, the words he or she especially hears, and the cues of body, tone, or position he or she perceives, person B "hears" what person A said. Person B thinks he or she hears the "full message" completely and objectively.

Person B then reacts to what he or she heard. What did person A say about him or her? Was it positive or negative? Was it expected or unexpected? Further, does person B agree or disagree? As these thoughts register, person B begins to frame a response. Or, person B may decide he or she knows what is being said. He or she may even have begun to frame a response after the first sentence of person A's talk, and may have tuned out all of the messages after that.

A major false assumption we make is that interpersonal communication is a simple process that is a meeting of the minds. Minds cannot meet; intent cannot be the same as the behavior expressed nor the impact of the message received by the other. There are too many intervening, ambiguous steps, each open to personal interpretation.

Another false assumption is that communication happens naturally. People express what they want to say in words and assume that those words are automatically understood by another in the same way one person walks across a room and another sees that someone has gotten up from a chair and moved to the window. They both believe they understand the behavior.

People assume that verbal communication is straightforward. They think that all they need to do is to express what they want to say in words, and the message sent is the message received. This is not so. The meaning one person has is never identical to that which another person has, because meanings are in people's minds, not in the words they use. Some people readily say, "I love you." They love their friends, their dog, their school, a movie, and their favorite recording stars. For them to say "I love you" to a date is an indicator of having had a pleasant evening. To another, who has been going with one woman exclusively for over a year, the words are hard to come by. She has regularly been asking, "Do you love me?" Yet he is highly resistant to replying. To him, these words reflect an intent to marry, and he is not sure he is ready for such a commitment.

READER ACTIVITY

Words evoke meaning in us in a special, very individual way. What is difficult for you to express? ______________________________

For those feelings difficult to express, how do you do it? What is your special way? (A way that those who know you well may understand, but that others may not.)

Anger ______________________________

Jealousy ______________________________

Unfairness ______________________________

Gratitude ______________________________

Resentment ______________________________

A terrific idea ______________________________

A ridiculous idea ______________________________

Affection ______________________________

Sadness ______________________________

What words do some people use that you think are

Phony ______________________________

Overbearing ______________________________

That stick in your throat (you don't seem to be able to say them, but you wish you could) ______________________________

____

Total accuracy in communication would require that two persons have an identical history of shared experiences. Only then could they perceive exactly the same meaning for a given message. Given the reality of different life experiences, such a situation is impossible (Chartier, 1976).

We have other false assumptions about communication (Coan, 1968; Luft, 1969; Watson, 1967):

1. That persons respond to each other objectively, listening only to the information conveyed. The key to what is happening in a group or between persons is not what is happening objectively but what is going on subjectively, what each person's feelings are. Subjective factors such as attitudes and values tell how people see themselves and others and how they order their world. The prime aim of most people is to survive in the group and, if possible, to enhance themselves in their own eyes and in the eyes of others. Each of us is forever bound up in the issue of our own personal needs and goals within the group, but unless the group can provide a means of personal self-fulfillment, we will move from the group either psychologically, with reduced participation, or physically.
2. What happens in a group is rational and easily understood as the group proceeds in an orderly, sequential manner to solve a problem or convey and receive information. Though some of the events in groups can be viewed as being orderly and making good sense, behavior is influenced more by emotions and by largely irrational strivings; logic and reason play relatively minor roles in human

interaction. There are questions of identity, "Who am I to be in this group? What image am I going to project to these people?" and "What roles will I undertake to project this image?" In some cases, our reponse to these questions is very natural, but in others it is strategic.

Other questions relate to power, control, and influence. Who has it, how much will be shared? Whether behavior will be facilitative or destructive to the group process will depend on an individual's particular needs and the realities of that particular group.

3. That the individual, like the group of which he or she is a part, is fully aware of the sources of his or her behavior and of the effects of his or her behavior on others. Parts of our behavior are unknown to us; it is often a surprise to find ourselves doing things that are difficult for us to understand or to make sense of. (We may have fantasized about being invited to become an officer of a professional group, yet when we are asked, we hear ourselves making an excuse about being too busy. Afterwards, it was difficult to understand how that had happened.)

We want to be accepted by other members, yet we have limited information on how they perceive us. We may be shy and frightened; they may perceive us as snobbish and aloof. We may want to be involved and see ourselves as offering suggestions that are helpful; we may be perceived as behaving in a highly dictatorial manner. We have very limited understanding of how our behavior influences others.

These false assumptions greatly reduce our ability to communicate in a group, or even understand what seems to be happening around us.

UNDERSTANDING COMMUNICATION IN GROUPS

The Johari Window

Joe Luft and Harry Ingram (hence, the name Johari, from their first names) developed a graphic model of behavior in groups known popularly as the Johari Window (Luft, 1969). They explain how communication proceeds in groups, recognizing the false assumptions and moving to a theory beyond them. The Johari Window[2] consists of four quadrants, depicted as follows:

[2] From *Of Human Interaction* by Joseph Luft. Reprinted by permission of Mayfield Publishing Company. © 1969 by National Press.

	Known to Self	**Not Known to Self**
Known to Others	1 *Open*	2 *Blind*
Not Known to Others	3 *Hidden*	4 *Unknown*

The four quadrants represent the total person in relation to other persons. The basis for the division into quadrants is an awareness of behavior, feelings, and motivation. Sometimes awareness is shared, sometimes it is not. An act, a feeling, or a motive is assigned to a particular quadrant based on who knows about it. As awareness changes, the quadrant to which the psychological state is assigned changes.

Quadrant 1 is the open quadrant; behaviors, feelings, and motivations here are known to the self and others. This is defined as the area of free activity; it is the window raised on the world; it is our public self. Quadrant 1 behavior is how we present ourselves, that we know and that others know. A person, new in a group, asks that others introduce themselves, as an initial action. He knows, and they know, that he has made an initial action. He knows, and they know, that he has taken the initiative and wants members to introduce themselves.

Quadrant 2, the blind quadrant, refers to behavior, feelings, and motivation known to others but not to the self. It is often referred to as the "bad breath" area; others may know that we have bad breath, but we don't know.

Quadrant 3, the hidden quadrant, refers to behavior, feelings, and motivation known to the self but not to others. This is often known as the "secret" area; we know all kinds of things about ourselves that we are not telling to the group. It is our private information.

Quadrant 4, the unknown quadrant, refers to behavior, feelings, and motivation known neither to ourselves nor to others. It is the area of "unconscious" activity, least accessible to us or to others.

According to the model, communication in a group moves according to certain principles—a change in any one quadrant will affect all other quadrants. Initially, members begin with a small quadrant 1; they retain their masks and public images and communicate in their style—being reticent to answer questions, or rarely speaking, or surreptitiously trying to find out the rules for behavior. The smaller the first quadrant, the more limited the participation and the poorer the communication.

As the group continues, people may make expressions of self-disclosure. ("I'm an undergraduate who got special permission to be here in this graduate course; I don't know what's happening and I'm terrified.") Now some of the secret, private information has become public. Quadrant 1 is enlarged and quadrant 3 is smaller. As more secret information is shared, the person has a larger area for communication; he or she is freer to express more. The person is freer because it takes energy to hide or deny behavior that is involved in interaction.

As the group continues, there may be feedback (information from others on behavior witnessed). With feedback, the blind area, quadrant 2, decreases. For example, "That's the third time you interrupted me today. You usually don't; are you angry with me?" It takes energy to deny feedback that occurs in interaction. As the person replies, the first quadrant is enlarged.

As the group continues, feedback may occur that results in self-disclosure and a reduction in quadrant 3 (secret or private area) and a reduction is quadrant 2 (the blind area), constantly increasing the open area, quadrant 1. In the process, some of quadrant 4 may even become known (Ah ha), and the open first quadrant is even further increased.

Working with others is facilitated by a large area of free activity (quadrant 1). It means more of the resources and skills of the persons involved can be applied to the task at hand.

According to the model of the Johari Window, our initial behavior in a group, based on our previous experience, reflects a constricted quadrant 1; we are careful about what we say, do, or reveal. Some people continue with a constricted behavioral repertoire. For others, their experience in a group leads to interpersonal learning. As the threat decreases and as mutual trust increases, they may appropriately disclose information or ask for feedback and increase their openness in the group. As members increase their open quadrants, there is not only greater variability in individual behaviors but also greater openness in interpersonal relationships in the group.

T-group members (an unstructured laboratory group) often comment on the difference between the group in the initial sessions, when everyone was guarded and careful about what they said, and the group in later sessions. At later sessions, people knew whom to tease and how, they knew who was sensitive and to what subjects; they knew who would respond spontaneously and who might be reluctant to reveal a feeling or opinion. There was so much more to talk about; people commented with wonder that it was not only easier to talk, but that they weren't anxious, no matter what was said. Not only were there no pauses, but there were so many areas in which they could talk that they had to monitor each other not to get off on tangents. After the group, they congregated to further process what had happened.

According to the Johari Awareness Model, miscommunication occurs frequently because our open window is so limited; we are not in touch with what we feel and cannot respond to what we hear. Luft gives the example of a young minister who came into a group, and through changes in quadrant 2 and quadrant 3, developed a much more open quadrant 1 and an increased understanding of himself.

> A young minister said that it bothered him not at all to be seen in the group as a yes-man. He claimed he had his own opinions and that he did not think of himself as overly agreeable. However, the others could see that the criticism did hurt, that he was more anxious and tense under the criticism. Yet, he denied these feelings because they were in fact walled off from him. He was in no position to see and feel what others were perceiving. His communication disturbed others because they were at such variance with what they saw in his blind area. Eventually he revealed (Q3 to Q1) problems he had in his church and his sense of failure and increasing isolation. He felt that such a disclosure would be devastating—but it wasn't, and he felt greatly relieved. In time he was able to express his own reactions about what was going on by acknowledging his own distress and by disclosing the ways he found to punish himself

when things went wrong. He discovered that by withholding all negative feelings and showing only the positive to his congregation and to his friends, he had been signaling to them to withhold too, and in the process, to keep their distance. His communication problems were bound up with difficulties in appropriate disclosure, resulting in bland and distant relationships.

He had become increasingly self-critical while at the same time, presenting himself as confident and self-accepting. He thought of himself as a failure but tried to act as if he felt he were doing rather well. It was as if he had painted himself into a corner and was now behaving in a way which said, "This is just where I wanted to be all along and haven't I done a fine painting job on this floor" (Luft, 1969, pp. 49–50).

How Tension and Defensiveness Arise in the Communication Process

If individuals desired to create problems in communication, there seem to be certain tried and true behaviors that will most assuredly help them on their way (Gibb, 1961; Rogers and Roethlisberger, 1952). A first step would be to keep other people from expressing their own ideas. People have a simple need to be heard, to have their ideas made visible. To have them accepted is desirable, yet not always possible. However, not to be even recognized or heard is an intolerable situation for most of us. Thus, there will be tension in a group if it is dominated by a few vociferous individuals while others listen passively. Whenever participants feel that a group is out of their control and in the possession of others, the atmosphere is likely to deteriorate.

Closely linked to this situation is one in which individuals respond with such certainty and force that only a full-scale verbal war would change their opinions. People naturally don't like to be pushed. Sometimes individuals in a group will attack a position in which they basically believe merely because one person has taken a "too certain" opposing viewpoint.

Even though we spend much time and energy evaluating people and events, if there is one thing that puts us on guard, it is the feeling that *we* are being evaluated by *others*. We are so used to judging the person along with the idea that we tend to become supersensitive to the same treatment. It is such a short step from the words: "Do you really believe that idea?" to the translation in one's mind: "How could you possibly be so stupid as to believe that idea?" In a group where our need for acceptance increases, the feeling that we are being personally judged is a sure way of developing internal friction. Similarly, if we feel someone is placing himself or herself above us in some sort of superior position, an immediate response is to prove to the world that this individual is "not that good." Quite often we find ourselves responding on the inside to the sharp but barely perceptible cues of superiority from another person.

Communication is also damaged when individuals do not trust the group enough to share what they really feel or think. The problem, of course, is that when we fail to express our feelings, others tend to read into this lack of expression what they believe we are feeling or thinking. More often than not the small streak of paranoia in each of us translates this neutral behavior into a negative perception. This behavior as well as other strategies that hide one's real self from the group will predictably result in defensive reactions by those on the receiving end. The diagram on page 35 reveals the subtleties involved in this complex process.

Bill's response to John's statement is partly the result of the selection of words, the context in which they are spoken, and his image of John as well as all the nonverbal cues he gets from the tone of voice, gestures, and posture. Also, the nature of the statement is partly the result of John's response to Bill's particular behavioral strategy with him (in this case neutrality). The result is a predictable increase in what Jack Gibb (1961) would call "defensive communication." Part of the problem obviously lies in John's insensitivity to Bill and in Bill's tendency to read more into the words than was actually intended. Worse than this is the fact that the underlying issues build in a cumulative fashion, which results in increased tension, deteriorated communication, greater polarization among group members, and less inclination to resolve these emotional roadblocks.

Gibb found that defensiveness increases when a person feels that he or she is being evaluated, controlled, or is the butt of a strategy (a plan or maneuver to accomplish an unknown outcome). A large part of the adverse reaction to much of the human relations training is a feeling against what are perceived as gimmicks or tricks to involve people and have them think they are really participating in a decision, or to make listeners think someone is really interested in them as people. Defensiveness also arises when a person feels that another is reacting to him or her "clinically" or as a case (neutrally); or from a superior position; or from a position of certainty. Arousing defensiveness interferes with communication and then makes it difficult, and sometimes impossible, for anyone to convey ideas clearly and to move toward the solution sought.

Statement by John: "Yes, but Bill, that's impossible. I've been here for five years, and I've never known that to work. Have you thought of the fact that. . ."

John feels:		*John as perceived by Bill:*
Reasonable	⟶	Evaluative, judging
Correct	⟶	Superior
Having heard Bill	⟶	Certain
I've got him backing up	⟶	Controlling
The group is with me		
He's probably angry; you never know with Bill		

Bill feels:		*Bill's eventual response:*
What's he mean "yes"?	⟶	Withdrawal
He never even heard me.	⟶	Neutrality
If it's impossible, I	⟶	Passive hostility
wouldn't have said it.		
Big deal—five years—he doesn't know everything.		
It's impossible to be right against him.		
He can sure make a person feel stupid.		
Obviously I've thought about it.		
It's always a fact coming from him.		
I'll bet everyone agrees with him.		
I should just keep quiet—that's better than looking stupid.		

Impact on Group: Unresolved hostility
Other members afraid to venture out against John
Those sympathetic to Bill strengthen their protective subgroup.

Feedback: A Means of Reducing Distortions in the Communication Process

The more group communication is allowed to be spontaneous and open the more the participants will be willing to recognize the perceptual distortions that develop and the behaviors that cause defensive responses. At any given time we see ourselves communicating a particular content message as well as an image of ourselves. Just as we often fail to communicate the verbal message of our intent, so too we fail to communicate the person or image we wish to bring to the group. In most groups, however, neither of these two levels is clarified for us and we can only assume the degree to which we were effective. We like to think that we are effective in our efforts, and it can be quite threatening to discover how often we are not. We are torn between a real desire to confront ourselves with how we are actually perceived (both at content and image levels) and our desire to live with the image we think or would like to think we are projecting.

Feedback is the process by which we find out whether the message intended is the message actually received. In the simplest sense, feedback refers to the return to you of behavior you have generated. A mirror gives one kind of feedback, as does a tape recorder, or a camera, or a videotape machine. A look in the morning mirror may indicate that we really didn't get enough sleep, and our voice is constantly a surprise as we listen to ourselves on the playback of a tape.

When photographs record us as beautiful, we are pleased; when they show more wrinkles than we were aware of, they are another kind of reminder. One of the most effective ways to train anyone, from a neophyte swimmer to a trial lawyer, is through feedback via videotape.

But the most powerful form of feedback is the human response. People can be excellent mirrors, cameras, and tape recorders. Optimal learning, however, requires sensitivity and judgment in the feedback process, and for this reason human response remains the most powerful instrument. Machines are limited to interaction that has been programmed into them and people have limitation problems also. A person is always faced with a choice of behavior from which to extract pertinent messages. As Laing (1972) and Gibb (1961) emphasize, even the simplest communication may be misinterpreted and misunderstood. The same may be said for every silence.

Since the simplest exchange may be misunderstood and thus quickly escalate to interactional difficulties, it is useful to be able to check with the other person when necessary. The manner is direct, but relatively unobtrusive. It should be obvious that even the simplest question can be transformed into a challenge or an attack. Leavitt and Mueller (1951) have done some classic work demonstrating the increased accuracy and confidence that come from being able to ask if what was heard was accurate, and also from asking questions for information.

Feedback can increase accuracy, instill a sense of being understood, and promote closeness and a sense of confidence. It can also increase defensive communication and the level of guardedness.

Feedback has been most effective when it is asked for (in contrast to the unsolicited, "I'm telling you for your own good."); when it is descriptive rather than evaluative; when it is behavioral rather than global; when it occurs as soon after the behavior occurs rather than after a long time lapse; when it is positive rather than negative (Campbell and Dunnette, 1968; Yalom, 1970; Jacobs et al., 1973).

Five varieties of feedback can be distinguished (Luft, 1969):

1. *Information.* The person giving feedback (G) repeats to the other person receiving feedback (R) what the other said. For example, "Did you say you didn't get a call?" Often this kind of feedback is prefaced by remarks like, "Am I correct in saying . . .", or "Are you saying . . . ." When the receiver hears the verbatim feedback, he or she has a chance to modify or confirm the essentials of his or her message.

2. *Personal reaction.* Here, the person giving feedback (G) informs the receiver (R) the effect the message is having on him or her. For example, "I've been angry at you since last week when you didn't come and didn't call. Now you act as if nothing happened." Because the person receiving the feedback is often unaware of the reaction of the person who is giving feedback (information from the third quadrant), conveying personal reaction is one of the most significant events in group interaction. The individual is informed of the specific impact of his or her behavior on another person. Because emotions are aroused, the group increases its positive interaction (Meulemans, 1973–1974).

Too frequently, these reactions are censored in everyday interchanges because we are afraid to hurt someone's feelings, or we are afraid that we won't be liked or that we will rock the boat. It is easier to be polite and to censor ourselves and perhaps take it out on someone on whom we can more easily displace our anger.

3. *Judgmental reaction.* In this kind of feedback, a person evaluates the behavior of another and delivers an opinion. For example, "You need to have a couple of more years experience before you can lead a group; you just don't have it." In this example, the person giving the feedback is presumably doing it for the receiver's own good. However, judgmental reactions tend to be resisted. Even when a group has worked to a high level of trust, such reactions still tend to be resisted, although they may be examined more thoughtfully. This is among the least desirable kinds of feedback for influencing awareness and change (Rosenwein, 1971).

4. *Forced feedback.* The person giving feedback calls attention to another's "blind" area (quadrant 2 in the Johari Model). For example, "Can't you see you attack anyone in a position of leadership?" The giver assumes that the receiver is unaware of his or her own feelings and motivations; it may even be true.

Another frequent example is prefixed by statements like, "I'm only telling you this for your own good . . ."; or, "I don't want to hurt your feelings, but . . . ." This kind of feedback usually does not produce awareness or change; it usually creates a situation in which the receiver feels he or she has an enemy or that he or she has been embarrassed in public. It more frequently produces anger and resistance. In situations in which an authority gives the forced feedback, the receivers may accept it too readily without seeing it themselves.

5. *Interpretation.* A variant of forced feedback, interpretation means explaining what happens by relating behavior to a reason or motive. For example, "You don't talk because you're an only child, and aren't used to talking in a group." Interpretation is a complex and subtle kind of feedback. Much depends on tone and phrasing. Much borders on pseudopsychotherapy, such as "You have trouble trusting us because you couldn't trust in your home as a child." In general, feedback that is interpretation is not helpful. It forces a past explanation on present behavior, when the ingredients of learning and growth exist in the present.

Gibb (1961) notes that defensive climates are more likely to arise when there is a judgmental reaction, when there is control, when there is evaluation, and when there is superiority and certainty. These occur in feedback that is judgmental, forced, or interpretive. Communication is more open and there is a more supportive climate likely when the feedback is informational or personal, such as when it is descriptive, spontaneous, empathetic, comes from a position of equality, and is provisional.

READER ACTIVITY

What do you know about yourself now that you did not know five years ago?

____

____

____

How did you learn that new information?

What experiences helped you learn that? What people or comments?

As you think back, what feedback was helpful to you?

What feedback was harmful? Why?

Jacobs and her associates (1973) found that positive feedback was rated as more credible, more desirable, and having greater impact than negative feedback. She further found that negative feedback that is behavioral was more credible than negative feedback that was emotional. We change by hearing (and seeking out) positive information on ourselves. We are most likely to hear negative information when it is behavioral and we can think about and modify the behavior to which it refers; we protect ourselves from others' judgments of ourselves, even to the point of not believing those judgments.

In most groups, the feedback process can be used to best advantage as a means of clearing the air, providing the opportunity to shift course or procedures, and raising important issues that could not easily be explored during the give-and-take of the meeting. It is possible to begin the process gently. For example, after a meeting, the participants can spend a few minutes discussing what might be handled differently the next time in order to insure a more effective meeting. In this way the process can focus on future behavior and events and not just on the behavior that hindered the present meeting. It requires the participants to develop effective modes of future behavior and a constructive attitude toward their own efforts. Similarly, without becoming too personal, the participants might each jot down on a piece of paper a specific type of behavior they feel was facilitative in the meeting as well as one that inhibited the progress of the group. The use of such immediate information can prepare a group to accept more readily specific information relating to individual behaviors. It can also increase the members' desire to solicit information about their own effectiveness. It is out of this search for personal learning and improvement that a climate of increasing support and openness will develop. Eventually it is possible for the group to develop enough trust so that the feedback process becomes an integral and unobtrusive part of the entire meeting, with members responding at both a feeling and content level and checking out their own perceptions with others in the group.

There is, of course, the possibility that feedback can become of greater importance to the members than the task facing the group. There is no doubt that the process can itself become distorted, inappropriately personal, and an actual imposition if mishandled. One way to control this situation is occasionally to appoint a member of the group as observer and nonparticipant. A brief descriptive report after the meeting can provide a stimulus for the group to reassess its own working goals and priorities. Of greatest importance is that the use of feedback not be imposed because it will inevitably create even more tensions and divisiveness and actually inhibit the very communication channels that the group is attempting to open.

Poor Communication: The Rule, Not the Exception

If one were to take a cross-section of American institutions, it is likely that breakdowns in communication are one of the primary sources of internal conflict and stress. Spend a day in a mental health institution and it soon becomes clear that the administration communicates badly with the staff, psychiatrists with psychologists, doctors with nurses, and nurses with day-care workers. Somewhere in the labyrinth of status, roles, job descriptions, and the multitude of internal conflicts that exist, help is given to the resident patient. There are, of course, those exceptional institutions where hierarchical power struggles are minimized, where role differentiation in terms of status is limited, and where, as a result, communication channels remain relatively uncluttered.

The same can be said in much of industry, although industry has generally provided leadership in areas of communication, specifically, and organizational development in general. At what appears to be the other end of the continuum, there are huge communication problems in organized religion. Between priests and bishops, religious and lay people, and even between religious orders, there are often fragmented lines of communication. In government, for example, the State Department seems to have a special language of its own, where words are used to imply meanings indirectly; it is part of the high art of diplomacy not to say what is meant, but to speak through innuendo and hidden meaning.

And then there are the schools. Perhaps nowhere is there a better example of tensions that exist because of the communication process. It touches every level, but most pointedly it is present in the classroom. Most of us have been part of classroom groups for between ten and twelve years of our lives. How well the lesson is learned, how well the tradition is carried into other groups. And what a price is paid in terms of efficiency, motivation, and personal identity. There are some exceptions, but they remain exceptions. The following conditions in education do exist in other institutions, but the school system is something we have all experienced firsthand.

1. Communication is one way—from a source of information to the receiver who can ask for clarification, but he or she is seldom in a position to transfer his or her learnings to others, be they younger students or age peers.
2. In the classroom group, the goals of learning are seldom established by the participants or even with them, but, instead, by an outside power source.
3. Rather than being shared, leadership is usually held tightly in the hands of the "responsible" person.
4. The participants are held accountable only for content information—usually in the form of an evaluative examination that labels individuals according to performance in terms of discrete letters or numbers.
5. Although held accountable in content areas, the participants are seldom held accountable in other areas, such as discipline and decision making, which are relevant to them.
6. The faculty are not held accountable for their performance in terms of the student participants. This lack of a two-way evaluation increases distance between student and teacher.
7. Rather than being perceived as an important resource to be used effectively by the classroom group, the teacher is established as *the* resource person.
8. Often the internal climate is highly competitive and sets student against student rather than stressing the educational venture as a cooperative one.

9. The communication of information from the students to the teacher is usually, for a variety of reasons, through a relatively small number of students.

With a very slight shift in titles and certain terms, the situation described here could easily be transferred to small groups within a variety of institutions. Obviously, it would be simplistic to say that by changing communication patterns, all these conditions would change. Nevertheless, research in the area of small groups suggests a variety of logical alternatives that could make for a more open communication.

FACTORS INFLUENCING COMMUNICATION

Group leaders, for example, are often hesitant to spend (waste) time developing interpersonal relations in a group where the goals are clearly defined around specific tasks (Slater, 1955; Grace, 1956). Thus, a program director may have a regularly scheduled 3-hour meeting every week (150 hours over a year) for his or her staff and never spend any time strengthening the communication process or exploring ways to improve interpersonal relations within the group. Similarly, a high school history teacher may spend from 3 to 5 hours a week with the same students for an entire year.

Almost any new group will be charged with tension (Crook, 1961) as individuals test out their environment and observe the various personalities involved. It is in this early period when most communication patterns develop. Among adolescents, an enormous amount of energy is spent worrying about or defining their self-images so that time spent early in opening lines of communication would help reduce initial personal anxiety and develop an atmosphere of greater support later. Groups that are attractive to their members have been shown to be better able to control deviant behavior by their members and to stay focused more directly on the task than less satisfied groups (Schachter, 1951; Back, 1951; Festinger and Thibaut, 1951; Gerard, 1953). Furthermore, there is a tendency to submerge negative feelings and a willingness to deal with them directly and aboveboard. Finally, as members feel increasingly satisfied with the group there is increased participation and more flexibility in the decision-making process. Thus, it appears that time spent initially and periodically in improving the communication process will pay dividends in terms of greater work efficiency (see also decision making and problem solving in Chapter Seven).

Also relevant is the notion of competition and its place in a group. There is little doubt that a group can be drawn closer together when competing with another group. However, there is little evidence that the seeking of mutually exclusive rewards even in the attainment of more general group goals is effective in producing an efficient environment for work (Deutsch, 1960; Wheeler and Ryan, 1973; Workie, 1974). It seems that when people enter into a task with a predefined need to be cooperative and interdependent there is more listening, more acceptance of ideas, less possessiveness of ideas and, in general, more communication. Within this atmosphere, the group will also tend to create achievement pressure upon itself. Furthermore, there seems to be more attentiveness to members' ideas and a friendlier climate than in groups where interpersonal competition is stressed to a greater degree. As suggested earlier, all these conditions help to make a group more attractive to its participants and will generally lead to greater group productivity.

Size of Group

There is no exact specification of how large a group may be before one no longer feels it is appropriate to call it a small group. The usefulness of the designation rests on the fact that size is a limiting condition on the amount and quality of communication that can take place among members as individual persons. This then affects the interpersonal relations that members have toward each other (Hare, 1976).

Size is a factor in group relationships since, as the size increases, the number of relationships possible between members increases even more rapidly. Kephart (1950) has demonstrated how that increase in the number of relationships becomes almost astronomical with the addition of a few more people. Notice in Table 1.1 how

Table 1.1 Increase in Potential Relationships with an Increase in Group Size[3]

Size of Group	Number of Relationships
2	1
3	6
4	25
5	90
6	301
7	966

[3]Reprinted from "A Quantitative Analysis of Intragroup Relationships" by William M. Kephart, *American Journal of Sociology* 60 (1950), by permission of The University of Chicago Press.

the addition of a person vastly increases the number of possible relationships.

Since the number of potential relationships between group members increases rapidly as a group grows large, the larger group tends to break into subgroups. Communication then takes on another dimension as subgroups relate to each other. Take the family as an example. Two brothers represent a relationship at an individual level, parents and children illustrate a relationship at a subgroup level, and grandfather and grandchildren illustrate relationships at both the individual and the subgroup level. The addition of an in-law to a household of five members, which includes mother, father, and three children means that 211 potential relationships have been added (Kephart, 1950).

In a work group, only a few of these possible relationships exist. Yet, it is easy to understand that in a discussion group, when time is limited, the average member has fewer chances to speak and intermember communication becomes difficult. Morale declines, since the former intimate contact between members is no longer possible. With a larger group, there are greater member resources for the accomplishment of problem solving, the average contribution of each member diminishes, and it becomes more difficult to reach agreement on a group solution.

Size influences communication and behavior. Two-person groups often result in considerable tension, because a dominant-submissive relationship inevitably develops. When one member does not feel that he or she has power over the other, he or she will tend either to fight the other person and his or her ideas or to withdraw into a passive pattern of behavior. Each member will use whatever behavior is required to balance the control component within the group. However, in the dyad, the possibility also exists for the greatest degree of intimacy (Wolff, 1950).

A three-person group, on the other hand, may have less tension, but only because two people will usually join forces and push their ideas into acceptance. The recognition of power through numbers decreases the resistance of the third member and allows a quicker resolution of the problem under consideration. The person in the minority may not feel good about it, but is better able to rationalize away his or her own impotence, given the obvious power of the opposition. Similarly, communication in odd-number groups tends to be smoother because the possibility of an equal split of opinion and the resulting struggle for power does not exist.

Above the size of five, members complain that a group is too large, and this may be due to the restriction on the amount of participation. A five-person group eliminates the possibility of a strict deadlock because of the odd number of members; the group tends to split into a majority and a minority, but the minority of two does not isolate an individual; the group is large enough for persons to be able to shift roles.

There appears to be no magic number for a successful working group. However, in general, as the size of the group increases, the affectional tie among members decreases (Schellenberg, 1959; Kinney, 1953; Berelson and Steiner, 1964). So much depends on the topic, individual personalities, motivations, and the past experience of its members. Nevertheless, a group of five seems to be optimal in a number of situations, because the group is large to allow for diversity of opinions and ideas, yet small enough to allow everyone to be heard (Hackman and Vidmar, 1970).

Physical Attractiveness

What factor does physical attractiveness play in interaction? The question is regularly asked (Heilman, 1980). One set of beliefs is that attractiveness makes no difference whatever. Others argue that such objectivity is naive, that the beautiful are a privileged group, consistently enjoying benefits solely because of how they look.

Snyder, Tanke and Berscheid (1977) have hypothesized that being attractive elicits certain behavior from others. They say that we have expectations of what a beautiful person is, and that we react to that image. They use the term *behavioral confirmation* to refer to how we may unknowingly induce others to treat us in such a way that our initial impressions of them are confirmed. For example, if we believe certain people are cold and hostile, we may be less friendly than if we believed they were warm and generous. They may be mildly offended by our cool and aloof approach and reciprocate in kind. Thus, our initial impression of their lack of warmth is confirmed though the confirmation is brought about by our own behavior. To see whether this sort of self-fulfilling prophecy occurs with stereotypes about the physically attractive, they designed an experiment in which male college students were to try to get acquainted with a female college student over the telephone.

The male-female pairs of students who participated in the research were not allowed to see each other before the telephone conversation. However, each male student was given some information about the woman he would be talking with and shown a photograph that had supposedly just been taken with a Polaroid camera. The female phone partners were in another room. All of the information furnished the males was supplied by the actual person they were to get acquainted with over the phone. The photographs, however, were selected from one of two pools of photos: attractive or relatively unattractive females.

Each male-female pair engaged in a 10-minute phone conversation. The conversations were recorded, and raters were asked to listen to the tape tracks that contained the female voices and answer a number of questions about the females. For example, they rated each female participant on a number of bipolar scales and answered questions such as "How much is she enjoying herself?" and "How animated and enthusiastic is this person?" Remember that the raters heard only the females, who were just being themselves but who were believed to be either attractive or unattractive by the person to whom they were talking.

Before the phone conversations took place, the male students had indicated their initial impressions of their partners. Those who saw an attractive photo, for example, expected their partners to be poised, humorous, and sociable. These expectations, which initially existed only in their minds, became reality! The raters listening to the phone conversations rated the females whose partners thought they were attractive as sounding more poised, more humorous, more sociable, and generally more socially adept than the females whose partners thought they were unattractive. Those believed to be attractive actually came to behave in a more friendly, likable, and pleasant manner.

The effect of our expectations may be one of the reasons research on the role played by physical attractiveness in communication is so mixed. We respond in kind to the images we conjure up that relate to attractiveness. In a recent Yale study (Heilman, 1980) participants were asked to decide who should be hired for a management job. Attractive male applicants were considered to be strong and competent, and they were more likely to be hired. Attractive female applicants were judged to be more feminine, associated with helplessness and high emotionality, and were not named to the position. The issue raises all kinds of questions about our objectivity and the relationship between physical attractiveness and our communication patterns.

Time for Communication

As group size increases, the time for overt communication during a meeting of any given length decreases. Each member has a more complicated set of social relationships to maintain and more restricted resources with which to do it (Huff and Piantianida, 1968). In larger groups, a few members do most of the talking (Zimet and Schneider, 1969). Members of groups are aware of this, and an increased number of members of discussion groups report feelings of threat and inhibition to participate as group size increases. The effect of increasing size is to reduce the amount of participation per member. As the group size increases, a larger and larger proportion of members have less than their share of participation time, that is, under the group mean (Bales et al., 1951).

Crowding

Crowding is another factor in communication, as is heat and cold. Crowding is not just a function of the number of people in physical space; it is also a psychological factor. For example, students studying alone in a library resent it when an "intruder" sits beside them at the library table (Patterson, Mullens, and Romano, 1971), although for some types of people, tasks in the presence of many other people in the room have no noticeable effect (Freedman, Klevansky, and Ehrlich, 1971). Sex seems to make a difference in the perception of crowding: females find smaller rooms more comfortable and are more likely to engage in intimate positive conversation; males prefer larger rooms (Freedman, 1971).

Raising the temperature of a room tends to create the effect of crowding. Under conditions of crowding and increased (or varying temperatures), people tend to react negatively to each other (Griffitt, 1970; Griffitt and Veitch, 1971).

Other Factors

There are some factors that can be stated as generalizations (Shaw, 1964; Bavelas, 1950):

1. Total group morale will be higher in groups in which there is more access to participation among those involved—the more open the participation, the higher the morale.
2. Efficiency tends to be lowest among groups that are the most open. Since more wrong ideas need to be sifted out, more extraneous material is generated and more time is "wasted" listening to individuals even when a point has been made.
3. Groups that are most efficient tend to be those in which all members have access to a central leadership figure who can act as an expediter and clarifier as well as keep the group on the right track in working through the problem.
4. Positions that individuals take can have a definite influence upon leadership in the group as well as on potential conflict among group members.

5. Groups with centralized leadership (see 3 above) tend to organize more rapidly, be more stable in performance, and show greater efficiency. However, morale also tends to drop and this, in the long run, could influence its stability and even productivity (Hearn, 1957; Glazer and Glazer, 1961).

It is clear that communication and the decision-making process are influenced by the physical structure (communication nets) that develop in any group, and it, in turn, helps to determine the success of the group (efficiency) as well as how satisfied the participants are in the experience. This communication pattern influences behavior, accuracy, satisfaction, and the potential emergence of leadership (Slater, 1958; Heise and Miller, 1951).

If one selects a circle that insures a relatively high degree of openness (access to participation), behavior will tend to be more erratic, participation will be greater, organization will be slow, and errors greater. It is possible, however, that the satisfaction gained by the participants from being heard and having their ideas made visible may prove beneficial in the long run as members develop interest, trust, more attraction for the group, and, eventually, greater efficiency in their work efforts.

Groups where the lines of communication are clear from the beginning and where relations with authority are specified tend to be more productive in terms of completing task objectives. The price for this, of course, is a reduction in the amount of information shared and the subsequent increase in the dependency upon the person(s) in authority. In the short run, it is doubtful that tension and resentment in such groups would be inhibiting when the concern for completing a task is greatest. However, in the long run, such communication patterns may well create numerous problems as individual frustrations build up with no legitimate ways for venting them.

Thus, a dilemma. One may choose greater leader control and efficiency at the price of lower morale and participation. Or, it is possible to choose higher morale and group satisfaction at the price of efficiency. The answer would seem to be a combination of the two, but it seems to be the rare person who is able to encourage sharing and full participation and still impose the restrictions desired to help maximize the operation of the group. Such a role is by no means impossible if the individual is aware of the many difficulties and traps when, for example, some individuals demand more structure and guidance, while others seek absolute freedom from restrictions. That there are no easy answers is suggested in the following example.

A meeting was planned for about 40 individuals to help introduce them to one another as well as orient them for a large convention involving 20,000 people the next day. The room was capable of seating nearly 250 people and chairs were arranged in rows. In order to alter the sterile environment, the program director changed the chairs around in a manner that would be more conducive for informal talking and getting to know one another. Thus, the chairs in the front of the room were rearranged to form loosely grouped circles of about five chairs each. Barriers were then arranged so that the chairs in the back of the room could not be reached.

When the director arrived 5 minutes before the meeting was scheduled to begin, he found his efforts to no avail. No one was sitting in the front of the room, and, as a result of much effort on the part of a number of individuals, the chairs in the back were now accessible and occupied. It was clear that the participants came to be talked to, that they felt more comfortable in straight rows and with a minimum of contact with one another, regardless of the publicized nature of the meeting. To meet and listen in straight rows, to not interact, to remain strangers in the group until drawn together by the force of a task, and to remain "comfortable" while being fed information are all the result of past conditioning. For the director to have allowed the situation to remain as it was would probably have resulted in many of the participants' leaving dissatisfied with the formal and structured nature of the program. To have moved them out of their security would have risked incurring a negative reaction as individuals became less secure and more dependent on themselves and not the authority.

The Influence of Status and Power on Communication[4]

A meeting is about to begin and we look around and say to ourselves, "Say, isn't this an interesting-looking group. I believe everyone from the janitor on up is here." And then the chairperson of the meeting says: "This should be a great experience today having so many different people with so many diverse points of view to talk together. I hope everyone feels as free as I do to express exactly what's on his or her mind. After all, that's why we're here, isn't it?"

The uncomfortable silence that follows is unnerving, but it captures the mood of impending failure. Unless extraordinary precautions are woven into a carefully planned program, the expected will occur. And the expected is that communication among those present will be dramatically influenced by the perceived status of individuals in the group. It has been shown that when high-status individuals are present in a group, both high- and low-status individuals

[4] See Hurwitz et al., 1968; Pepitone and Reichling, 1955; Mills, 1967.

direct their communication to them (Hurwitz et al., 1968). It is the high-status individuals who will tend to be accepted more, and they will find it easier and to their advantage to speak more. Similarly, because low-status participants don't value acceptance by their own status peers, they will often avoid association with one another during the meeting. Rather, they will wait until later to express their own feelings and attitudes concerning the proceedings. Also, since there is a general fear of evaluation by those with power (Kelley, 1951), it can be expected that those lacking it will take few risks, generally speak inconsequentially, and avoid candidness in their statements. Because of this expected trend in behavior, it becomes even more difficult to contribute if one lacks power, since considerably more attention will be given to each contribution by a low-status person. This in turn will only increase his or her fears of intimidation and critical evaluation. The cycle is further extended by the probability that those with influence will hesitate to reveal any of their limitations or personal vulnerability among those with lesser influence, thus lending an artificial quality to the whole proceeding.

So it is that status and power talk to status and power, with others tending to become observers in the process. What appears to be voluntary silence may be subtly imposed by the group. Unless it is legitimate to "draw in" those pushed to the periphery by the sheer power present, they will tend to feel an increasing sense of impotence. This may not occur if the individual is able to share vicariously the ideas and influence of a person with high status. But even the individual who participates least in the group has feelings about what is going on, has ideas that could be a contribution to the discussion, and, most of all, a desire to feel worthwhile. However, just as certainly he or she may also lack the skills, trust, and energy to overcome the obstacles to his or her communication.

It is not forced participation that is essential, but an atmosphere in which an individual really feels his or her contribution is desired and his or her ideas perceived not as "interesting" but rather *with real interest*. It does not occur when a high-status person is suddenly aware that he or she has obviously avoided John throughout the discussion and now wishes to recoup with a condescending question like, "Well, John, what do you think?" And then John, called into action with no real choice, can only stammer (as expected) a few simple comments, more because he wants to get the group off his back than because the words really represent his opinion.

It is always amazing when a group of ten or so is involved in a heated discussion among three or four individuals with the greatest status and overt involvement and someone suggests that it might be helpful to spend a few minutes in twos and threes discussing progress up to that point. Usually the noise level in the group goes up tenfold, as individuals who have not talked but suddenly realize they have a brief time to express themselves in a less inhibiting situation, pour forth ideas and feelings. It is not that they had nothing to say previously, but the atmosphere in the group simply did not allow a free expression of their ideas. Even among individuals skilled in working with groups, it takes a concerted effort to push beyond immediate needs and sources of gratification, and, instead, seek and actually cultivate opportunities for participation. This process does not seem to evolve naturally—there are too many personal needs in the way.

The purpose of this chapter has not been to impart great wisdom concerning perception and communication in a group. Rather, it has been to focus upon the painfully obvious: an increased awareness of the *process* underlying group endeavors. It is not necessary to print great lists of "how tos" since the awareness of what is happening is the first and largest step in correcting some of the obvious problems that exist. It is an awareness of ourselves at both a feeling and behavioral level, and then an awareness of what is happening among other group members that is most important. Thus, the prescriptions are missing and so too, hopefully, is the moralizing. Communication problems do exist, but they are merely reflective of our own fallibility and the extent of our needs whenever we get together with others. For that very reason communication is difficult but very possible to improve.

EXERCISE 1 The Three-Stage Rocket: An Exercise in Listening and Speaking Percisely

Objectives

To stimulate participants to listen more carefully

To develop skills in the feedback of verbal content

To help in the clear and succinct expression of ideas

To increase one's awareness of nonverbal cues in the communication process

Rationale

We are forever in a hurry to say what must be said and to be listened to in return. We expect instant attention on the part of others and an alert response to *our* responses. However, we are so busy formulating our own ideas, preparing rebuttals, and thinking beyond the person who is speaking (with the same expectations) that we often fail to hear the message he or she is sending our way. As a result, some people feel that expending energy in the conversation is senseless, and withdraw; others try to make their point by overwhelming the other person with words. Neither response is very effective.

Setting

The group is divided into sets of three persons. The facilitator may want this group to be with participants who are unfamiliar with one another, or with individuals who communicate regularly. There is a tendency for a structured activity to be more effective if the individuals who are working together in the skill session are not the best of friends. Among strangers the norm is usually to participate, and among friends it is easy to become sidetracked. Individuals within the three-person groups are labeled *A*, *B*, and *C*. The three stages of this session can easily take between 45 minutes and an hour including discussion. It is assumed that the participants have been having some difficulty communicating or in some other way have been readied for the exercise. This might be nothing more than talking with the group about the factors that make simple verbal communication such a difficult task (for example, poor speech, saying too much, not listening). The facilitator then asks for two participants to demonstrate an activity that will help focus more directly on the problem, and a topic is selected in which they can comfortably take opposite sides. He or she then establishes the rule that each individual must recapitulate what the other has said to his or her satisfaction before he or she is able to express an idea or opinion of his or her own. Thus, person *A* opens with a statement and person *B* must capture the essence of the message and feed it back to *A* to *A*'s satisfaction (a nod of the head is sufficient). If the feedback is not satisfactory, *B* must try again until *A* is certain that he or she has grasped his or her message. *C* (in this case the facilitator) acts as a moderator to make sure both participants are listening and recapitulating before injecting their own ideas into the conversation.

Action

Stage One *A*s and *B*s in all the trios now begin talking (it may facilitate things if a common topic is selected) with *C* as the moderator. The facilitator should float and see that the instructions are understood. After 5 minutes, *B* talks to *C* and *A* becomes the moderator. This continues for another 5 minutes (if there is time, the facilitator may want *A* and *C* to have a chance with *B* as the moderator).

Stage Two Another rule is imposed on the participants. The process is to continue, but a time limit is added. *A* makes a statement to *B*. Now *B* must reflect the essence of the message to *A*'s satisfaction and introduce his or her own idea in no more than 25 seconds. If he or she is unable to do so, *B* forfeits the chance to add his or her own idea. The aim is to sharpen listening and recapitulating and, at the same time, to reduce *B*'s input to just what is essential. It is important that *C* be a strict referee or else this stage of the exercise will prove ineffective. After 5 minutes or so *B* and *C* interchange in this manner and *A* is the time referee. It is possible during this second interchange (and possibly a third—between *C* and *A*) for the facilitator to have the participants sit on their hands, thus adding another restriction.

At this point, it is possible to have a brief discussion among the members of the trio concerning what has occurred up to that moment. Some facilitators, however, do not like to break up the sequence of the activity and hold off the discussion until after the third stage.

Stage Three Now the time restriction is removed and all three participants take part equally. They are all to discuss a particular topic (by this time they will easily select a topic of interest by themselves). They must, however, still reflect upon what the speaker has said and recapitulate to the speaker's satisfaction. The new restriction is that all the participants must keep their eyes closed during this entire stage. Thus, the conversation should be relatively natural since the recapitulating will be fairly natural by then. The rationale for this stage is to make the participants aware of their dependency upon many nonverbal cues in the process of normal conversation.

Discussion

The participants may talk together within their own trios about their learnings and the implications of the Three-Stage Rocket in their efforts to communicate. Or, it is sometimes helpful to establish new trios for the discussion with members from different groups helping them to focus on specific questions such as:

At which point in the exercise did you feel least comfortable? Why should this be?

What did you learn about yourself from this exercise that may have implications for you in your future efforts to communicate?

Did the exercise prove annoying to you at any time? Why?

What did the time restriction do to you? Was it helpful?

What did you learn about yourself with your eyes closed and how you listen and how you communicate?

With these types of questions in mind, the discussion should be profitable. It may be helpful to have the groups report specific findings to the entire group in an effort to begin closure to the exercise. Also, although feedback is held strictly to the verbal content level, it may be worthwhile to use this exercise as a first step in readying the group for a more in-depth analysis of the subject.

EXERCISE 2 Communication Role Reversal

Objectives

To make group members aware of how easily they tune out one another

To force people into a position where listening becomes expected at both an emotional (affective) level and at a content level

Setting

The large group is divided into subgroups or sets of four or five participants. Each person is given a rather large name tag that is pinned on him or her or placed in his or her lap so that every other person can clearly see it. This is to be done even if members know one another's names. Again, it is probably helpful if the members are not too familiar with one another, since it usually requires greater concentration and insures greater involvement (unless the members are voluntarily together where familiarity is one of the aims of that particular program).

Action

The subgroups are given a topic that insures some involvement on the part of all the participants. If possible, the topic should be of such a nature that opposing views will be presented. After 5 or 10 minutes—at a point when the discussion has developed to a considerable degree—the facilitator requests that each individual give his or her name tag to the person across from him or her. They are then asked to continue the discussion as if they were the person whose name tag they now have. After another 5 minutes, the facilitator asks the group members to begin expressing the views of the person on their right (another exchange of tags).

Discussion

If the participants have really been listening to one another and the discussion is moving with most individuals participating, the exercise will not prove difficult. However, if a person has not been participating for some reason, this poses questions for the person playing him or her, and it poses questions for the group. Why was he or she not involved in a topic about which he or she must have ideas? The group is also asked to discuss whether the switch made them uncomfortable or made the task particularly difficult. Also, what was present in this situation that is present in most group communication? Was it easy to pick up the emotional as well as the content information of the person you were playing? Did the learnings from the first switch carry over into the second switch?

EXERCISE 3 The Blind Builder: A Task in Interdependent Communication

Objectives

To observe how different individuals give direction

To observe how different individuals receive direction

To gain a better understanding of what happens when communication occurs under stress conditions

Setting

This exercise is limited to some extent by the availability of materials and space. Groups consisting of four participants are established. Two of the four are to be observers. Each group must have the following materials:

A blindfold

A backsaw or other small handsaw

Odd-sized pieces of wood board (perhaps 5 or 6)

A hammer

Between 10 and 20 tacks or small nails

A 3-foot piece of rope or twine

The participants are told the following story:

> Two people flying across a group of islands in the South Pacific were forced to crash-land on a small, uninhabited island. The temperature is extremely hot and, although the land is quite arid, it appears that there will be rain sometime during the day. The two fully expect to be rescued in a few days and have enough food for 5 days. However, their water supply was lost in the crash and they haven't even got a container for holding water should it rain. Because of the heat and their fear of dehydration, the two survivors feel it is essential to build a water container and then wait for what looks like an inevitable thunderstorm. They find some wood and tools and are ready to set about their task. The only problem is that during the crash one of the two received a heavy blow on the head and is now both blind and mute. The second person burned both hands while pulling the tools from the burning wreckage and is not able to use them at all. But, together they must build the container—and before the rain comes. A few drops begin to fall.

Thus, the hands of one individual are to be tied with the twine securely behind his or her back, and the eyes of the other are to be blindfolded.

Observer Roles

Both observers are to take notes on the more general aspects of the activity, but each is responsible for a more detailed observational report concerning one of the people in the activity. How does the second person give directions? Is it possible for the blind person to understand him or her? Do they establish a basic nonverbal system of communication so that the blind person can communicate? What signs does each of the people reveal as the task becomes increasingly difficult? How is this frustration communicated to the other, and what is the other's response? What could have been done to facilitate their communication?

Action

The task will take between 15 and 30 minutes. The observers may wish to observe other groups for the sake of contrast. When the task is completed (or not completed), the facilitator consolidates the two groups of four into one group of eight for a discussion of (*a*) the observers' data and (*b*) the feelings of the two participants. He or she then has the group of eight try to draw together some general learnings and implications to be shared with the large group.

EXERCISE 4 One-Way versus Two-Way Communication[5]

Objectives

To illuminate problems inherent in any communication between one person and another or a group when the aim is to give specific directions or information

To explore the feelings generated from two very different approaches of transferring information to others

Action

It will probably facilitate the experience if this particular exercise is carefully outlined step by step since there are many phases which, if discussed in a general format, might prove confusing.

1. The facilitator makes a few warm-up comments relating to the problems facing individuals who must transfer information or directions to others. The exercise provides some insights into this process as well as some possible behavioral alternatives.
2. It is important that each individual is facing the person who is to be designated to give him or her certain information. This person or the *sender* should be selected from volunteers. It is also important that the person selected have a clear and distinct speaking voice. Each member of the group should have a pencil and paper.
3. Then the facilitator informs the volunter that he or she is to give the group members all the verbal information necessary for them to reproduce the figures in Diagram I exactly as they are drawn on the page. He or she is to begin with the top figure and then describe each figure in succession as he or she moves down the page. The group members must reproduce each figure in the same relationship to the one above it, as shown in the diagram. The volunteer should have his or her own copy of Diagram I. It is important that the sender follow certain specific rules in his or her presentation. They are:
   a. Sit with your back to the group. You are not to face them during this phase of the exercise.

[5]Design is based on one developed by Harold J. Leavitt in *Managerial Psychology*, Chicago: University of Chicago Press, 1958, pp. 118–128.

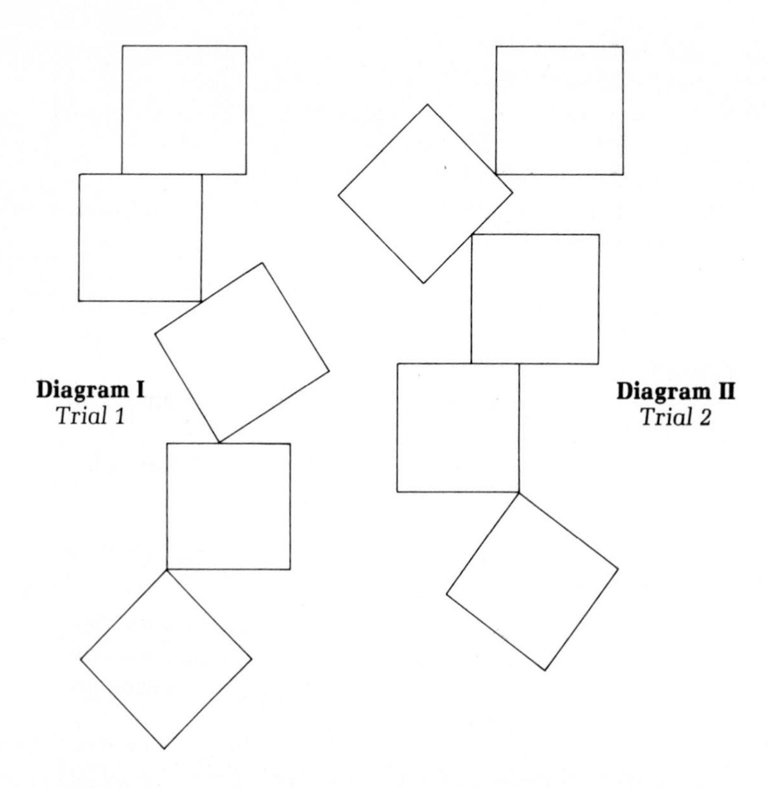

b. There will be no questions. Simply give the group as much information as you feel necessary to transmit the information necessary for them to reproduce the Diagram I.

4. A large chart similar to the one on page 59 should be prepared. The facilitator may not wish to put it up until after the final phase. However, it will be necessary to gather some data in the meantime. First, the facilitator notes the time it takes for the sender to give the group the information necessary for reproducing Diagram I. Next, he or she asks the group to write down the answers to the following questions:

a. How many squares in the diagram do you believe you have correctly reproduced on your paper?

b. Note your feelings as you took part in the exercise. Be as specific as you can, using descriptive adjectives.

The responses to these questions will be collected and posted on the chart following the final phase of the exercise.

5. Now, the sender is given a copy of Diagram II and is told that his or her task is the same. He or she is to study the figures in the diagram and then describe the diagram to the group so they can reproduce it on their paper. Again, he or she is to begin with the top figure and then describe each figure in succession as he or she

moves down the page. The group members must reproduce each figure in the same relationship to the one above it, as shown in the diagram. The sender should follow certain specific rules during this phase of the exercise.

a. He or she must sit facing the group.

b. He or she may answer any questions that will help clarify the task to the group. The facilitator may help them verbally in any way but of course should not show them the diagram.

6. Again, the facilitator should note the elapsed time during this second phase of the exercise. When it is completed, the group is asked the following questions:

a. How many squares in the second diagram do you believe you have correctly reproduced on your paper?

b. Note your feelings as you took part in the exercise during the reproduction of the second diagram. Be as specific as you can, using descriptive adjectives.

It should be noted here that the sender is also to respond to the questions after each of the phases. But, he or she is to guess how many correct reproductions the group (or average person in the group) had for the first diagram and later for the second diagram. He or she is also to note his or her feelings during each presentation and record them.

7. Then the data are collected. On the large chart, the facilitator is to note the following information: First, he or she records the elapsed time for the completion of each presentation. Second, he or she asks the group to show by raising their hands how many believe they were able to reproduce zero (0) figures correct during the first presentation; how many believe they had one (1) correctly reproduced;

	Elapsed Time	0	1	2	3	4	5	**Feelings**
		Guessed Correct Trial 1						
Trial 1 *Diagram I*								
		Actual Correct Trial 1						
Trial 2 *Diagram II*		0	1	2	3	4	5	
		Guessed Correct Trial 2						
		Actual Correct Trial 2						

and so on through five (5). He or she repeats these same questions for the second phase or trial in Diagram II.

8. At this point, Diagram I is presented to the group and each person calculates how many he or she actually had correct. A show of hands again allows the facilitator to record these responses on the chart. The same is then done with Diagram II and the data recorded. A cross-sampling of the feelings generated during the first trial is then posted, followed by a posting of feelings from the second trial. (It is possible to actually collect these data and make a more accurate tabulation, although this will probably not influence the impact of the information.) Now, the guesses of the sender during each trial are noted alongside the other data as well as his or her feelings during each presentation.

9. In analyzing the data and discussion, there are many possible procedures for drawing significant information. The facilitator could lead a general discussion of the results, asking key questions in relation to the data. For the most part the data will lead to some natural conclusions relating to time, accuracy, and feeling dimensions. One possible design would be to:

a. Present the data in an objective fashion—drawing no conclusions.

b. In groups of about five, have the participants analyze the data and draw specific conclusions in terms of the communication process.

c. Post these conclusions in a session with the total group supplementing the implications they have drawn if necessary. Total time set: 1½ hours.

EXERCISE 5 An Introductory Micro-Laboratory Experience

Objectives

To hasten the development of interpersonal communication within a group

To help establish a norm of openness within a group in which both content and feeling level statements are appropriate

To begin exploring perceptions of one another among the participants

Rationale

A micro-experience is designed to move group members through a variety of activities in a relatively short period of time (1 or 2 hours).

By reducing the amount of time spent on any one activity, a wide range of ideas and experiences can be developed and then built upon later. The objectives of each micro-lab will differ according to the particular aims of the facilitator and the needs of the group. This particular design is directed at helping to open channels of communication in a new group of individuals or in a group where channels of communication have been restricted or narrowly defined. Of course, a key in this approach is the knowledge by the participants that they are free to communicate at any level that is comfortable for them and that the various activities are merely meant to help provide a structure within which to relate. This approach has met with considerable success, since people basically desire to be known and enjoy talking about themselves and others. An opportunity to do this in an atmosphere of fun and goodwill can be most satisfying and may build innumerable personal bridges among individuals in the group.

For the facilitator, it is important that the activities build on one another in a sequential fashion and make sense in terms of the stated objectives. Too often the approach is misused and individuals are submerged too rapidly into personal areas that bring on later feelings of guilt or anger.

The following design sequence lasts about 90 minutes but could easily be extended to 2 or 3 hours by adding other activities. (Many of the exercises in the section on Selective Perception could be integrated with little difficulty, if these proved appropriate in terms of the aims of the program.)

1. Getting to Know You

(Time: 15–20 min.) The group divides itself into pairs, in each case with people they do not know. They have 5 minutes to get to know each other in any way they wish. The facilitator has a blackboard or newsprint available and after 5 minutes asks members of the group to call out the topical areas that were being discussed. He or she posts these quickly.

He or she then says something like this to the group: "In each of our lives there are usually a handful of individuals that we can call true friends. These are the people in whom we could have complete and ultimate trust. If this person to whom you are now talking were to become one of these very special friends, what would you have to know about him? What would be the most important thing to discover before you could have a relationship with the kind of trust necessary to sustain it?" The group members discuss this for about 5 minutes with their "potential" friend.

Again, after 5 minutes or a bit longer, the facilitator asks the group to share the topical areas they were exploring together. He or she lists these next to the original list developed after the first talk. The difference should be apparent. The second list should be much less superficial, more personal, and with much more involvement. In this period of 10 minutes, the participants will probably have shared more about themselves than they ever have in many of the "friendly" relationships they now have. How simple it is to go a little deeper, if only the two participants agree that it is all right to venture into these areas. Both usually like the experience. There is no need for much discussion of the data outside, only perhaps a few expressions of feelings from the group. The point is too self-evident to lose in a superficial discussion.

2. Discussion of Communication with a Person Like Yourself

(Time: 15–20 min.) The facilitator has the members of the group look around and find someone that, for some reason, they believe is like themselves. He or she has them sit with that person. This may take a little encouragement and pushing on the part of the facilitator, but once individuals see others doing it, it becomes much easier. These new pairs are to spend a few minutes discussing what it was about themselves that caused them to get together and what else they have in common. (5–10 min.)

At this point, the group is asked to listen to a record carefully (for example, "Sounds of Silence" or "Dangling Conversations") and discuss the implications it has for their own participation in a group. The aim is to get the individuals involved at a content level, thus using different behaviors than when they first began talking. (5–10 min.)

The pairs are to discuss some apparent dissimilarities they have now noticed about themselves, things that were not apparent when they first began to talk, and also whether the original similarities were just as apparent at this point. (5–10 min.)

3. Discussing a Critical Event in Life with a Person Unlike Yourself

(Time: 15–20 min.) The facilitator now has the members of the group look around and find someone that, for some reason, they believe is unlike themselves. Again, the new pairs are to spend 5 or 10 minutes discussing what it is about themselves that they perceive as probably different.

Then the facilitator asks the two individuals to share an experience in their lives that proved to be absolutely crucial in shaping "who they are" today. The experience can be positive or negative but must have been essential in their development as a person.

The participants discuss once more the initial differences they perceived. Are they the same? Are there others they have since noticed during the course of their conversation?

4. Childhood Fantasy with a Person You Would Like to Know

(Time: 20–30 min.) This particular experience can be rich in a variety of learnings and, if taken naturally along with the other events, will not be perceived as threatening to the participants. Evaluation or analysis is not the key. Rather it is the pulling together of a wide variety of verbal and nonverbal cues that can communicate to each one of us an enormous amount of information about the other person.

The facilitator might introduce this section by giving everyone a final chance to meet someone else in the group. He or she has everybody seek out another person they would simply like to get to know better and gives them 5 minutes or so to talk about the kinds of things that usually keep people from getting to know others. What keeps people from being personal with one another and allowing much more information with which we can all relate and bring to the surface? (5 min.)

The facilitator continues by saying: "Now, knowing just what you have been able to learn in these few minutes, let's see how much more we can discover about the person with us. Think for a few minutes about this person and imagine him or her when he or she was an eight-year-old. Try to draw a picture of him or her at that time in his or her life. For example, try to imagine him or her at play, how he or she played, with whom, his or her types of friends in school, how he or she enjoyed it, what he or she liked about it, his or her family, wealthy or poor, how many children, oldest, youngest, his or her relations with his or her family, how he or she expressed anger, and so forth."

The facilitator can close by saying: "There are a thousand possible avenues you might want to focus upon. Try to be as thorough as you can so as to obtain the most complete picture possible. Now, after thinking about all this, share it with the individual. Then let him or her share his or her picture to you. After that tell him or her how accurate he or she was and what the *real* picture was. You may want to know why certain things were pictured."

It should be stressed that the most a person can be is absolutely wrong. After at least 15 minutes or when it appears that most of the pairs are through conversing, it might be useful to have two groups of pairs come together and share why they believe they were able to be so accurate or why they were so inaccurate. Also, why go through this exercise in the first place among relative strangers? It is always interesting to have a show of hands among the total group to discover how many individuals were very accurate in their descriptions. Often as many as 80 percent feel their partner hit the mark. Why? Because we are the product of our past. For most of us, our behaviors at 30 are remarkably similar to those at 8. We are open books to the world, though we seldom realize it.

A general discussion of the significance of the micro-experience for the participants and its relation to the rest of the program can provide a useful conclusion to the total experience.

BIBLIOGRAPHY

The American Heritage Dictionary of the English Language. Ed. W. Morris. New York: American Heritage Publishing Co., and Houghton Mifflin, 1969, p. 600.

Asch, S. E. "Studies of independence and conformity: I. A minority of one against a unanimous majority." *Psychological Monographs 70,* 9, Whole No. 416 (1956).

Back, K. "Influence through social communication." *Journal of Abnormal and Social Psychology, 46,* No. 9 (1951).

Bavelas, A. "Communication patterns in task oriented groups." *Journal of the Acoustical Society of America, 22,* No. 725 (1950).

Bales, R. F., F. L. Strodtbeck, T. M. Mills, and M. E. Roseborough. "Channels of communication in small groups." *American Sociological Review,* 16 (1951), 461–468.

Becker, F. D., R. Sommer, J. Bee, and B. Oxley. "College classroom ecology." *Sociometry,* 36, No. 4 (1973), 514–525.

Berelson, B. and C. A. Steiner. *Human Behavior.* Short ed. New York: Harcourt, Brace and World, 1964.

Berne, E. *Beyond Games and Scripts.* New York: Grove Press, 1976, pp. 44–53, 123–135.

Boszormenyi-Nagy, I. and G. M. Spark. *Invisible Loyalities: Reciprocity in Intergenerational Family Therapy.* New York: Harper & Row, 1973.

Campbell, H. P. and M. D. Dunnette, "Effectiveness of T-group experiences in managerial training and development." *Psychological Bulletin,* 70 (1968), 73–104.

Chartier, M. R. "Clarity of expression in interpersonal communication." *The 1976 Annual Handbook for Group Facilitators.* La Jolla, Calif.: University Associates, 1976, pp. 149–156.

Coan, R. W. "Dimensions of psychological theory." *American Psychologist,* 23 (1968), 715–722.

Codol, J. P. "On the system of representations in a group situation." *European Journal of Social Psychology,* 4, No. 3 (1974), 343–365.

Crook, R. "Communication and group structure." *Journal of Communication,* 11 (1961), 136.

Culbert, S. A. "The interpersonal process of self-disclosure: it takes two to know one." In *New Directions in Client-Centered Therapy.* Ed. J. T. Hart and T. M. Tomlinson. Boston: Houghton Mifflin, 1967.

Davison, W. P. "On the effects of communication." *Public Opinion Quarterly,* 23 (1959), 342.

Deutsch, M. "The effects of cooperation and competition upon group process." In *Group dynamics.* 2nd ed. Ed. D. Cartwright and A. Zander. New York: Harper and Row, Publishers, 1960, pp. 414–448.

Ehrlich, H. J. "Affective style as a variable in person perception." *Journal of Personality,* 37 (1969), 522–539.

Fabun, D., ed. "Communications." *Kaiser Aluminum, 23,* No. 3 (1965).

Ferreira, A. J. "Family myth and homeostasis." *Archives of General Psychiatry,* 9 (November 1963), 457–463.

Fertig E. S. and C. Mayo. "Impression formation as a function of trait consistency and cognitive complexity." *Proceedings of the 77th Annual Convention of the APA,* 4, Part 1, 345–346.

Festinger, L. "Informal social communication." *Psychological Review,* 57 (1950), 271.

Festinger, L. and J. Thibaut. "Interpersonal communication in small groups." *Journal of Abnormal and Social Psychology,* 56 (1951), 92.

Frauenfelder, K. J. "A cognitive determinant of favorability of impression." *Journal of Social Psychology,* 94, No. 1 (1974), 71–81.

Freedman, J. L. "The crowd: Maybe not so madding after all." *Psychology Today,* 5, No. 4 (September 1971), 58–61.

Freedman, J. L., S. Klevansky, and P. R. Ehrlich. "The effect of crowding on human task performance." *Journal of Applied Social Psychology,* 1, No. 1 (January 1971), 7–25.

Gerard, H. "The effect of different dimensions of disagreement on the communication process in small groups." *Human Relations,* 6 (1953), 249.

Gibb, J. "Defensive communication." *Journal of Communication,* 11 (September 1961), 141–148.

Gibb, J. R. "The effects of human relations training." In *Handbook of Psychotherapy and Behavior Change: An Empirical Analysis.* Ed. A. E. Bergin and S. L. Garfield. New York: Wiley, 1971.

Glazer, M. and R. Glazer. "Techniques for the study of group structure and behavior: Empirical studies of the effects of structure in small groups." *Psychological Bulletin,* 58 (1961).

Goffman, Erving. *Asylums.* Garden City, New York: Doubleday-Anchor, 1961.

Goffman, Erving. *The Presentation of Self in Everyday Life.* Garden City, New York: Doubleday-Anchor, 1959.

Goffman, Erving. *Stigman: Notes on the Management of Spoiled Identity.* Englewood Cliffs, N.J.: Prentice-Hall, 1963.

Grace, H. "Confidence, redundancy and the purpose of communication." *Journal of Communication,* 6 (1956), 16.

Greenspoon, L. "The reinforcing effect of two spoken sounds on the frequency of two responses." *American Journal of Psychology,* 68 (1955), 409–416.

Griffitt, W. B. "Environmental effects on interpersonal affective behavior: Ambient effective temperature and attraction." *Journal of Personality and Social Psychology,* 15, No. 3 (1970), 240–244.

Griffitt, W. J. and R. Veitch, "Hot and crowded: Influence of population density and temperature on interpersonal affective behavior." *Journal of Personality and Social Psychology,* 17, No. 1 (January 1971), 92–98.

Hackman, J. R., and N. Vidmar. "Effects of size and task characteristics on group performance and member reactions." *Sociometry*, 33, No. 1 (March 1970), 37–54.

Hare, P. *Handbook of Small Group Research*. New York: Free Press, 1976, p. 214.

Harrison, A. A. *Individuals and Groups*. Monterey, Calif.: Brooks/Cole, 1976, pp. 100–107.

Hearn, G. "Leadership and the spatial factor in small groups." *Journal of Abnormal and Social Psychology*, 54 (1957), 219–272.

Heilman, M. E., "Sometimes Beauty Can be Beastly," *New York Times*, Business and Finance, June 22, 1980, p. 8.

Heise, G. A. and G. Miller. "Problem solving by small groups using various communication nets." *Journal of Abnormal and Social Psychology*, 46 (1951), 327.

Huff, F. W., and T. P. Piantianida. "The effect of group size on group information transmitted." *Psychonomic Science*, 11, No. 10 (1968), 365–366.

Hurwitz, J., A. Zander, and B. Hymovitch. "Some effects of power on the relations among group members." In *Group dynamics*. 3rd ed. Ed. D. Cartwright and A. Zander. New York: Harper and Row, 1968, pp. 291–297.

Jacobs, M., A. Jacobs, G. Feldman, and N. Cavior, "Feedback II—the credibility gap: delivery of positive and negative and emotional and behavioral feedback in groups," *Journal of Consulting and Clinical Psychology*, 41, No. 2 (1973), 215–223.

Johnson, M. P. and W. L. Ewens. "Power relations and affective style as determinants of confidence in impression formation in a game situation." *Journal of Experimental Social Psychology*, 78, No. 1 (1971), 98–110.

Jourard, S. M. *The Transparent Self: Self-Disclosure and Well-Being*, New York: Van Nostrand, 1964.

Kelley, H. "Communication in experimentally created hierarchies." *Human Relations*, 4 (1951), 39.

Kephart, W. M. "A quantitative analysis of intragroup relationships," *American Journal of Sociology*, 60 (1950), 544–549.

Kinney, E. E., "A study of peer group social acceptability at the fifth-grade level in a public school." *Journal of Educational Research*, 47 (1953), 57–64.

Klausner, S. "Now if the government only used a community

approach." *The Pennsylvania Gazette*, December 1978, pp. 10–11.

Kohler, W. *Gestalt Psychology*. New York: A Mentor Book—The New American Library, 1947.

Laing, R. D. *The Politics of the Family and Other Essays*. New York: Vintage Books, 1972.

Leavitt, H. J. "Some effects of certain communication patterns on group performance." *Journal of Abnormal and Social Psychology*, 46 (1951), 40–41.

Leavitt, H. and R. Mueller, "Some effects of feedback on communication." *Human Relations*, 4, (1951), 401.

Livesley, W. J., and D. B. Bromley. *Person Perception in Childhood and Adolescence*. New York: John Wiley & Sons, 1973.

Loomis, J. "Communication, the development of trust and cooperative behavior." *Human Relations*, 12 (1959), 305.

Luft, J. *Of Human Interaction*. Palo Alto, Calif.: Mayfield Publishing Company, 1969.

Maslow, A. H. *The Psychology of Science*. New York: Harper and Row, 1966, p. 19.

Mellinger, G. "Interpersonal trust as a factor in communication." *Journal of Abnormal and Social Psychology*, 61, No. 52 (1956), 304.

Meulemans, G. "Social interaction and emotional arousal in two training groups." *Bulletin du C.E.R.P.*, 22, No. 3 (1973–1974), 153–160.

Mills, T. M. *The Sociology of Small Groups*. Englewood Cliffs, N.J.: Prentice-Hall, 1967.

Mowrer, O. H. *The New Group Therapy*. New York: Van Nostrand, 1964.

Nidorf, L. J. "Information seeking strategies in person perception." *Perceptual and Motor Skills*, 26 No. 2 (1968), 355–365.

Nidorf, L. J. and W. H. Crockett. "Some factors affecting the amount of information sought by others." *Journal of Abnormal and Social Psychology* 69, No. 1 (1964), 98–101.

Nisbett, R. E. and T. D. Wilson. Institute of Social Research, University of Michigan, *Newsletter*, (Summer 1978), 4.

North, A. "Language and communication in group functioning." *Group Psychotherapy*, 10, No. 4 (1957), 300.

Nottingham, J. A. "Attitude extremity and the process of judging others as related to information-seeking behavior." *Proceedings of the 78th Annual Convention of the APA,* 5, Part 1, 409–410.

NTL Institute of Applied Behavioral Science. *Laboratories in Human Relations Training.* Washington, D.C: NTL Institute of Applied Behavioral Science, 1969, p. 23.

Nylen, D., R. Mitchell, and A. Stout. *Handbook of Staff Development and Human Relations Training.* Washington, D.C.: NTL Institute for Applied Behavioral Science associated with NEA, 1967, pp. 71–74.

Patterson, M. L., S. Mullens, and J. Romano. "Compensatory reactions to spatial intrusion." *Sociometry,* 34, No. 1 (March 1971), 114–121.

Pepitone, A. and G. Reichling, "Group cohesiveness and expression of hostility." *Human Relations,* 8 (1955), 327–337.

Reid, D. W. and E. E. Ware. "Affective style and impression formation: Reliability, validity, and some inconsistencies." *Journal of Personality,* 40, No. 3 (1972), 436–450.

Rogers, C., and F. J. Roethlisberger. "Barriers and gateways to communication." *Harvard Business Review,* 30, No. 4 (1952), 46.

Rosenthal, R., D. Archer, M. R. DiMatteo, J. H. Koivumaki, and P. L. Rogers. "The language without words." *Psychology Today.* (September 1974), 64–68.

Rosenwein, R. E. "Determinants of low verbal activity rates in small groups: A study of the silent person." *Dissertation Abstracts International,* 31, No. 12B (June 1971), 7578–7579.

Schachter, S. "Deviation, rejection and communication." *Journal of Abnormal and Social Psychology,* 46 (1951), 190.

Schellenberg, J. A. "Group size as a factor in success of academic discussion groups." *Journal of Educational Sociology,* 33 (1959), 73–79.

Sebald, H. "Limitations of communication mechanisms of image maintenance in the form of selective perception, selective memory, and selective distortion." *Journal of Communication,* 12 (1962), 42.

Shaw, M. "Communication networks." In *Advances in Experimental Social Psychology.* Vol. I. Ed. L. Berkowitz. New York: Academy Press, 1964.

Slater, P. "Contrasting correlates of group size." *Sociometry,* 211 (1958), 129.

Slater, P. "Role differentiation in small groups." *American Sociological Review,* 20 (1955), 300–310.

Snyder, M., E. D. Tanke, and E. Berscheid. "Social perception and interpersonal behavior. On the self-fulfilling nature of social stereotypes." *Journal of Personality and Social Psychology,* 35 (1977), 656–666.

Sorensen, G. and J. C. McCroskey, "The prediction of interaction behavior in small groups: Zero history vs. intact groups." *Monographs,* 44, No. 1 (March 1977), 73–80.

Strodtback, F. et al. "Social status in jury deliberations." *American Sociological Review,* 22 (1957), 713–719.

Verplanck, W. "The control of the content of conversation." *Journal of Abnormal and Social Psychology,* 51 (1955), 668–676.

Watson, R. I. "Psychology: a prescriptive science." *American Psychologist,* 22 (1967), 435–443.

Watzlawick, P. *How Real Is Real? Confusion, Disinformation, Communication: An Anecdotal Introduction to Communications Theory.* New York: Vintage Books, 1977.

Watzlawick, P., J. H. Beavin, and D. D. Jackson. *Pragmatics of Human Communication: A Study of Interactional Patterns, Pathologies, and Paradoxes.* New York: W. W. Norton & Co., 1967.

Wheeler, R. and F. L. Ryan. "Effects of cooperative and competitive classroom environments on the attitudes and achievement of elementary school students engaged in social studies inquiry activities." *Journal of Educational Psychology,* 65, No. 3 (1973), 402–407.

Winthrop, H. "Focus on the human condition: Interpersonal and interactional processes as extinguishers of structured communication. *Journal of Human Relations,* 19, No. 3 (1971), 418–438.

Wolff, K. H. *The Sociology of Georg Simmel.* Glencoe, Ill.: Free Press, 1950.

Woollams, S. and M. Brown. *Transactional Analysis.* Dexter, Michigan: Huron Valley Institute Press, 1978, pp. 118–120.

Workie, Abaineh. "The relative productivity of cooperation and competition." *Journal of Social Psychology,* 92, No. 2 (1974), 225–230.

Wrench, D. "The perception of two-sided messages." *Human Relations* 17 (1964), 227.

Yalom, I. D. *The Theory and Practice of Group Psychotherapy*. New York: Basic Books, 1970.

Zimet, C.N. and C. Schneider. "Effects of group size on interaction in small groups." *Journal of Social Psychology*, 77, No. 2 (1969), 177–187.

Two

Membership

Membership is a central concept in thinking about ourselves, from birth with our initiation into our first group—a family membership—through the myriad memberships we will experience in the course of our lives. Membership can be thought of as a ticket of admission. What gets you in? How much does it cost? Is it a first-class ticket, or steerage? Do you even have a ticket? Consider the following situation, in which an individual attempts to become a member of a group.

Tony grew up as a minimally educated immigrant in a working-class neighborhood. He had worked since he was eleven, that he was sure of, and maybe even before that. One summer, he worked at

Whitemarsh Country Club, and as he watched the cars, the clothes, the style, and the assuredness of the members, he had a dream. Someday, someday, he would be middle class, and be one of them. Membership in Whitemarsh would signify proof that he had transcended being poor and a foreigner, lower class and a nobody. The dream was always there. Dues were $1,000 per year, and the initiation fee, triple that. He saved, he cultivated two members who might sponsor him, he was accepted for membership—and he made it, at last!

The first Sunday, he dressed carefully in his expensive, casual-looking sport outfit. He entered, heart pounding at the moment. Others in the foyer were coming in and being greeted by the house committee chairperson; still others were smiling at each other, waving, casually acknowledging each other. He moved into the card room. Several groups were playing, and as he walked in a couple of people looked up and then returned to their cards in nonrecognition. He watched some of the players, and listened to their talk (he knew none of the people about whom they were talking, nor the events they described). He moved to another table and watched. No one came near him, asked him a question, or invited him in. He walked over to the bar, where the bartender was engaged in conversation at the other end. In due time, he came over and asked Tony what he wanted to drink, and not another word.

Tony looked at his drink, the bar, the card room, the easy ambiance the others had—the kidding, sharing, joking, talking about people they knew and a world he didn't. Suddenly—strange that it should be sudden—Tony realized that he didn't have a ticket to membership. He might belong for twenty years, and he still wouldn't have a ticket to membership.

It goes like that. We can move into a small town where families have been for generations, and thirty years after moving in, still be treated as a stranger and newcomer, not to be fully trusted—like "summer people" in Vermont, who come to the same house, year in and year out, but who are never invited to a "regular's" house or considered worthy of town membership. Religion, social class, or ethnic origin can be an obstacle to full membership in the family of one's marriage partner. Almost anything can be a factor, since the family determines whether you will ever be a member.

In a small group, the situation is the same. Each of us enters, not knowing and fearful. How will our appearance measure up? Our sex? Our status? How we speak? What we say? What are the expectations they have for us? And the question we ask internally is, will I be accepted as a member of this group?

The feeling never goes away. Because membership is so important to our sense of ourselves, whenever we enter a new group, the old feelings return. Most of us hope we will be accepted, and that membership will represent a supportive and caring experience. Some, of course, join for reasons of status, access to power, opportunities for self-development, or a hundred other reasons.

READER ACTIVITY

Consider what are the tickets of admission

In your classroom ____________________

In your club ____________________

Among your social peers ____________________

In your family ____________________

Some of the tickets are behavior, wealth, appearance, degrees, religion, sex, style.

Where don't you feel accepted where you would like to feel accepted?" ______

Why? ____________________

Do you really lack the ticket, or have you simply assumed you wouldn't be accepted? ____________________

Sometimes we assume we don't have the ticket, but we may be wrong.

RELATIONSHIP OF GROUPS AND MEMBERSHIP

A group is seen as a frame of reference and an environment in which an individual moves. Membership describes the quality of the relationship between the individual and the group. In fact, concepts of membership and groups are so related that groups are often defined in terms of members. Some of the properties of groups are:

Membership is defined.
(It is known who is, and who is not a member.)

Members think of themselves as composing a group.
(They have a shared image, thinking of themselves as a group.)

There is a sense of shared purpose among members.
(Members can give reasons for being in the group.)

There is a feeling of greater ease of communication among members, than between members and nonmembers.
(There is a lot to talk about, and being "in" makes it easier to talk than with "outsiders.")

There is a sense of approval or disapproval of members, and there is feedback from others in the group.
(They let you know how you are coming across.)

Members have expectations for certain ways of behaving in various situations in which the group finds itself.

There are leadership policies and roles.
(Members need to coordinate efforts and maintain conditions for problem solving.)

A status system emerges among members.
(There is a hierarchy of worth of the individual to the group: members know where they fit on the scale.)

No two groups are alike on these dimensions. Membership is the perception by the individual of the quality of his or her relationship to the group (Thelen, 1954).

TYPES OF MEMBERSHIPS

Formal Membership

When we say membership in a group is clearly defined, we are talking about a boundary condition. Who is in, and who is not. Tony, who had paid his membership dues, was a member—he would appear on the rolls, receive regular bills and announcements, and sign his name on meal checks. Dues-paying, card-carrying, computer printout people are formally recognized as members of a group, and defined as within the boundary condition. Their affiliation places them within the boundary, they perceive themselves as part of the club. They have an image of the group and its purposes that is generally shared by the other members.

Aspiring Membership

Another kind of relationship is that of an *aspiring* member. For years, Tony had been an aspiring member, noting how members dressed and emulating them, trying to talk in their style, practicing golf and tennis as he was "getting ready." Aspiring members are not formally within the boundary of the group. They are not members, but act as if they were. Or, to put it another way, although they don't have a ticket of admission, they act as if they might get one, and want to be prepared should the opportunity arise. Consider this example:

His father had gone to Penn, his mother had gone to Penn, three cousins and an older brother had gone to Penn, and as long as he could remember, he had planned to go to Penn. The family had season tickets to the football games, he always went to Alumni Day, and he followed the news of Penn's progress in sports more avidly than his high school scores. His books had Penn covers, he regularly wore his Penn sweatshirt, and already knew which were the best dorms on campus. He felt "his" school had made a serious mistake in cutting varsity ice hockey.

If asked, the boy would readily admit that he was not a member of Penn's student body, but Penn greatly influences his dress style, his activities, and his aspirations for the future. He frequently feels like a member when he walks around campus in his sweatshirt with his brother, and psychologically he is a member. In terms of formal membership, however, he is outside the boundary at least at the present.

Marginal Membership

In contrast, consider the following situation:

A senior at Penn is being interviewed for jobs for next year. He doesn't attend the football games or get involved in sports scores. He disdains Penn covers and sweatshirts, or for that matter, any Penn emblems. He dashes for the train to New York immediately after his last class on Friday. He belongs, he is a member of the University of Pennsylvania student body, but neither the school nor its members influence him to any degree. He identifies much more with people making it in New York; college is already in his past. He is defined as a marginal member.

Membership might be thought of as if the group were a drawn circle. Those who belong are within the circle. Those who are actively involved and influenced by members can be viewed as being in the center of the circle. Conceptually, they are centrally involved—usually as the president or high-status members. The marginal member may be seen as within the circle (the boundary), but close to the edge (for example, the Penn senior). The aspiring member is outside the circle. Full psychological membership occurs when a person is positively attracted to membership and is positively accepted as a member. In our earlier illustration, Tony was not a full psychological member because he was not positively accepted as a member, despite his positive attraction toward membership. The others he enviously watched were full psychological members.

Being in a group does not influence a person's entire behavior. Membership may involve only a limited investment, such as membership in a country club, a union, a church group, or a neighborhood association. However, the degree to which people are involved will affect both the functioning of the group and its significance for them as members.

Membership in the Formal or Informal Organization

There is yet another type of membership. It is membership in the formal organization and in the informal group within the organization.

According to its organization chart, the director of a state hospital, the two assistant directors, the medical director, and the business manager make up the director's cabinet. The director invites the others to meet with him weekly for staff meetings. The meetings consist of routine reports of actions taken during the week, and scheduling or treatment plans for the coming week.

Only the stranger thinks that decisions are made in this group. Real decisions—on strategies for dealing with various inspection boards, on promotions, on personnel problems, on training, on budget—are made by a different group. This group, which cannot be located as an entity on the organization chart, consists of the director, the director of personnel, a psychiatrist, and a physician (internist). The meetings are held in the office of the director of personnel, usually on Tuesdays in the late afternoon. How did this informal group form; how did they obtain their power? How does the cabinet feel about this "other" group?

Within organizations, there are often two memberships. There is membership in the formal organization where criteria are usually known. Concurrently, there is the informal organization, which also has a membership. Criteria for membership here are often not stated, and membership rules may support or contradict the rules of the formal organization.

Voluntary/Nonvoluntary Membership

In speaking of memberships, we are usually considering participation of a voluntary nature. Should we join the consumer's union, a film-making group, or a woman's group? Often we forget that we are members of many groups in which we have no choice—nonvoluntary groups. A nonvoluntary group is described below.

A difficulty with many programs, not only governmental ones, is that they often have a different, even opposite, effect from the one intended. What sounded like a good cost-effective plan doesn't work, and only afterwards do we understand why.

An example of this ineffectiveness is the "Work Incentive Program." The goals of the program were to reduce the welfare rolls, solidify the family unit, and bolster the competence of the head of the household. The program identified people who were the heads of households, with limited job skills, and on welfare. They were then targeted as the beneficiary of all services. The targeting approach was viewed as being an efficient utilization of funds—the rationale being that if one helps one person (the target) the benefits will radiate out to the entire household.

Klausner (1978, 10–11), who evaluated the program for over six years, found it largely ineffective. The prime reason for its failure was the targeted approach. He says, facetiously, that targeting is a good idea if you are on an artillery range but not in a human services program. "The whole program is one to one; the individual and the government, and what is forgotten is that the head of a household is part of a household with roles assigned, prescribed, desired, and expected." The household did not recognize her significant membership and this nonunderstanding was a major gap in the functioning of the program. "Influencing the individual is insufficient, and detrimental to changing the household or ameliorating the conditions of poverty."

One involuntary membership is a household (although with divorce it need not remain as it was—for the adults or for the children as they grow up). Aside from birth, there are a variety of reasons for membership in nonvoluntary groups. These may be age and neighborhood (as in tenth-grade English class at West High), a court order (as in a drug group in a rehabilitation program or a prison group), or political turmoil (as a group of Americans taking refuge in the American Embassy prior to a hasty departure related to a governmental coup d'etat). Compulsory memberships may be nurturing and supporting, or they may be horrendous and terrifying, as in prisons.

WHY PEOPLE JOIN GROUPS

There is little we can do about compulsory memberships. The real question is: Why do people join the groups they do join?

Some people seem to have a need to belong: they are joiners. It is said that they join because it is good business, and that memberships produce exposure and contacts. Others, however, are quick to point out that the person has a variety of interests, and membership is an indication of a willingness to work for each of these groups.

This controversy over why a person joins a large number of groups was a significant political issue in a mayorality campaign in a large Eastern city. One candidate was a member of over forty organizations. The question was raised about whether these memberships were for political show or represented personal interests. In the campaign, his detractors utilized the memberships to point him out as a dilettante, and derisively commented that his running for office was just another passing fancy to which he would not commit the time nor priority necessary. His advocates used the same memberships to refer to him as a Renaissance Man, a man of broad interests and knowledge, whose familiarity with such a diversity of groups and organizations especially qualified him for the post.

Others, not so outgoing, also join. A quiet woman, middle-aged and alone, not given to small talk nor interests in families or children, used to find herself alone on weekends and holidays—until someone convinced her that collecting stamps was a good hedge against inflation. Stamps changed her world. To be knowledgeable, she joined stamp groups. First local, then regional, then national, and international. There are meetings to go to, friends with whom to talk, conventions and programs on weekends, and international conventions to consider as sites for vacations.

An understanding of why people join groups is complicated (Quey, 1971). However, there seem to be three major reasons:

1. They like the task or activity of the group. People join a stamp club because they like talking about stamps and buying or selling stamps; they join a tennis club because they like playing tennis and having courts and partners readily available; they join a consumer's cooperative because they would like to pay lower prices for food.
2. They like the people in the group. Being in an activity with friends (or people who are liked) is a powerful inducer for joining. Most social activities fall in this category. Sororities, fraternities,

luncheon clubs, and country clubs are all characterized by certain types of people; persons seeking membership investigate for the one where they will be comfortable and find their type.

Being with people you like is not only a reason for joining social groups, but seems to be the major factor in determining whether a person finds a group experience significant. Stiles, in attempting to construct a theory of group process experiences, found that "positive values attributed by a group member to his group experiences depends upon the presence of a significant other person for him in the group" (Stiles, 1973). She found that the "significant other" was not psychologically healthier than the person picking, and that being a significant other was not a function of being a "star" (someone who is chosen often). Rather, her major finding was that when a member of a group found a significant other in the group (someone special to them), they found the group experience a positive one.

Frequently, people join because they like both the task and the people. They may want to join a church group, yet travel some distance to a church where they feel they will get along, even though a church of the same denomination may be closer. Students join college houses with little concern about the activities or purpose of that house, but special regard for who is in it, knowing that with the right people they will find the activities they want and people with whom they want to be.

People may join an organization initially because of an interest in the task or activity, and then find they enjoy the people as well. Often, they will maintain their membership long after their task interests have waned in order to continue to participate in the pleasant personal associations. The reverse situation also occurs. A person may originally join an organization only to please a good friend who is already a member, but that person may later become genuinely interested in the project.

3. A third reason for joining a group is that the group can satisfy needs lying outside the group. The group itself does not satisfy the person's needs directly, but is a *means* to satisfying his or her needs. Some early research illuminates the reason of group membership as a means to external satisfactions. Willerman and Swanson (1953) found that girls joined sororities primarily because of the prestige accorded them as sorority members. Rose (1952) found that people joined unions not because they enjoyed union activities or working with other union members, but because unions got them higher wages and job security. Schachter (1959), in his classic experiments, found that under conditions of experimentally induced anxiety, people wanted to be with other people.

Brief observations produce numerous examples. The fledgling lawyer joins an expensive, prestigious luncheon club because it is a good place to meet prospective clients. Funeral directors are well-known for their lengthy memberships, because people feel better calling in "one of their own" rather than a stranger at times of sorrow. Other examples include joining a popular activity to make socially desirable contacts and joining the PTA as a newcomer in order to meet others about the same age in the neighborhood.

READER ACTIVITY

How about you?

List three groups to which you belong:

1. ______________________

2. ______________________

3. ______________________

For each, list why you joined, and now, why you continue.
What are your main reasons for joining a group?

The Back Experiments

Kurt Back (1951) was intrigued by the following questions: Does it affect the group as to *why* a person joins? Does it make any difference whether the person joins because he or she is interested in the task, or likes the other people, or because the group is a means to meeting his or her needs? If it makes a difference, how? What kind of difference?

Back designed a series of experiments to get answers to these questions. He arbitrarily paired subjects, but told them that, on the basis of previous tests of personality and other measures, they had a special relationship to each other. He told some pairs that their personalities were similar and that they would have a great deal in common (the liking condition). He told another grouping of pairs that they had common goal interests in the project (interest in task condition). In the third grouping, he told each of the partners that the other would be an important person to know and could be influential (the group as a "means to" condition).

The results indicated that why a person joins does make a difference on the functioning of the group. Those primarily attracted as friends interacted at a personal level—they had long conversations, were pleasant to each other, and expected to influence each other. In those groups attracted primarily by the task, members wanted to complete it quickly and efficiently. They discussed only those matters they thought relevant to achieving their goals. In those groups attracted by potential prestige from membership, members acted cautiously, concentrated on their own actions, and in general were careful not to risk their status. When there were none of these bases of attraction, members of a pair acted independently and with little consideration of each other.

The Back experiments lead to the generalization that the nature of group life will vary with different sources of attraction (Lang, 1977).

MULTIPLE MEMBERSHIPS

The complexities of membership are compounded when it is remembered that individuals belong to a wide variety of groups. Some they may join for task reasons, some because of the people, some to meet needs beyond the group; in some they are an involuntary member and in others they have sought membership—and to each they bring their unique selves, and behavior to enhance their "position" in that situation. If each membership produces a tension within an individual, and a unique mix of perceptions, communication patterns, and desired goals within the group, imagine what happens when we think that an individual belongs to a multiplicity of groups.

That multiplicity can carry with it a number of assets. There are advantages in the variety of contacts. At least every group is not laden with strangers; some of the members can be known and produce support. There is the acquisition of transferable experiences based on skills obtained in different kinds of groups; for example, sitting on one university committee allows one to learn of the complicated procedures required to arrange for use of a room for a program—procedures which become simpler when making arrangements for another event for another committee. There is the possibility of expediting what one needs to do; for example, knowing someone from one context allows one to make a phone call to check for information or possibilities for action, rather than write a formal letter. (The often cynical comment, "It's not what you know, but who you know" is based, in large part, on the access we have to persons in positions of power based on knowing them in another context.) A group can often accomplish its objectives more rapidly because of those contacts. There have been a number of studies that point out that memberships on prestigious community boards are held by a small number of the citizens of the community (Klein, 1968).

Members of a board who make up the nominating committee recommend members of other boards whom they know and like, to serve. Elected, and later themselves on the nominating committee, those new members continue the process. At one time, prestigious community boards were divided up by families. Everyone knew that certain families were interested in particular activities and served on those boards. Gershenfeld (1964) found that, in tracing one community organization back twenty years, 50 percent of board members had been preceded by one member of their family. Presently, however, among most recently appointed members, this is not true. When boards are made up of families or friends, the members find it easy to conduct business as personal relationships. Pervasive fears of acceptance have been dealt with previously and satisfactorily resolved. Multiple memberships can even be a source of creativity and innovation when a diverse group comes together to search for a mutually agreeable solution to a problem, as in the following example:

After decades of redevelopment, residents of a large metropolitan area stayed away in droves from a restored neighborhood with housing and quaint shops. Both the business associations and the city were frantic for a gimmick to bring people (and business) to the redeveloped historic area. One idea was to bring together representatives of the mother churches in the area, and have them plan a heritage week or festival. Each church and synagogue had spawned others in the metropolitan area, and that might be a method for bringing suburbanites in for the activities. It was hoped the special events (dancing, cooking demonstrations) might also induce the nonreligious populations to have a look. Although the idea sounded promising, the initial enthusiasm palled at the prospect of bringing together what were perceived as "prima donnas" from all of those seventeen competitive, barely surviving historic churches. There were images of constant bickering and upstaging, endless philosophical filibusters, and impossible decision making.

So much for perceptions. It didn't happen that way. Each of the struggling groups was so delighted for the opportunity for publicity for their special quality, was so encouraged by the funding for a festival, and was so revitalized at the prospect of a heritage festival, which had been beyond their provincial dreams, they vowed (individually and collectively) to produce a festival that would long be remembered. The first meeting of two representatives of each of the religious organizations began with most of them not knowing each other, but committed to the prospect of creating an event. Enthusiasm ran high. Diversity was the keynote for excitement. Ideas tumbled around ideas as the initial meeting ran until two in the morning. There were suggestions of sample refreshments to be tested, learning greetings in other languages so that committee members would be knowledgeable, and, despite the differences of the organizations, they represented a common bond of creation.

The diversity of memberships enhanced what they had to offer, and the heritage festival committee experienced themselves in new ways.

Multiple memberships coexist in each of us. We are members of a family, a neighborhood association, a religious organization, a P.T.A., a raquet ball group, a professional association, and a faculty journal club. These coexist in a person's life with little conflict. There may be an occasional time problem—sometimes three meetings are scheduled for the same Tuesday night—but this conflict is minor and resolved with little tension.

Conflicts of Multiple Membership

Sometimes, multiple memberships can be the source of serious conflicts. *The Lonely African* (Turnbull, 1963) gives an example not soon forgotten. In Africa today, many young men, reared within their families and the sharing tribal community, leave them to enhance their futures in the city. They learn, with great difficulty, to accommodate to new standards. First, they are alone, they are individuals—not part of a tribe. If they are to experience success and achievement, it is the individual who must earn it. Painfully he learns to accept the loneliness, to strive for these new standards, to become successful. His new success brings with it increased pain, loneliness, and conflict. In time, tribal members learn of his success and come to visit their kinsman. They remind him of the rules of the tribe—of sharing his wealth and his possessions with all of them as they would with him in the tribe. The conflict between his memberships in the tribe and in an individual-oriented society are unbearable. Many decide to totally cut themselves off from their tribal community. They change their names, move to a different address, and become the "lonely African."

Generally, people do not join organizations or groups with conflicting norms or values. A person is not likely to be a member of the National Association of Manufacturers and also a member of the

Socialist Party; nor of the Catholic Church and the Abortion Rights Lobby; nor the Ku Klux Klan and the Southern Baptist Christian Leadership. Although a person will attempt to avoid membership conflicts, sometimes they cannot be avoided. A businessperson who believes that survival and eventual success depend on taking advantage of every situation may have difficulties on Sundays in church when the minister preaches about ethical behavior. He or she may resolve this discomfort by not going to church, changing business practices, or compartmentalizing memberships—engaging in one kind of behavior that is appropriate in the business world and another kind of behavior on Sunday that is compatible with Christian ethics.

Some multiple memberships result in conflicts that present serious problems, where the dilemmas created cannot be resolved with satisfaction and conflict continues to rage. Frequently, we feel whichever choice we make will induce negative consequences. A common area here is the conflict between membership in a family group and in a career/business group. Long hours required for career achievement mean distance from one's family and being an outsider to the daily events that build closeness. But being there at home for those daily events may mean that the company views one as a nine-to-fiver not committed to the company and not considered as promotable as others. Many law firms have as a norm that young lawyers routinely work long hours for at least ten years; then they will be evaluated for elevation to partnership. The standards are clear—at what cost to the family. With the spiraling increase of women in the labor force also committed to careers, the tensions of multiple memberships are escalating.

Such conflicts become exacerbated as we have fewer traditional male-female duties, we value family and parenting relationships more, and we have more interests and activities. Consider the busy doctor who was called out on an emergency, when she had promised to take her daughter to the movies; the salesman father who travels during the week and is only home on weekends; the scout leader whose family resents his spending a week of his precious two-week vacation taking his troop camping; the woman student who finds herself spending many hours in the library in addition to time at classes—hours she is no longer spending with her family.

Yet another kind of multiple membership conflict is common when a committee is composed of representatives of subordinate groups. The conflict creates intrapersonal dilemmas as the individual vacillates between representing the subordinate group, acting in a way about which he or she can report back proudly, and also being open to the present discussion taking place and acting in a way that fits that situation. It is difficult to make a decision or take an action without fearing the consequences of questioning whether it is the right one. One group of researchers (Schwartz, Eberle, Moscato, 1973) found that "individuals with high group awareness tended to be less successful in problem solving in an *ad hoc* problem solving environment than groups of individuals with low-group awareness." That is, maintaining the links took energy from the present situation.

The resolution of conflicts of multiple membership is often attained at great personal cost and with much anxiety. Often these conflicts are resolved in accord with the standards of the group that are most salient at the moment.

READER ACTIVITY

List ten groups in which you have membership. List at least one conflict each membership can produce for another membership.

How do you resolve these differences?

REFERENCE GROUPS

Of the many groups to which people belong, which are most salient to them? Which influence how they typically feel about things? To which do their attitudes most closely relate? Membership in a group implies the right to influence others in the group; it also implies an agreement to accept influence from others, as in the following example:

An Eagle Scout certainly considers the Boy Scouts a reference group. He has attended patrol and troop meetings for years, accepted varying degrees of responsibility, and influenced boys younger than himself as well as his peers. He has risen through the ranks in advancement programs, acquiring numerous merit badges to become an Eagle. Scouts have obviously been an important reference group for him, as years later he has a bedroom wall displaying his scouting memorabilia, a sash with his badges, and annual OA meetings to relive his scouting memories. In contrast to the Eagle, there are also scouting candidates who attend a few meetings and never achieve the beginning rank of Tenderfoot. The weekly meetings are too frequent, they don't like the other boys, they may not be as agile or competitive as others, and for them, the Boy Scouts never becomes a referent group.

Those groups an individual selects as his or her reference groups are the ones whose influence he or she is willing to accept. Consider the following illustration:

A handsome black college student, well dressed and with a styled hair cut, has pledged for fifteen weeks prior to admission to a prestigious black national fraternity. The final ordeal, which determines whether he is serious in desiring membership, requires that he submit to having his head shaved, as a reminder of what slaves endured. There are no exceptions.

Reference groups have been described as serving two distinct functions. The first is a comparison function. "A group functions as a comparison reference group for an individual to the extent that the behavior, attitudes, circumstances, or other characteristics of its members represent standards or comparison points which he uses in making judgments and evaluations" (Kelley, 1952, p. 413). This is the concept of social influence on an individual's perception, cognition, and level of aspiration (Sherif, 1936; Festinger, 1957; Asch, 1960).

Second, reference groups serve a normative function. "A group functions as a normative reference group for a person to the extent that its evaluations of him are based upon the degree of his conformity to certain standards of behavior or attitude and to the extent that the delivery of rewards or punishments is conditional upon these evaluations" (Kelley, 1952, p. 413). This concept says that reference groups invoke social pressure (Mirande, 1968). Both of these functions may be served by the same groups. People's reference groups may greatly influence their attitudes toward themselves and will affect their relationships with other groups. There are a variety of groups in which individuals may hold membership and to which they address their behavior (Theland, 1954; Boszormenyi-Nagy and Spark, 1973; Berne, 1963).

The Actual Group

It is the actual group in which individuals can interact, test ideas, hear the responses of others, get feedback on their own behavior, and learn. It is only in this actual group where they can have the experience on which to build their understanding of their behavior and their attitudes. In a given group however, there are only some people who are referents, those whom they attempt to influence and who in turn influence them. The effective group is determined subjectively for each member- those who "make sense" to the member, those who seem to be in touch with reality as the member sees it, and those with whom the member identifies. These few, perhaps only eight to ten in a group of thirty (and even their composition may change over time), are the total group for that person. It is they who influence him or her, and it is they to whom he or she can relate.

The Group We Represent

There are two levels of reference groups in this concept. At one level are the groups that appoint or elect a member to represent them in another group. A faculty member is appointed by his or her department to represent them on the university senate; representatives of neighborhoods are elected to sit on the community mental health advisory board; condominium owners select their representatives to the resident council. Here, the group the person represents is thought of as "his" or "her" group, whom he or she must speak for, fight for, and defend. In defending his or her group, the representative believes or comes to believe that their approach and perspective are his or hers. The representative strives for goals that favor his or her group—these are his or her vested interests. A member relates to the group he or she represents, and this becomes a reference group.

At another level are the groups a member represents when he or she is not officially designated to do so. For example, in a classroom discussion belittling the value of sororities, a sorority member may feel compelled to enter the discussion in heated terms as she defends the worth of sororities. Some say that when people defend their vested interests almost blindly, it is because these groups are important reference groups in their lives. Another theory holds that the defense occurs when people are anxious about their membership in that group; they defend the group to assure themselves that they do belong, and that it is worthwhile belonging.

In both cases, people react to the actual group, as well as the group they represent (formally or subjectively). The group they represent influences their behavior.

Who are those subjective groups who serve as our reference groups beyond those we formally represent? They are all those past and present groups of which we have been a part, where we "rubber-band" or flash back to a former incident or relationship or moment and respond in the present as if we were still in that other situation. Influences from our families, an early school experience, our racial and ethnic heritage, the events of our lives, even areas of activity continue to influence us in present groups. "Big sisters" may come into groups constantly giving orders and trying to control others, all the while feeling like martyrs who aren't appreciated. Even some Catholics, who have not been in church for years, may blanch and try to avoid being involved in a discussion of abortion. All of those groups from our past are there within us. Some are readily identifiable in color, sex, language, and dress, and others are revealed as a discussion or topic collides and triggers one of those other groups. These groups may be classified as abstracted groups, hangover groups, and fantasied groups.

READER ACTIVITY

Listed below are some of your hidden reference groups. As you think about it, how do they influence you in groups?

A. Birth order in family: which were you?

________ Oldest ________ Middle ________ Youngest

How was that for you (who were you expected to be, contrasted with what you wanted to be)? ________________

____

How does that influence you in groups today? ________________

____

B. Sex

________ Male ________ Female

How were you expected to be in the family? How were you supposed to act? ________________

____

How does that influence you now in groups? ________________

C. School

If you phrased what you learned in school as a motto (or proverb) what would it be? ________________

____

How does that influence you in groups now? ______________________

____

D. What is one activity that you do now that you're good at, or that makes you special? (for example, marathon runner, gourmet cook, expert on Robert's Rules)? ______________________

____

How do you let people in a group know that? ______________________

____

How does it influence your behavior in a group? ______________________

____

The Abstracted Group

Here people are influenced by groups from their past, although that influence is not specifically remembered. They might have found that a stern, "no nonsense" teacher accomplished the most in a classroom; they might have worked for a direct, authoritarian employer whom they found to be very efficient. They may be influenced by these dimly remembered models, and feel that the ultimate goal of all groups is to function in this way. Their behaviors are influenced by this abstracted group.

Difficulties emerge when values abstracted from these forgotten groups come into conflict with the values of the current group. It is then that there needs to be a reexamination—how was the situation then similar to the situation at present; how is the situation now different from that earlier one? What were the objectives then; what are they now? How was the group composed then? Who is in it now?

The Hangover Group

Unresolved membership anxieties and problems in important reference groups may be continuously dealt with in other groups. For example, the child who was the "little brother" in his family—the last to be heard, the most frequently disregarded, the one allowed almost no responsibility because he was considered "young" even at voting age—may in other groups strive continually for leadership as proof of his competence. Or, conversely, he may be very antileadership, rebelling against any authority in retaliation for all of the times he could not "talk back" as a child.

Some feel that many of the problems of leadership are not legitimate problems of the actual group, but are unresolved "hangovers" from previous groups—unresolved family relationships that influence behavior in the present group. Or, consider this example of another kind of hangover group:

A young woman from a workingclass family bought into the great American dream—she could become anything she wanted if she applied herself and worked. She had. She had judiciously selected models to learn how to study and do well at school; she had scrutinized well-dressed classmates and fashion magazines to learn how to have a "classy" image; she had learned how to wear her hair and use make-up. She had learned, like Eliza Doolittle, how to make "small talk" and how to express her ideas. She had been accepted into a prestigious college.

Only a couple of months into her freshman year, she was being bombarded by sorority invitations. She had what they wanted. She was flattered by the attention, and wanted desperately to be accepted as a member by the most elite sorority. They continued to "rush" her to the end; they liked her, and she was ecstatic about being one of them. As a final stage in the process, she filled in standard forms asking routine information on her background and agreement to be responsible for the financial obligations of membership if accepted. She wasn't accepted. She has listed her father's occupation honestly; he was a bartender. Because of her social class, she was dropped immediately.

Consider who was the reference group of the committee who voted to drop her. (In some groups one will never get accepted, because their decisions are based on reference to "some other group.")

Decisions are not made in the present, based on present circumstances. Rather, old reference groups continue to dominate the discussion and decision. It might be that the committee felt that once they knew her father was a bartender she suddenly was not "our type," or that her sophistication was a thin veneer and that she was not really "one of us". Thus, a hangover group can influence the behavior of members consciously or unconsciously.

The Fantasied Group

This is the kind of group that may give a person the emotional support that he or she definitely needs but is not receiving in the present actual group.

A person who has read some legal materials on an issue may fantasize himself or herself presenting the data not as a lay member to fellow members but logically and convincingly to a panel of three judges. Consultants who may know nothing about an organization frequently come in and fantasize that they are the ultimate authorities on any problems of organization, because they have taught organizational theory in their classes for years. With just a few minutes of background, they handle a serious problem with the aplomb of a professor answering a student question. Other consultants fantasize themselves speaking with graduate students rather than with clients; they theorize or discuss situations in ivory-tower terms rather than discuss the practical realities of the situation.

When people do not accept and address their behavior to the actual group, they will use some other group or mixture of groups, or even a constructed group, to meet their needs for emotional support. Some people, when gently reminded or prodded, may recognize that they lapse and that they are indeed addressing their behavior to groups other than the actual persons present; to others these groups remain hidden in their subconscious.

In summary, it is evident that there are actual groups that influence us and to which we address our behavior. However, when there is a perplexing situation in the actual group, it is as if we are caught in the web of overlapping memberships. Which membership (real, past, abstracted, fantasied) should guide our behavior? Which one seems the one we should use in this instance? Memberships that overlap for us at a particular time are the ones that have special salience for us in that situation. Those cues in the present situation that remind us of another, so that we respond in that way, give us clues about membership roles in which we are anxious.

It even seems that there is a hierarchy of reference groups, since whenever a decision is reached in an overlapping or conflicted membership situation, membership in some groups is enhanced at the expense of other memberships.

FACTORS INCREASING ATTRACTIVENESS OF MEMBERSHIP

The life blood of a group are its members; they are the resources through which accomplishment occurs. The satisfaction of those members, the degree to which they feel accepted, the degree to which they want to return are critical to the survival of the group. Recognizing this importance, one of the objectives of a group is to create cohesiveness. Cohesion is defined as the attraction of the group for its members; the greater the attractiveness, the higher the cohesion.

Excerpts from an older graduate student's paper on her experience in a group may serve as an illustration of the growth of cohesivensss in a group:

> My first encounter with the group made me feel uncomfortable. I realized that I was older than most of the group members, and felt that I did not have too much in common with them. Because of my previous life experiences, I felt I had already dealt with conflict, decision making, and not being a part of the "in group" more frequently than most of my peers in the class. I kept in mind, first and foremost, my purpose for being in the group—it was to successfully complete the course requirement of being in a group, nothing else. . . .
>
> After five weeks of class, I felt much more comfortable in the situation. It was a small group (five of us) and I felt much more at ease. The group was quite mature, and it was apparent that different individual interests were not going to interfere with accomplishing a group project. Surprisingly each group member became totally involved in the goal of the group. A special moment occurred in our third group session. While brainstorming for an idea for a group project, I was aware that each member had input that was recognized and discussed. The atmosphere was friendly, informal, and free. Each idea was discussed pro and con about the feasibility of undertaking the project. The decision to create our special "olympics" in class was mutually agreed to by the group. . . .
>
> My relationship with the group grew from one of skepticism in the beginning to total involvement with the group in a very short time. The skepticism was the result of fear of the unknown. Was I going to be accepted by my peers in the group? Quite early in our group relations it became apparent to me that most of the group had the same fears. It also became apparent quite early that the people in the group were warm, friendly people, although initially they didn't look that way. I found some group members more appealing to me than others, but I can say with pride that I know each member and can work with them. I also believe each of us would also say it—not one person in our group was absent all term, and several times when class ended we were the last to leave. Who would think that I would ever say I love that group; it has been one of the best experiences of my life.

Generally, we know that the attractiveness of a group can be increased if members (or potential members) are aware that they can fulfill their needs by belonging to that group. Since it is difficult to change the members' needs, it is a more feasible approach to emphasize the properties that meet members' needs or the gains to be derived from belonging. Some of the properties that increase attractiveness are as follows:

Prestige The more prestige a person has within a group, or the more that appears to be obtainable, the more he or she will be attracted to the group (Kelley, 1951; Aronson and Linder, 1965). People who are placed in a position of authority over others are more attracted to the group than those low in authority. This is especially true of those in authority who expect to remain in that position. For example, the group or organization will be attractive to a principal who is appointed "for life," army officers who may be promoted but are rarely demoted, or chief executives who are appointed and remain in that position until a bigger executive position is available.

However, those in a position of high authority who may be demoted to one of low authority are attracted to a lesser degree. Those of low authority who expect to remain in that position are not attracted to the group. The shipping clerk, the member of the telephone squad, and the "envelope licker" are not attracted. Yet those of low authority who envision being moved up can also find the group attractive; for example telephone squad members who see themselves as potential officers or committee chairpersons.

Those most attracted of all are members of high prestige who see themselves remaining in that position or those of low prestige who perceive themselves as rising in the group. High authority members who can be demoted (the president after a year of office) and low authority members who cannot rise (frequently women in a business enterprise) are not attracted to membership. In addition, persons who are valued members are more likely to be attracted to a group than those who do not have much social worth (Jackson, 1959; Snoek, 1959; Lott and Lott, 1969).

Imagine a questionnaire being passed around to a group in which members are asked to rank members on who is most important to the group. Suppose also that a questionnaire is distributed in which each member is asked how attracted he or she is on a scale from 0 (not attracted) to 10 (highly attracted). It is likely that when the data are tabulated and analyzed, the results will indicate that those seen as most valued or most important in the group will be those who are most attracted to it. Thus, when we feel that our ideas are listened to and acted upon, we are more attracted to the group.

Group Climate A cooperative relationship is more attractive than one that is competitive (Deutsch, 1959; 1967). If a group works together as a team to develop a product (or outcome), and if it will be evaluated on the basis of a team effort, the members will be more friendly toward one another than in a competitive situation. However, when members are rated on the basis of individual performance, there are fewer interpersonal relationships, more withholding of information or not volunteering information, and fewer influence attempts.

The Deutsch cooperation studies are classic; more recently, Worchel (Worchel et al., 1977) directly tested the hypothesis that cooperation leads to increased attractiveness of members for the group. In the first phase of the study, groups were led to believe they were either competing, cooperating, or having no interaction. In this phase, competition led to the least group interaction. (In competition, secretiveness is a central strategy.) In the second phase, two groups were combined and worked cooperatively on two tasks. They received feedback that their combined effort had either succeeded (success condition) or failed (failure condition). Following the feedback, members of the groups were asked how attractive the group was to them and whether they would like to continue to work with this group. The findings were significant for understanding cohesiveness. Among the groups that had previously competed, those who were successful found the group more attractive, and those who had been part of the "failure" groups reported decreased attraction to the group. For groups who had previously cooperated, both success and failure increased intergroup attraction.

Degree of Interaction Among Members Increased interaction among members may increase the attractiveness of the group (Homans, 1950; Good, 1971). Participating in the give and take with members, getting to know some of the others, making some good friends—these by-products of membership make the group more attractive. Being a member increases the contact with people who are liked, offers an opportunity to know them better, and especially, allows for real chances for clarification in influencing and being influenced. While pleasing interaction increases attractiveness of membership, if the interaction is unpleasant (if members disregard each other, bore each other, or if there are members who are considered repulsive), attraction to membership will be decreased (Festinger, 1957; Aronson, 1970; Amir, 1969).

Size As most of us have experienced (Seashore, 1954), the size of a group greatly influences our attraction to it. Smaller groups are likely to be more attractive than large ones (Wicker, 1969). In a small group it is easier to get to know the other members, to discover similar interests, to have dedication to the cause, and to have a sense of being a significant participant in the group. As the group increases in membership there is a corresponding heterogeneity of interests. Feelings toward each other become less personal, concern with the "cause" is often less intense, and there is a reduction in the degree of individual participation, intimacy, and involvement (Tsouderos, 1955).

Relationship with Other Groups Relationships with other groups are also a factor in attractiveness. Groups are more attractive if their position is improved with respect to other groups. A group that had been all but disregarded in its efforts to create parent awareness of alienated adolescents suddenly became valued when the Governor invited them to participate with him at a televised news conference. Not only was the group deemed more prestigious, but membership increased in attractiveness for the individual members.

Success The maxim that nothing succeeds like success applies to groups also. Members are more inclined to join groups or continue in groups that are successful (Jackson, 1959; Shelley, 1954; Martens and Peterson, 1971; Nixon, 1976).

When a sports team, in competition with other teams, wins more games, membership on that team is more attractive. One need only briefly watch the pandemonium of the winning team after a close game to understand what winning does to enhance membership. And after a big game, the hugging, champagne dousing, the embracing and dancing are all signs of being part of the winners. Not only a place to be, but a place to be next year too.

It isn't only sports teams who enjoy winning. A charity ball that was successful and raised the money anticipated has little difficulty retaining its members to "do even better" next year. They are able to recount how they were a terrific team who pulled it off against great odds. As they congratulate each other on how well they did and what each contributed to the success, they decide again that they will continue to work for the organization; the cause is even more important than it ever was.

Because they had been successful, they agreed that they were winners. They then become more satisfied with other members and with the task; they know they have the resources in members and the skills to be successful, and consequently are even more attracted to the group. They trust the group to the extent of being open to the group's solution to a problem rather than their own solution (Shelley, 1954).

Even if, in an experimental condition, members are told that their unit is *potentially* successful, the group takes on new prestige and members are more attracted to it.

It is interesting that if a person desires membership and it is difficult to obtain, he or she will value the membership more than if it were easy (Aronson and Mills, 1959). As if to reduce an internal dissonance he or she says, "This membership had to be worth it for me to go through such an ordeal—of course it's worth it—it's a great group—I'm lucky to be a member" (Festinger, 1957).

There is no evidence, however, to indicate that the same situation prevails if a person does not desire membership. Then the difficulty of the ordeal simply becomes another reason the membership is unattractive.

Fear Under conditions in which subjects (in experimental conditions) were in a situation of fear, they preferred to be with others, and the opportunity to be with others rather than alone increased the attractiveness of the group. When it was appropriate, subjects were more likely to choose to be with the others who were also anxious (or fearful) about the same thing, thus providing an opportunity for "social comparison" of their feelings (Smith et al., 1972; Buck and Parke, 1972; Dutton and Aron, 1974; Morris, 1976).

Children in border settlements in Israel report rushing to the bomb shelters at the first sound of attack, and being comforted by seeing the familiar faces of adults and other children at the entrance. They already feel less frightened; being with the others feels good. Londoners report similar experiences in the underground during the bombings of World War II.

And it isn't just wars and bombings. The research evidence indicates that subjects aroused by hunger, sexual stimuli, danger, or negative evaluations, seek out others to be with.

In a recent discussion on crowding and overpopulation, it was hypothesized that those who formed into groups would experience less crowding and stress in a high-density residential condition where there would be increased regulation and control of social interactions; being in a group would be a desired method for reducing stress due to overcrowding (Baum, Harpin, and Valins, 1975).

FACTORS DECREASING ATTRACTIVENESS OF MEMBERSHIP

When does a group become less cohesive? When is membership less attractive so that people prefer to leave? Members will leave when the reasons for their initial attraction no longer exist (the people they joined to be with have left); when their own needs or satisfactions are reduced (a member of the scout parents committee until a child leaves scouts); or when the group becomes less suitable as a means for satisfying existing needs (a young lawyer joins a political party committee and later is appointed to the District Attorney's staff; she resigns from the party committee because such membership is frowned upon). Members may also leave when the group acquires unpleasant properties, such as a diminished reputation, constant fighting among members, or an activist stand with which they disagree.

It is possible to be less and less attracted to the group, but not leave. A group can retain its members when attraction is at the zero or near zero level. In such situations, the group is inactive and provides little influence over its members; the members in turn provide little internal support for each other or to the organization. Members in this category—in conceptual terms, borderline members—are pushed over (and out) when the precarious balance is disturbed—as when the meeting time is changed or dues are raised even a small amount.

Research findings indicate that groups can lose their attractiveness for several reasons:

1. A group disagrees on how to solve a group problem. Some will walk away from the discussion, or not attend the meetings at which such a problem is on the agenda; others withdraw by working on private problems or become "turned off." Members may sense real personal frustration in such instances and the group is viewed as a source for precipitating feelings of personal inadequacy and impotency.
2. If the group make unreasonable or excessive demands on people, or people feel inadequate in the group situation, the group is less attractive and they will leave (Horowitz et al., 1953). If people are assigned a job that is too difficult for them (like arranging a program) or if people feel inadequate in the group situation (which requires their giving verbal reports at meetings when they feel inadequate as speakers), the group will be less attractive and they will leave, often without stating or being fully aware of their reasons.
3. Groups that have members who are too dominating or who have other unpleasant behaviors reduce attractiveness of the group.
4. Staff conferences in which there was a high degree of self-oriented behavior were viewed as less attractive to the staff (Fouriezos et al., 1950). Members who dominate the discussion and severely limit the opportunities for participation by others reduce attractiveness.

5. Some memberships may limit the satisfactions a person can receive from activities outside the group. For example, women in some religious groups are not permitted to drink, dance, wear make-up, or wear short skirts. Membership in such groups clearly limits satisfactions which might be derived from going to dances, or being "stylish." Police officers on rotating weekly shifts or nurses on night duty are also limited in their outside activities.
6. Negative evaluation placed upon membership in a group by people outside the group (which gives the group low status) also reduces attractiveness of the group. Teenage boys who enjoy scouting and become Eagle Scouts rarely tell their schoolmates of their scouting activities, nor even consider wearing their uniforms to school during Boy Scout Week because of the derisive comments anticipated. Being a member of the school discipline committee is not a sought-after appointment because of the reactions of peers to members of such committees.
7. Competition among groups also reduces attractiveness unless people have reason to believe they will be with the "winners."
8. People will leave one group to join another if the second is better able to meet their needs or if they have limited time for participation. For example, members may belong to an organization and then move to another part of the same city. They may join a branch of the organization closer to their new home. Or they may instead join a similar, but different, organization since they are moving and planning new relationships.
9. If individuals are "scapegoated", if they are blamed for negative events, they come to view those who attribute responsibility for failure to them less favorably. They come to see them as less competent and being less cooperative. They identify less with the group; and come to feel themselves outsiders. The psychological process soon leads to the actual process of separation (Shaw and Breed, 1970).

ATTRACTIVENESS OF MEMBERSHIP AND GROUP SUCCESS

What difference does cohesiveness make? How a group functions does depend on how attractive it is to its members. This will be reflected in the energy members expend on reaching their goal, how easily they attain it, and how satisfying the outcome will be. There is evidence to suggest that if people are attracted to membership, they are more likely to accept the responsibilities of membership (Dion, Miller, and Magnan, 1970). They are likely to attend meetings more regularly (Sagi, 1955), they are more apt to participate readily in meetings (Back, 1951), and they are more likely to accept responsibilities in the organization (Larson, 1953). In addition, they will be more persevering in working toward long or difficult goals (Horowitz et al., 1953), and are likely to work harder regardless of outside supervision. Cohesive groups will be especially productive if they are also motivated to do the task well (Hall, 1971; Landers and Crum, 1971).

If the group is attractive, members are more open to interpersonal influence. It has been found that when members are attracted to the group, they are more willing to listen to others, they are more flexible in accepting other opinions (Rasmussen and Zander, 1954), and they will attempt to influence others more (Schachter, 1951). It is especially worthy of note that attracted members will change their minds more often to take the view of fellow members.

In conclusion, membership issues are important to each of us in our experience in groups, and are one of the critical components of groups' existence and effectiveness. Increased understanding of membership factors can help us be more aware of them in groups, and thereby increase our opportunities and abilities to be more effective. It also increases the possibilities for the successes of the groups with which we are concerned.

EXERCISE 1 Multiple Memberships: Representative Group—Member Role

Objectives

To experience the conflicts of multiple memberships

To understand the conflicts of representative memberships in a familiar community situation

Setting

A table with five chairs is placed in the center of the room.

The facilitator asks for five volunteers to play various roles in a discussion of sex education.

The roles assigned are:

A parent, a representative of the P.T.A.

A conservative businessperson, a representative of the Chamber of Commerce

A middle-aged lawyer, a representative of the Citizens' Association

The principal of the high school, a representative of the faculty
The executive director, representing the Mental Health Association

Situation

The setting is a suburban community in the Midwest; there are 30,000 inhabitants. The community is conservative and incomes are above the national median. The facilitator describes the situation as follows:

An anonymous benefactor feels there is a need for sex education in this community. Such education has never been part of the budget; the benefactor will donate $25,000 which would be sufficient for an excellent program—the community will determine the program. He would like to see such a program instituted this year, but there is a stipulation that we must let him know our decision at the end of the month. If we agree to the program, we can have the funds immediately. If not, the benefactor has made plans to use the funds for a project in another community.

During the past month we have brought the situation to the attention of groups within the community. Opinion has been divided. Your groups are the major ones involved. We must make our decision promptly because today is the last of the month. You have been appointed by your groups to help arrive at the final decision; the other groups will follow your decision.

The following developments have taken place during the month: the P.T.A. has not made a commitment due to disagreement within the organization; the Chamber of Commerce was opposed; the Citizens Association unfortunately had their meeting canceled due to the elections; the faculty want to represent the wishes of the whole community although they themselves feel it is a desirable project and a special opportunity; the Mental Health Association strongly supports the project and might have been influential in finding the benefactor.

The role players are told to think about their roles and how they will act in the situation as a representative of their organizations. If possible, each player should have a "coach" with whom he or she can practice his or her role. The coach may help the individual to magnify certain aspects or modify others. The coaches would take about five minutes with role players.

The facilitator announces, "The meeting will begin with the representative of the Mental Health Association as chairperson."

Action

The meeting begins. Nonparticipants watch the role playing, the discussion, and the decision.

Discussion

To the nonparticipants:

At the beginning, how accurately did the representatives speak for their organizations?

Who changed from his or her organizational position? How?

What conflicts in membership did you see? For whom were they greater?

Who remained unchanged? How could you interpret this? (Observers might be asked to look for egocentric behaviors, for example, deflating others, sarcasm, building up of self, defensive replies, withdrawing or nonparticipating behaviors.)

To the role players:

What conflicts did you feel? When? What influenced your decision?

EXERCISE 2 Conflicting Memberships

Multiple memberships are situations in which one person has membership in a number of groups simultaneously. These may or may not be in conflict. In conflicting memberships or conflicting roles, there is conflict within the person as to which membership or role should be a determinant of his or her behavior at that moment.

The conflicting memberships can be experienced readily in a number of situations. Take, for example, the family situation wherein a woman plays a dual role of wife and mother.

Setting

The facilitator asks for volunteers to be a family,—a mother, father, oldest child, middle child, youngest child. The facilitator assigns each volunteer a role, and explains the situation. A number of situations will be presented. For example, one situation might begin with the wife-mother saying: "My husband and children pull at me constantly. My husband wants me to go downtown with him, and my children want me to be home when they come out of school." (The issue of conflicting memberships—her two roles—is evident.)

The situation is staged with a woman in the center with her arms stretched out at shoulder height. On one side her "husband" will be pulling her in his direction. On the other side her children will be lined up pulling in their direction. The youngest might be pulling the mother's knees or the bottom of her skirt. The middle child might be pushing the youngest away and pulling the mother at the waist, the oldest might be pulling the mother's shoulder or arm in his or her direction. Once the family knows its roles, the facilitator says, "Act out." The father pulls and entreats, "Come on, honey." The youngest pulls and may say, "I want my mommy," the middle one may pull with determination silently, the oldest may plead, "Please, listen to me." Each is pulling, and the mother is feeling pulled and swayed and harassed and inadequate. She may simply be moaning, or she may say, "Stop pulling so hard, you're hurting me. Please stop, I feel as though I'm caught in a vise." (In acting out the situation, participants are encouraged to actually pull, indicate their feelings nonverbally, or with just a few words.) After a few moments of this, the facilitator calls, "Stop." All relax their holds.

Discussion

Then the "family" is asked: How did it feel? How did you like your role? What were the relationships with the other members of your "family"?

Some questions are also put to the audience:

What did you notice?

How did the central person feel?

What behaviors might have helped reduce the conflicts?

What could have been done so that members coud be accepted or respected?

There are other possibilities for acting out role conflict in a family situation.

The "husband" might say: "My wife, my children are strangling me—I can't move, they're clutching me so tightly. I want to get ahead, but how can I if I have to be home every night for dinner at six, if I have to spend every free minute playing with the children?"

One of the children might say: "I'm being pulled in all directions—my parents want me to be one way, my boyfriend wants me to go another way, my best friend wants me to do other things."

After several possibilities are decided upon, the "family" may reenact these new situations, observe the effects, and then both actors and audience discuss the same questions posed the first time.

EXERCISE 3 Reference Groups: "Who Am I?" Composition

Objectives

An understanding of reference groups

An understanding that reference groups, groups to which a person can relate, are subjectively determined

An understanding that groups that may be referent groups for some are not referent groups for others

Setting

Members are divided into groups (four to six per group). Each member should have a pencil and paper.

Action

The facilitator asks each person to write a 100-word composition explaining who he or she is so that "people will know what you're like." The compositions are written, and each person reads his or hers to the group. Members are then asked to determine from the composition who that person's reference groups are.

Discussion

How were reference groups different among members?

What difference does it make to the person?

What will be the effect in the group, or in other groups?

EXERCISE 4 Increasing Attractiveness of Membership

Objectives

To understand that cooperative rather than competitive relationships can increase attractiveness of the group

To understand that increased interaction increases attractiveness of the group

To understand that interdependence can increase attractiveness of the group

Setting

Participants are divided into tens (approximately) and seated five at a table. Each participant is given a piece of paper. At one table each person is given a crayon. At the other table the entire group is given one crayon. The members at the second table are instructed that they can hold the crayon only 20 seconds and then must pass it to the next person. The groups are instructed to "draw something from your life which is characteristic of you" or "draw something characteristic of you in the group." The facilitator may call out at the end of each 20-second period or have a timekeeper at the table do it.

Action

The groups draw as instructed. In one group each person works with his or her crayon on the drawing; in the other, members pass a crayon to the next person every 20 seconds.

Discussion

How did you feel about the process?

How do you feel about your product?

How do you feel about your group? Talk about the experience.

Variation

This can be done as a group drawing. Each group is given one piece of paper. In one case each member is given a crayon; in the other group, one crayon is to be rotated. This variation heightens interdependence and cooperative relationships in the passed-crayon group.

The exercise is as described through the action phase. Then, before the discussion, individual evaluation sheets are distributed. These evaluation sheets permit those who worked individually to compare their evaluations with those who worked cooperatively. Some questions in the evaluation sheet might be:

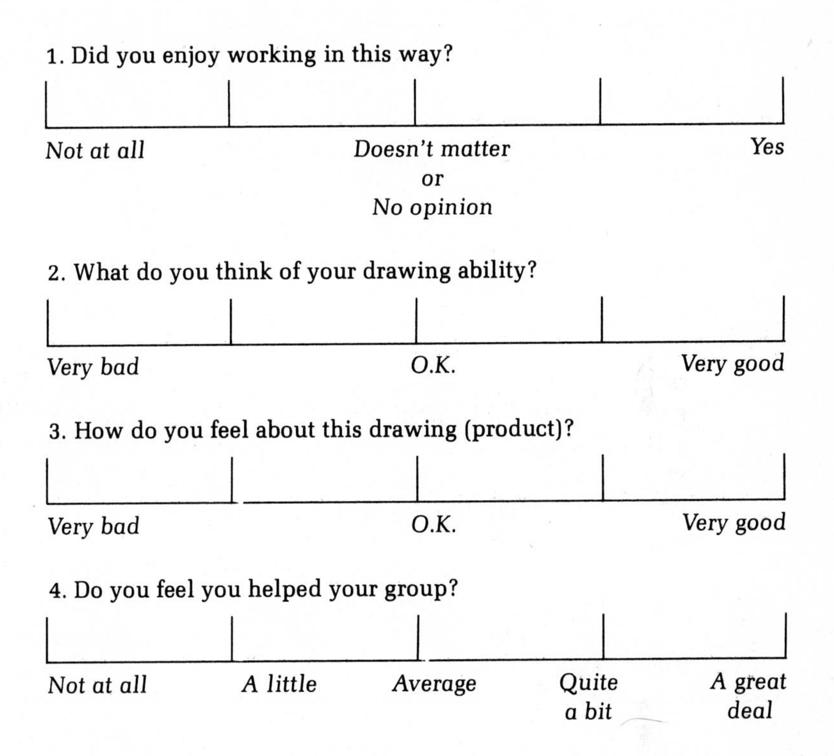

Sheets should be identified either by the color of the paper (white to groups who worked individually—colored to cooperative groups) or a coding designation (I, individual; II, cooperative). Rating sheets are frequently utilized to obtain opinions along a range. Sometimes they are numbered (1–5) and results tallied with presentation of the lowest, highest, and median scores for each question. Data should be tallied, comparisons made by questions, and results reported.

Discussion

See objectives as a basis for discussion.

EXERCISE 5 Factors Influencing Attractiveness of Membership

Objectives

To understand personal sources of attraction in a group

To become aware of and experience the problems of having new members in a group

To become aware that changed membership in a group influences not only relationships with new members but also the relationship of older members with one another

To understand that changed membership influences sources of attraction to the group

To understand that increasing the size of a group has advantages as well as problems

Action

The facilitator says, "Select a partner. Get acquainted. Be aware of how you feel in this pair." Pairs talk for 15 minutes. While they are talking, the facilitator assigns each pair a number. After 15 minutes he tells them to stop talking and asks, "How do you feel about this dyad? Would you like to continue in this group? *Think* about your answer." Then he says, "Would groups 2, 5, 8, 11, etc. [every third group] split? One member go to the group on one side of you; the other go to the group on the other side of you. There will now be a series of triads or three-person groups. Get to know one another."

Discussion

One person comes alone and must develop a whole new set of relationships. How does he or she feel? How is he or she treated?

Others were in a pair that was or was not satisfying. How do they feel toward one another with the newcomer? How do they respond to the newcomer?

How do members of the triad feel in comparison to how they felt in the dyad?

Variations

Discussion can be within triads rather than with the whole group, or two triads combined.

There can be progressions in the size of groups. Triads 2, 5, 8, 11, etc. disband. One person goes to the group at the left, the pair goes to the group at the right. The feelings of one person going into an existing group (newcomer joins trio) are examined; the same is done with a pair going into an existing group (pair joins trio).

EXERCISE 6 Increasing the Attractiveness of a Group: High Talkers—Low Talkers

Objectives

To present an opportunity for "high talkers" and "low talkers" to develop empathy for the other

To recognize the limitations of either behavior

To present low talkers an opportunity to recognize that their situation is not unique

To afford an opportunity for high talkers to practice listening and for low talkers to speak

To experience how group membership can be determined by certain behavioral patterns

Situation

This exercise is appropriate after the groups have been working together for some time and usual behavior patterns are known by the members. The facilitator asks each person to categorize himself or herself as a high talker or a low talker. (This can be written down, and there can be a perception check among the members.)

Action

A fishbowl is set up with the low talkers sitting in the center, and the high talkers standing around the outside observing. The low talkers discuss the problems of being a low talker. (8 minutes)

Then the groups switch. The high talkers go to the center and discuss why they talk a great deal, and the problems involved. (8 minutes)

Both groups come together and talk about their feelings and observations.

Variation

Instead of action in a fishbowl setting, action occurs in mixed quarters, that is, two low talkers and two high talkers in a group. In each, there is a discussion of the problems of being a low or high talker. (20 minutes)

The entire group comes together and discusses their feelings and observations.

EXERCISE 7 Conflicts of Multiple Memberships — Alter Ego Exercise

Objectives

To understand the conflicts of multiple memberships and the influences of the various memberships on behavior in a given situation

Setting

The exercise is a role-playing one. Five chairs are placed around a table, and behind each there are two or three others. The chair at the table represents the person; the chairs behind the person represent other memberships who speak out to try to influence him or her. The person then speaks in terms of hearing some influences and ignoring others. Fifteen volunteers are selected. There will be five participants, and two alter egos behind each participant (or more as the situation permits). Participants and alter egos should practice privately and build their cases.

The facilitator should establish a situation for discussion that involves counter influences on a person. An example might be a committee meeting to determine a new high school dress code. Members of the committee could be:

President of the Student Body (Alter Egos)	1. Peer representative—students want a more liberal dress code—jeans, no socks, and so forth.
	2. Football team member—coach thinks athletes should wear hair short; if longer hair passes, it will look more out of place.
Principal of School (Alter Egos)	1. Faculty representative—students who dress for school more likely to value schooling.
	2. Adviser of students—want to be popular; will this reduce effectiveness on other issues?
President of P.T.A. (Alter Egos)	1. Representative of parents—conservative, would prefer dresses, slacks, not jeans; they are for hay rides or working on lawn; miniskirts indecent.

2. Member of a parents' discussion group—parents must be open to new ideas, not impose their values on children if they don't make sense to children; each generation has its clothing fads—remember yours?

President of Eleventh Grade Class (Alter Egos)

1. Representative of class—better get a more liberal code—as liberal as possible, or students will think I have done nothing this year, and I won't get re-elected.
2. College aspirant—I need a good recommendation for college from the principal; my scores and grades aren't as good as they might be, and a favorable letter would help. If I antagonize him, it won't be so good.

Faculty Member (Alter Egos)

1. Faculty member—I'm pretty liberal about what students can wear, but there are some who always want to go further. If you say skirts are allowed two inches above the knees, they go five; if you say button shirts, they wear tee shirts. To preserve any semblance of dress you have to have a more severe code, but enforce it liberally.
2. Parent—kids have so many problems these days—draft, college competition, drugs, authority problems—let them have some feelings of power, of making their own decisions. What difference does it really make what kind of clothes they wear?

Action

The meeting takes place, and the alter egos speak as appropriate. The meeting goes on until a decision is made or it is evident that a decision is impossible.

Discussion

How did multiple memberships affect members' behavior?

Which memberships were influential? Why?

How do multiple memberships influence the productivity of a group? Its efficiency?

Variation

The setting and the action are the same as described previously, except that the facilitator notifies the alter egos that they may not speak the first five minutes of the discussion. At the end of the first time period, the alter egos speak to their counterparts and continue to do so for the remainder of the discussion. For this variation, observers can be appointed and asked to note changes in the participants' behavior when there are not alter egos and later when there are.

The participants report their feelings with and without alter egos and tell who they felt influenced their behavior. Reports are then taken from observers, and a discussion as in previous section follows.

BIBLIOGRAPHY

Amir, Y. "The effectiveness of the kibbutz-born soldier in the Israel defense forces." *Human Relations*, 22, No. 4 (1969), 333–344.

Aronson, E. and J. Mills. "The effect of severity of initiation on liking for a group." *Journal of Abnormal and Social Psychology*, 59, (1959), 177–181.

Aronson, E. and D. Linder. "Gain and loss of esteem as determinants of interpersonal attractiveness." *Journal of Experimental and Social Psychology*, 1, No. 2 (1965), 156–171.

Aronson, E. "Who likes whom and why." *Psychology Today*, 4, No. 3 (1970), 48–50, 74.

Asch, S. E. "Effects of group pressure upon the modification and distortion of judgments." In D. Cartwright and A. Zander, eds., *Group dynamics*. 2nd ed. Evanston, Ill.: Row, Peterson, 1960, 189–200.

Back K. "Influence through social communication." *Journal of Abnormal and Social Psychology*, 46, (1951), 9–23.

Back, K. W., S. R. Wilson, M. D. Bogdonoff, and W. G. Troyer. "Racial environment, cohesion, conformity, and stress." *Journal of Psychosomatic Research,* 13, No. 1 (1969), 27–36.

Baum, A., R. E. Harpin, and S. Valins. "The role of group phenomena in the experience of crowding." *Environment and Behavior,* 7, No. 2 (1975), 185–198.

Berne, Eric. *The Structure and Dynamics of Organizations and Groups.* New York: Grove Press, 1963.

Boszormenyi-Nagy, I. and G. Spark. *Invisible Loyalties: Reciprocity in Intergenerational Family Therapy.* N.Y.: Harper and Row, 1973.

Buck, Ross W. and Ross D. Parke. "Behavioral and physiological response to the presence of a friendly or neutral person in two types of stressful situations." *Journal of Personality and Social Psychology,* 24, No. 2 (November 1972), 143–153.

Deutsch, M., Y. Epstein, D. Canavan, and P. Gumpert. "Strategies of inducing cooperation: An experimental study." *Journal of Conflict Resolution,* 11, No. 3 (September, 1967), 345–360.

Deutsch, M. "Some factors affecting membership motivation and achievement motivation." *Human Relations,* 12, (1959), 81–85.

Dion, K. L., N. Miller, and M. Magnan. "Cohesiveness and social responsibility as determinants of group risk taking." *Proceedings of the annual convention, American Psychological Association,* 5, Part 1 (1970), 335–336.

Dutton, Donald G., and Arthur P. Aron. "Some evidence of heightened sexual attraction under conditions of high anxiety." *Journal of Personality and Social Psychology,* 30, No. 4 (1974), 510–517.

Feldman, R. A. "Group integration and intense interpersonal disliking." *Human Relations,* 22, No. 5 (1969), 405–413.

Festinger, L. "A theory of social comparison processes." *Human Relations,* 37 (1954), 184–200.

Festinger, L. *"A theory of cognitive dissonance.* Evanston, Ill.: Row, Peterson, 1957.

Festinger, L., and H. Kelley. *Changing attitudes through social contact.* Ann Arbor, Michigan: Research Center for Group Dynamics, 1951.

Fouriezos, N., M. Hutt, and H. Geutzkow. "Measurement of self-oriented needs in discussion groups." *Journal of Abnormal and Social Psychology,* 52 (1950), 296–300.

French, J. P. R., Jr. "The disruption and cohesion of groups." *Journal of Abnormal and Social Psychology,* 36 (1941), 361–377.

Gershenfeld, Matti K. "Leadership on Community Boards." Report for Federation of Jewish Agencies, Philadelphia, Pa., 1964.

Good, L. R., "Effects of intergroup and intragroup attitude similarity on perceived group attractiveness and cohesiveness." Dissertation Abstracts, 1971, 60–3, 3618–B.

Hall, J. "Decisions, decisons, decisions." *Psychology Today*, 5, No. 6 (November 1971), 51–54, 86–88.

Homans, G. *The human group*. New York: Harcourt, Brace, 1950.

Horowitz, M., R. Exline, M. Goldman, and R. Lee. "Motivation effects of alternative decision making process in groups." ONR Tech. Rep. 1953, Urbana, Ill.: University of Illinois, College of Education, Bureau of Educational Research.

Jackson, Jay M. "Reference group processes in a formal organization." *Sociometry*, 22 (1959), 307–327.

Jovick, R. L., "Cohesiveness-conformity relationship and conformity instrumentality." *Psychological Reports*, 30, No. 2 (April 1972), 404–406.

Kelley, H. H. "Communication in experimentally created hierarchies." *Human Relations*, 4 (1951), 39–56.

Kelley, H. H. "Two functions of reference groups." In G. E. Swanson, T. M. Newcomb, and E. L. Hartley, eds. *Readings in Social Psychology*. New York: Holt, Rinehart, and Winston, 1952, 410–414.

Klausner, Samuel. "Now if the government only used a community approach." *Pennsylvania Gazette* (December 1978), 10–11.

Klein, Donald C. *Community Dynamics and Mental Health*. New York: John Wiley and Sons, Inc., 1968, pp. 47–56.

Landers, D. M. and T. F. Crum. "The effects of team success and formal structure on interpersonal relations and spirit of baseball teams." *International Journal of Sport Psychology*, 2, No. 2 (1971), 88–96.

Lang, P. A. "Task group structuring, a technique: Comparison of the performances of groups led by trained versus nontrained facilitators." Dissertation Abstracts Int., 1977, 38 (6–A), 3186–3187.

Larson, C. "Participation in adult groups." Unpublished doctoral dissertation, University of Michigan, 1953.

Lott, A. J. and B. E. Lott. "Liked and disliked persons as reinforcing stimuli." *Journal of Personality and Social Psychology*, 11, No. 2 (1969), 129–137.

Martens, R. and J. A. Peterson. "Group cohesiveness as a determinant of success and members' satisfaction in team performance." *International Review of Sport Sociology*, 6 (1971), 49–61.

Mehrabian, A. and S. Ksionzky. "Some determinants of affiliation and conformity." *Psychological Reports*, 27, No. 1 (August 1970), 19–29.

Mirande, A. M. "Reference group theory and adolescent sexual behavior." *Journal of Marriage and the Family*, 30, No. 4 (1968), 572–577.

Morris, W. N. "Collective coping with stress: group reactions to fear, anxiety, and ambiguity." *Journal of Personality and Social Psychology*, 33, No. 6 (1976), 674–679.

Ninane, P. and F. E. Fiedler. "Member reactions to success and failure of task groups." *Human Relations*, 23, No. 1 (1970), 3–13.

Nixon, H. I. "Team orientations, interpersonal relations and team success." *Research Quarterly*, 47, No. 3 (October 1976), 429–435.

Quey, R. L. "Functions and dynamics of work groups." *American Psychologist*, 26, No. 10 (1971), 1081.

Rasmussen, G. and A. Zander. "Group membership and self-evaluation." *Human Relations*, 7 (1954), 239–251.

Rose, A. *Union solidarity*. Minneapolis: University of Minnesota Press, 1952.

Sagi, P., Olmstead, D., and F. Atlesk. "Predicting maintenance of membership in small groups." *Journal of Abnormal and Social Psychology*, 51 (1955), 308–331.

Schachter, S. "Deviation, rejection and communication." *Journal of Abnormal and Social Psychology*, 46 (1951), 190–207.

Schachter, S. *The psychology of affiliation*. Stanford, Calif.: Stanford University Press, 1959.

Schwartz, T. M., R. A. Eberle, and Donald R. Moscato. "Effects of awareness of individual group membership on group problem-solving under constrained communication." *Psychological Reports*, 33, No. 3 (1973), 823–827.

Seashore, S. *Group cohesiveness in the industrial work group*. Ann Arbor, Michigan: Institute for Social Research, 1954.

Shaw, M. E. and G. R. Breed. "Effects of attribution of responsibility for negative events on behavior in small groups." *Sociometry*, 33, No. 4 (1970), 382–393.

Shelley, H. P. "Level of aspiration phenomena in small groups." *Journal of Social Psychology*, XL (1954), 149–164.

Sherif, M. *The psychology of social norms.* New York: Harper and Row, 1936.

Smith, R. E., L. Smythe and D. Lien. "Inhibition of helping behavior by a similar or dissimilar nonreactive fellow bystander." *Journal of Personality and Social Psychology,* 23, No. 3 (1972), 414–419.

Snoek, J. D. Some effects of rejection upon attraction to the group. Unpublished doctoral dissertation, University of Michigan, 1959.

Stiles, D. B. "The significant other as a determinant of positive perceptions of group process experience." Doctoral dissertation abstract from *Dissertations in Education, Guidance and Counseling,* p. 51. University of Miami: 1973.

Stotland, E. "Determination of attraction to groups." *Journal of Social Psychology,* 49 (1959), 71–80.

Thelen, H. A. *Dynamics of groups at work.* Chicago: University of Chicago Press, 1954.

Tsouderos, J. "Organizational change in terms of a series of selected variables." *American Sociological Review,* 20 (1955), 207–210.

Turnbull, C. M. *The lonely African.* New York: Anchor Books Edition, 1963.

Wicker, A. W. "Size of church membership and members' support of church behavior settings." *Journal of Personality and Social Psychology,* 13, No. 3 (1969), 278–288.

Willerman, B. and L. Swanson. "Group prestige in voluntary organizations." *Human Relations,* 6 (1953), 57–77.

Worchel, S., V. V. Andreoli, and R. Folger. "Intergroup cooperation and intergroup attraction: The effect of previous interaction and outcome of combined effort." *Journal of Experimental Social Psychology,* 13, No. 2 (March 1977), 131–140.

Zander A. and A. Havelin. "Social comparison and intergroup attraction. *Human Relations,* 13 (1960), 21–32.

Three

Norms, Group Pressures, and Deviancy

How does the Supreme Court work? How do nine justices, presumably the most brilliant legal minds in the country, make decisions that can override a president and counter a Congress—and, simply in their decision to take or not take a case, determine which laws stand and which will be changed? How do these nine justices, wedded for life to one another, arrive at their decisions?

Bob Woodward, the Pulitzer Prize winning author of *The Final Days of the Nixon Administration*, and Scott Armstrong went about finding out how the Warren Burger court functioned between 1969 and 1976. In the course of their investigation, they interviewed more than 200 people, including several Supreme Court justices, over 170 of their law clerks, numerous employees, and assorted savants. They assembled 8 file drawers of thousands of pages of documents as they sought through interview, verification, and reverification from a further interview to piece together how the nine went about conducting their business. Their object was not to illuminate the problems of the Court or to press for better justices. Their object was to understand and report on the process[1] (and, as the subject had fascinated them, they hoped that it would also intrigue a curious nation and produce a best seller).

Information on the Supreme Court is sparse; Woodward is credited with a tenatious style that led to the resignation of a president. That combination of factors produced major reviews of *The Brethren*. One after another of the reviewers was incredulous. To quote from John Leonard's review in the New York Times[2],

> . . . the Supreme Court behaves like any other committee with which I've had any acquaintance. Its scruples are relative; its personalities clash; its many pairs of eyes are on the main chance, the good opinion of posterity, the boss, the clock and sometimes the Constitution. One imagines that, even in the Agora, Socrates was hustled . . . .
>
> An associate justice of the supreme court is allowed to be ordinary . . . . We are advised that Chief Justice Warren Burger is most ordinary. He delays voting until he can be in the majority or finagle the assignment of the writing of an opinion; that he is no stranger to tantrums . . . and that he holds a grudge.

Reviews continue, noting that one justice doesn't do his homework and delays decisions because he is ill-prepared; another, on the slightest provocation, launches into one of his favorite ideological sermons as others tune him out and impassively wait for him to finish. One justice was almost blind, and the group had to make special concessions in working with him; another, of high status as a liberal and a favorite of the press, was an "unpleasant man for whom and with whom to work."

One after another of the reviewers expressed shock that these men, esteemed for their fine logical minds and ability to think through enormously complicated legal issues, act like any other committee. It seems impossible that *they* are frustrated with a leader

[1] Bob Woodward and Scott Armstrong, *The Brethren. Inside the Supreme Court* (New York: Simon & Schuster, 1979).

[2] John Leonard, *New York Times Book Review*, December 16, 1979, p. 1.

who is not viewed as the most brilliant; that they subtly coerce and are coerced; that they know what they can and cannot expect of each of the others; and that their way of functioning is quite different from the Warren Court, which preceded them.

It is not only book reviewers who express their surprise. Despite the loftiness of the participants, the titles they hold, or the seriousness of the questions they decide, each group (committee, task force, or court) has its norms—its procedures for members' interactions with each other, its expectations for behavior, its processes for working and arriving at decisions. One of the fundamental properties of groups is that each group has norms that develop over the time of the group's interaction. Norms emerge from the participants who work together, at a given time, to accomplish a task. These norms emerge whether the groups are in high places or low ones. Norms are probably the most difficult group concept to convey and to understand. We somehow think that when *they* work together, that brilliance, that erudition produces some form of interaction that is of a rarified form, with which we can in no way identify. How can it be that *they* interact like other committees? How can it be?

This fundamental idea of group norms is a crucial one for understanding what is happening in a group. The prime reasons we don't "see" group norms is that we have never looked. We like to think that we act moment by moment, spontaneously and appropriately. We think any problems are somehow related to "difficult" personalities or some kind of unchangeable situation in which personalities as different as oil and water don't and won't mix. We think that we march to our own drummer and are not conformists to a group. Not true. For the most part, we are conformists to the group.

Unless we see the norms of that group—how members control each other, which behaviors are permitted and which are not—unless we see, we are victims. As an unseeing participant in a group, we can then only be swept with the tide, and have very limited options on how to gain acceptance or how to work for the outcomes we want. It is literally (actually, visually) the difference between looking at a group and seeing a group. The awareness will make that much of a difference. Even those who are highly trained psychologists can miss seeing the group, as in the following example:

A talented, well-trained Ph.D. clinical psychologist became a member of a university department that had a strong group focus. As part of his familiarization with the departmental group emphasis, he participated in a group led by another member of the faculty. He emerged from the experience shaken. In response to a casual, "How was it?" he raged, "She saw everything, everything. She is a witch;

that's the only way anybody can see all that. She saw all these things happening; she predicted what would happen—and it did. It was one of the scariest experiences of my life. There are only two explanations possible. Either I am totally uneducated, and I sure felt that way, or—the more plausible one—she is a witch."

That story is told with a special verve today. Because one year later, he ran that same group (course). "Today," he boasts, "I am a witch, a seer. After looking and becoming aware of the group in its interactions, understanding its norms and its environment, I can *see*. I can predict, and amaze people who are probably thinking, "How does he do it? Either I am terribly uneducated or he is a witch."

THE CONCEPT OF GROUP NORMS

Norms are the rules of behavior, the proper ways of acting in a group that have been accepted as legitimate by the members of that group. They are accepted as legitimate procedures of the group as a system as well as of each member within the system. Group norms regulate the performance of the group as an organized unit, keeping it on the course of its objectives.

Norms at the Individual Level

At an individual level, group norms are ideas in the minds of members about what should and should not be done by a specific member under certain specified circumstances (Homans, 1950; Hare, 1976). They are learned by members. Usually, norms provide one of the most important mechanisms of social control of behavior of individuals within society.

It is important to understand that norms are not only rules about behavior in the group but also ideas about the patterns of behavior. Rarely can the ideas be inferred directly from behavior, rather they must be learned. A fraternity member learns it is all right to complain about poorly conducted social events within the fraternity, but it is traitorous to suggest disappointment to outsiders because an important fraternity norm is image. A student in a group project may want to do a considerable amount of reading in preparation for a class presentation. However, he can see from the disapproving looks he receives from the others that it would be wiser to do a project that can be developed on class time so that other group members can do an equal amount of work. The group will now redefine its project to mean one that they can complete with only class time. Despite a speech by a dorm "counselor" that drugs are prohibited and any infraction of the rules will be reported to the police, the dorm member learns that if it is done in a room and not a lounge, and is not bragged about, the behavior is acceptable. In fact, it is the preferred mode of spending Friday evenings sociably. A scientific team member learns that attire is of little importance to other members. Dress can be as informal and casual as individually preferred. However, at monthly report meetings with the divisional managers, male members wear shirts, ties, and suits; women can continue in their usual dress style.

Thus, through these experiences, group members learn that the significance of an act is not the act itself, but the meaning the group gives to it. They learn that the meaning may change according to who performs the act and the circumstances under which it is performed. This experience results in what are called shared ideas among members about what should or should not be done by a specific member under certain circumstances. A shared idea means, in the previous examples, that an illusion of producing terrific social events must be maintained to the outside world; that while a class presentation should be well done, it should not occupy personal time; that drugs are permissible in that dorm despite the stern speech; that at monthly report meetings, men are expected to be part of the organization and dress accordingly in suits and ties, but that women need not conform to special dress expectations.

When the norms are expectations for the behavior of a particular person, these are called *role expectations.* For example, if someone asks who will take notes at a meeting, and all eyes turn to one member who has unofficially taken notes at several previous meetings, role expectations are readily visible.

READER ACTIVITY

Below are two activities to help you understand the concept of norms.

1. One way to understand norms is to understand the difference between the "new" you, and the "savvy" you. Select one of these situations:
   a. Being a freshman and being a senior.
   b. Being a new employee and being that employee a couple of years later.
   c. Being a member and being an officer in the same group. Using the situation you selected, list five things you found difficult (or were fearful of) as a new member.

____

____

As you became experienced in the group or organization, you learned what the real rules are. How do you handle these situations now?

For example, freshmen are very concerned with what to wear and often read magazines on the current college fashions; they are very apprehensive about looking too phony or too new and having the "right" look. Seniors ridicule reading such magazines; they know what to wear.

Consider dealing with registration, meeting people, making friends, having a "crowd," picking the right teachers and the right courses, and, with the college situation, even getting the right hours. Think of yourself then (not knowing the norms) and later (when you knew). The difference is understanding how the system operates; and when you know, how much less energy and anxiety are involved.

Norms at the Group Level

At a group level, norms (or more correctly, the *normative system*) are the organized and largely shared ideas about what members should do and feel, how these norms should be regulated, and what sanctions should be applied when behavior does not coincide with the norms (Mills, 1967). Group norms function to regulate the performance of a group as an organized unit, keeping it on the course of its objectives. They also regulate the differentiated but interrelated functions of individual members of a group. Norms in some cases specify particular behavior, and in other situations merely define the range of behaviors acceptable. In some cases, no transgressions are permitted; in others, wide variability may be practiced.

Similarly, norms range from those formally stated to those held by group members at an unconscious level, only recognized when they are somehow transgressed. At all levels, there are implicit or explicit standards that have important implications for the feelings and behaviors generated in the group.

READER ACTIVITY

Think of two groups with which you are involved. Consider your role in each. How are you different in each? List as many ways as you can.

Group A

____

____

____

____

____

Group B

____

____

____

____

____

Among the factors to consider: dress, amount of talking, to whom you talk, seating, responsibilities you take in the group, expectations for you in the group, your feelings of acceptance in the group.

____

____

How is the group culture different in each?

____

____

____

How did it happen?

____

The Invisibility of Group Norms

Whenever we enter a strange (synonym for new, but it feels strange) group, we are uncomfortable. We need to know how it operates, and how we will fit in. Until we know, until we have some lay of the land so that we can know how to navigate, we utilize all of our strategies from former groups that have had positive payoffs. We are dressed in the way we think they will find us acceptable, we are our most affable and charming, we hope to exude intelligence and status. We scan for clues—who is "in," who is "out"; what is the leader like; who is popular; how do people talk; how does one gain acceptance. For new members, there is constant strain in scanning, learning, watching, imitating. Until they learn how to operate in that group, they are bewildered, fearful, and feel unaccepted, or at best marginal.

To insiders, the long-term members, the idea that there are rules within the group is viewed as ridiculous. They will say that they just act naturally and are accepted. They feel acceptance; they are not under any recognizable tension. They can't understand what this talk of tension is about—it must be something that psychologists dreamed up. Insiders are so familiar that they cannot see the forest for the trees.

Thus, in real-life groups, norms are invisible, in that they are taken so much for granted that they are given little thought. The invisibility of norms is analogous to the classic figure/ground gestalts in that norms are often the "undifferentiated ground," rather than the figure that naturally emerges. We see the task, membership, problem-solving methods, but we don't see the process of interaction, the ways members conform or influence—these are the background. Too often, we ignore this background as a part of understanding the functioning and effectiveness of individuals and groups.

In many training groups, norms are a threatening topic for the members. Identifying small group norms implies taking responsibility for creating, building, or maintaining the group culture that seems to exert an external influence on one's own behavior. Thus, groups often collude to keep their own norms out of conscious awareness.

This collusion to keep norms invisible is most pronounced in family systems, where some norms or myths literally go unrecognized (and thereby unchallenged) for generations (Laing, 1972; Nagy and Spark, 1973; Ferreira, 1963.)

HOW GROUP NORMS DEVELOP

How is it that people who are so imbued with their own uniqueness, and, especially in our culture, where the rugged individualist is upheld as an ideal, become so attuned to what is happening in a group? How do norms develop, especially in what has been characterized as the "me generation"? A fascinating question: How do norms develop when we vigorously say we will not be bound by what "*they* want," and we will "do our own thing." There are a number of interpretations.

A Sociologist's View

Erving Goffman may be thought of as a sociologist and a minimalist. He believes that there are regulations that govern all social contacts. Without implicit obedience to these laws of contact command, orderly public contact would be imperiled. Every human contact involves rules that are learned, with the shadings and nuances of a fine theatrical performance. Goffman has devoted his life to seeing, cataloguing, and attempting to understand the interaction rituals performed by persons in the immediate physical presence of others. These contacts may be so fleeting and informal as to be unrecognizable as a social function—an elevator ride, a dash for a bus. They also include such major events as weddings and funerals. "More than to any family or club, more than to any class or sex, more than to any nation, the individual belongs to gatherings, and he had best show that he is a member in good standing. Just as we fill our jails with those who transgress the legal order, so we partly fill our asylums with those who act unsuitably—the first kind of institution being used to protect our lives and property; the second to protect our gatherings and occasions" (Goffman, 1963). Whether occurring in public or private places, the rule of behavior that seems to be common to all such encounters, and exclusive to them, is the rule obliging participants to "fit in."

In his classic *Presentation of Self in Everyday Life* (1959), Goffman views social contact in theatrical terms. Every scene develops as an interaction between the actor and the audience (observers). Goffman is especially interested in "thespian technique"—how is it that the actors develop their performances to look real? How do they present their behavior as acceptable to the audience while in the "front region" (on stage, before the observers) and act very differently in the "back region" (off stage, not before the observers)? Goffman notes the audience knows the actors are giving a performance that is not real, but colludes with the performers to act as if the performance was authentic, spontaneous behavior. Both the actors and the audience collude for a good show.

For Goffman, then, learning the rules of social contact—the rituals and then the more difficult "acts"—is a necessary condition to avoiding an asylum. It is a basic condition of social life and social survival, and occurs even in the most tenuous contacts. The basic rule in social contact is to fit in. Norms thus develop from an individual's learning to fit in.

A Behavioral Interpretation

Learning theorists have advanced relatively straightforward explanations for what appears to be conformity to group norms. They extend the law of effect; that is, people behave in ways that win them rewards and avoid or suppress behaviors that are punished.

Simply, people learn to identify cues that signal what behaviors will be reinforced, and, especially upon entering a new group, are extremely sensitive to cues signalling punishment. For the most part, establishing or retaining membership in a group (itself a reinforcing condition) is a consequence that only results from the choice to conform to these standards—deriving appropriate reinforcements from appropriate behavior. As a pattern of reciprocity appears in this exchange of behavior-for-reinforcement, the phenomenon that we call *group norms* takes shape.

Because individuals differ in the consequences that they find reinforcing, nonconforming behavior also can be found in groups. Yet, if individuals are going to retain membership (expulsion is the most severe form of punishment) they must either conform, find a way to change the norms so that deviant behavior becomes redefined as socially appropriate and reinforceable, or must be permitted to formulate a role that permits deviance and reinforces other members' tolerance of that deviance, such as an authority above the law or a lovable class clown.

This behavioral interpretation is most frequently applied to the practice of group psychotherapy, where it is hoped that clients will establish socially adaptive norms, and that their behavior can be shaped to the desired behavior. In such situations, norms are made explicit and highly visible, so that expectations are clear and coaching is directed to conformity (Bandura, 1977; Rose, 1977; Heckel and Salzberg, 1976; Liberman, 1970).

Therapeutic applications of social learning principles require explicit, highly visible group norms and consistent tracking of the clients' conformity to these norms. Yet, real-life groups operate very differently. In real life, group norms (the agreed upon criteria for receiving and giving rewards and punishments) are usually established through a process that is largely subliminal.

Behavioral research offers evidence for such a subliminal, though potent, social learning process:

1. Stimuli that are outside of conscious awareness (such as tachistocopically presented pictures of popcorn in a movie theatre) can influence behavior (getting up to buy popcorn). Visual perception studies have shown that pornographic pictures presented too quickly to be consciously recognized nonetheless can cause male college students to become sexually aroused.
2. Reinforcement can increase the performance of desired behavior and decrease undesired behavior, even though the target individual (whose behavior is being modified) remains consciously unaware of the manipulation. Two examples can be cited:
    1. In many cases, when school teachers apply behavior modification principles to reinforce sitting still and to decrease the incidence of acting out inappropriately, such manipulations can be successful without the student even being informed or consciously recognizing the teacher's intervention (Martin and Lauridsen, 1974).
    2. In individual psychotherapy studies, active listening responses such as "Um Hm" or smiling increased the client's tendency to talk about certain subjects or themes, or even show certain emotions, without being aware of or being able to verbalize the therapist's subtle manipulation.

Thus, one could argue that group norms develop through a process of *subliminal conditioning*, as, through trial and error guided by past experience and preconceptions, one learns to identify the criteria by which reinforcements and punishments are meted out in the present group. Social behavior, as the norms to which one conforms, is thus situation specific; it is based on the distinctive features and reinforcement criteria of the specific group.

Communications Theory

Communications theorists build a somewhat different rationale for the formation and acceptance of group norms. The major principles of the "Palo Alto Group" (Watzlawick, Beavin, and Jackson, 1967) follow:

All behavior is communication.

A person cannot not communicate (because even the act of choosing not to communicate conveys messages).

All communication (and, hence, all behavior) consists of two components, namely, information and (an implicit) command. The information and command components of communication have a figure-ground relationship. Usually, it is the information that emerges as figure, what we hear, and the command recedes into the background, forming the context of information.

The give-and-take of the command components, which are often implicit and unconscious, define the ground rules of the interpersonal relationship.

The give-and-take of implicit commands, if successful, ends with the adoption of a mutually agreeable quid pro quo, or this-for-that exchange of complementary behaviors. This would be a typical sequence of events leading to the establishment of an acceptable quid pro quo:

A. The command component implicit in any act of communication is the communicator's proffered definition of the relationship between self and other(s) involved in the act of communication.
B. In response to the first person's act of communication, the second communicates acceptance of the protagonist's definition of the relationship between them—or offers a counter-proposal, an alternative definition of the profferred relationship.
C. Successive acts of communication represent a negotiation or bargaining process, until, in accordance with social learning principles, a mutually acceptable quid pro quo—an acceptable exchange of social reinforcements from complementary points of view—is established.
D. If the negotiation process is not successful, the participants will terminate their relationship. Thus, the quid pro quo agreed upon will be the termination of further communication between them. This bargaining process is largely subliminal, as the participants learn of each other's criteria for social reinforcement and preferred reinforcers as givers and receivers. A vivid illustration of this bargaining process occurs in the movie *Annie Hall*, when Woody Allen and Diane Keaton are on the balcony of her apartment exchanging information as new acquaintances, and the subtitles reveal negotiation of a quid pro quo.

Especially when dealing with groups that have a history, the give-and-take that leads to the establishment of a quid pro quo is ongoing. The beginning or end of an interaction episode is an arbitrary marker, an act of punctuation that determines how the participants will understand (or misunderstand) what has transpired during the course of their interaction. (Watzlawick gives the example of the experimental rat who "has the experimenter trained to issue food pellets whenever he presses the bar in the cage. . . ."—a form of punctuation in their interaction that differs from the experimenter's understanding of the relationship.)

Once established, the quid pro quo pattern of the relationship is resistant to change. *It is in the exchange of command components and the subsequent quid quo pro that social reinforcers come.* Information rarely, if ever, has this strong interpersonal impact.

In summary, norms exist in all social contacts. We pick them up in the course of living, even as we go about learning content or information. Norms develop through our communications with others, but not directly and straightforwardly for the most part. Rather, norms develop by subtle, subliminal, beyond-awareness processes of inference from noting raised eyebrows, or hearing supportive "uh-hums," or watching how others gain approval. They may evolve through an interpersonal process of negotiation as we attempt to follow the rule of fitting in. Within each group there is a history of what is and what is not acceptable behavior, which has developed over time in that situation; and which members learn and understand.

CLASSIFICATION OF NORMS

If it is understood that norms are a set of standards that groups develop for themselves, is there a way to classify these standards? Are there dimensions along which these norms develop? Is there a way to understand a group in terms of its norms, and can this method also be a means to contrast groups? Sociologist Talcott Parsons (Parsons and Shils, 1951) believes there are. He thinks that norms in any society or group must provide answers to questions relating to at least four dimensions:

1. *Affective Relationships.* How personal are the relationships? Are relations among members to be based upon expression of feelings they have toward each other, or are feelings to be suppressed and controlled? Is an expression of emotion considered legitimate and appropriate, or is it understood that any expression of emotion is personal and will hamper the movement of the group? For example, in a school situation, teachers are not supposed to express their

feelings of dislike for a particular student or even another teacher. Hospitals often set aside a separate section in the cafeteria for doctors, to insure their not being too personal with other members of the staff or patients. In a family, on the other hand, the norm is for eating together, sharing personal experiences, being expressive of feelings, approving and disapproving of members.
2. *Control, Decision Making, Authority Relationships.* Is the involvement with another to be total and unbounded (as with a parent and a child), or is it to be restricted and specific (as with a swimming instructor and a pupil)? Parents have almost total decision-making control over their children, yet the swimming instructor's control is limited to the pupil at the time of the lesson.
3. *Status–Acceptance Relationships.* Is the relationship with the other due to the fact that he or she represents a type, or a class (a servant, a client, a teacher) or is it due to the uniqueness of the relationship with the other (a brother, a cousin, a friend)? In some groups, the norm is to leave the minute the session is over, without even a goodbye. In others, a personal relationship among colleagues develops as they become friends and depart for lunch together.
4. *Achievement–Success Relationships.* Is the significance of the other due to his or her personal qualities (intelligent, trustworthy), or is it due to his or her professional skills (as a researcher, as an athlete)? Within the faculty of a college department, some are respected for their professional contributions, but others are evaluated by varying criteria for what is considered success. To all, rank is a significant factor.

One way to compare norms across groups might be with regard to answers members give to the questions above. A great variety of combinations is possible. It may prove a simple, meaningful way to contrast the difference between primary and secondary groups. It may also clarify the difference between members' wishes for norms and the actual existing norms. It may also make it possible to reduce stress for the members or to precipitate action toward changing the norms.

Prior to entering a university, students frequently believe they can meet and discuss ideas and issues with their dynamic professors. They expect that being in their classes should give them special access to those professors who sound exciting, or are involved in interesting research, or who consult with important clients. Students who attempt to arrange such exchanges are often disappointed. Some faculty rush out immediately as class ends to avoid being trapped by a student, others are never in their offices to answer the phone, or do not return calls. It is not only difficult to make appointments, but often they are casually kept. At the appointment, faculty members may not even recognize students when they see them, and may only be willing to be involved in discussions related to their course. They are not receptive to discussions of career opportunities in the field, hearing a student perspective on the university, or personal matters. That intellectual dialogue so touted in the catalogue is not a source of contact beyond the classroom; and at some institutions this can be the norm.

Understanding these norms may help the student clarify the difference between his or her wish level and the reality level. There can be a consequent reduction in stress and an increased ability to function at a newly understood reality level. And, those who would like further contact with faculty can devise some strategies to modify the norms, at least with some faculty members.

KINDS OF NORMS

Learning norms is not a simple matter at best; there are so many, and it is difficult to determine which take precedence over others; which are time specific and which are general; which apply to all members and which to some members; which are to be strictly adhered to and which are to be totally ignored.

Written Rules

Some norms are codified as in bylaws and code books. They may be formal, written statements intended to be taken literally as group rules, and they are enforced by organizational sanctions (that is, actions to insure compliance). They are stated and presumably available to members who are willing to examine the constitution or corporate policies. But, there are complications.

Sometimes statements in code books are not adhered to as stated. For example, it may be stated with regard to procedures for promotion for university faculty that teaching skill, service, and publication will be weighted equally. However, since publication is easier to measure and more prestigious, it takes on greater weight. The norm then is that publication is most important; teaching skill and service become secondary factors.

Sometimes there is a tacit understanding that formal laws can be ignored, much like old statutes that remain on the books but are not enforced, a classic example being the old Connecticut statute that makes taking a bath on Sunday unlawful.

This distinction between norms and written documents is an important one, for, as some of the formal rules are weighted differently than perceived, as some are adhered to and others are not, a new set of rules is established. Frequently, those unfamiliar with this distinction will examine a copy of the group's by-laws and believe these are the procedures by which that group functions; yet there may be little similarity between such official pronouncements and what actually occurs in practice.

Explicitly Stated Norms

Some norms do not appear in the codifications or in formal written form, but may be explicitly stated verbally or may easily be recognized by members. In being hired, an employee may be told, "Everyone gets here by 9:30" (the explicitly stated norm being that you're late if you arrive after 9:30). Although nothing may be said about attire at work, a new employee may notice that all the men he has encountered in the office are wearing suit jackets and ties; there is the easily recognized norm that he will be expected to wear a suit jacket and tie in his work. However, as previously stated, explicitly stated norms may not be the actual norms practiced.

Nonexplicit, Informal Norms

Within each group there may be nonexplicit or informal norms that influence member behavior. For example, a stated norm may be that all members of a team are expected to be present at weekly hospital rehabilitation staff meetings. However, what happens is that the physician, nominally the team leader, calls just before the meeting stating that she is tied up and asking that they meet without her. The nonexplicit, informal norm that develops is that the entire team is to be present except the physician, who will call to say she won't be there. After weeks of this practice, the team makes assignments to members knowing they do not have to clear them with the physician. She is informed by reading the weekly staff-meeting minutes.

Or, in another illustration, although there may be no assigned seats at committee meetings, the chairperson generally sits at the left end of the room, flanked by the vice-chairpeople.

Sometimes, norms become known only when they are violated. A minister may preach about justice and racial equality and may urge his congregation to live according to these principles, all of which they accept from him. However, when he marches in a picket line, the congregation may rebuke him for transcending his position. He may be sanctioned with a statement to the effect that ministers may preach about justice and equality, but action on social issues is for others. In this situation, neither the minister nor the congregation knew the norm existed until an action took place that was contrary to the norm; the congregation then made known the violation by the threat of sanctions.

Norms Beyond Awareness

There are also norms that are created as if by osmosis, in a gradual, unconscious pattern. We conform without even knowing that we conform to these pressures. We automatically raise our hands when we want to be recognized; we say hello to those we know when entering; we expect a certain order at a meeting: an opening, the minutes, the treasurer's report, old business, then new business. We expect paid-up members to be notified of meetings.

When we are supposed to be factual and objective, we rate ourselves and our group as best (Codol, 1973 a, b). Two researchers set up teams of undergraduates in a competitive situation. It was their hypothesis that there would be an evaluative bias, and that it was a social norm of which participants would be unaware. Results supported their expectations: that persons were expected to evaluate their own group's product more favorably than that of the other group and that the group judged a member who showed evaluative bias more favorably than one who did not. (Dustin and Davis, 1970). An illustration of how norms develop, and the incredible craziness that sometimes ensues, may help explain what rationally is unexplainable.

A large, sprawling psychiatric hospital is situated on over a thousand acres, with ten patient buildings. There are over three thousand patients and they are treated by a wide array of professional staff: physicians, nurses, rehabilitation counsellors, social workers, psychologists, neurologists, psychiatrists, physical therapists, and occupational therapists. Many of the patients are on medications for chronic illnesses (such as diabetes), as part of psychiatric treatment (anti-depressants, tranquilizers, or muscle relaxants), for allergies, or for infections. Medication is a central aspect of patient care. In addition to medications, there are a variety of routine hospital-care items ordered through the pharmacy (for example, thermometers, bandages).

The policy developed by the pharmacy for ordering was simple and easily understood. Pharmacy picked up the list of medications and other items from each unit on set days, twice a week. Staff members who needed to order medications submitted their requirements to the nursing staff, who were responsible for ordering the medications and making up the lists. Although the procedures and the order methods were clear and known, there were indications, in terms of quantities ordered, that some buildings were stockpiling medications. However, since each building had a different kind of patient, admissions fluctuated and sometimes patients were transferred from one building to another, the pharmacy orders of each building varied.

When someone from pharmacy questioned a large order, there were a variety of evasive answers, such as, "I'm not the head nurse and I didn't compile the list," or, "I didn't order it, maybe the second shift needed it," or, "Some of the patients from building 6 are being transferred here next week, and we need their medications." The pharmacy staff was exasperated.

In an organization development program, the chief pharmacist presented his case for study. What emerged was a situation without controls, a craziness that had developed as a method to cope with what had become an impossible system.

Because sometimes the nursing staff was forced to wait a month for certain often-used medications, and serious problems developed for patients who did not have them, a norm developed to increase the order of those medications. In that way, the staff felt they would always have some on hand and would never be out.

When the pharmacy thought an order was excessive, they did not fill it. They would stall by saying they had ordered it and that it would be in soon.

When none of the medication ordered was delivered, requiring an even longer wait, the nursing staff then ordered even more on their next list.

As the order grew, the pharmacy decided to reduce the amount ordered by 80 percent. The informal norm that then developed was that only 20 percent of an order would be delivered.

The pharmacy conjured up an excuse about why the order was short, and the nursing staff repeatedly complained that they needed the medications ordered and that they were always being shortchanged.

The lunacy of the situation was illustrated when the chief pharmacist was out for a week, and someone from another institution briefly replaced him. In accord with the normative system that had evolved, one nurse wrote a high order for a medication rarely used, assuming the usual 80 percent cutback. Instead, the new pharmacist (unfamiliar with the system) filled the order as written. The nurse receiving the order was in shock; she now had enough of the medication to last for three years. She was in a real quandary. Should she find storage space for the bonanza, or should she return the excess? Either way there would be problems, but she decided to store the medications.

FORCES THAT INDUCE ACCEPTANCE OF GROUP NORMS

The process by which a group brings pressure on its members to conform to its norms, or by which a member manipulates the behavior of others, is the process of social control. It is recognized that some groups may legitimately exert pressures for uniformity of behavior and attitudes among members, for example, church groups, political parties, or professional societies. Others exert influence on members without their awareness that it is happening, such as office associates, teachers in a given school, a lunch group—those who interact frequently, though they have not created any formalized structure. They exert influence through their informal group standards and may have an important effect on members' behavior.

If the norms of the group are compatible with an individual's norms and goals, that person will conform to the norms of the group. However, if an individual finds that his or her behavior deviates from the group norms, he or she has four choices: to conform, to change the norms, to remain a deviant, or to leave the group.

Why do people bother to learn the norms of a group so that they can conform? Is it worth the effort? How do members induce other members to conform?

Basically, there are two kinds of forces that induce an individual to conform. One set of forces can be thought of as internal forces, those based on intrapersonal conflict, the second set of forces are external forces, in which others attempt to influence the person directly.

Internal Forces Based on Intrapersonal Conflict

One of the major early studies in group dynamics sought to demonstrate how a group influences individuals to conform. In the classic Sherif experiments (Sherif, 1935, 1936, 1961), each subject is placed in a darkened room and asked to judge how far a dot of light moves. (Although the light appears to move, it actually does not. The phenomenon is known as the autokinetic effect.) The subject sees the dot of light and makes a series of individual judgments. Then, the subjects are brought together in twos and threes to again judge how far the light moves. In this situation, their judgments tend to converge to a group standard. Later, when they view the light again as individuals, they retain the group standard and give that answer.

This experiment has been replicated with variations for almost four decades, and the results are so predictable that it is even conducted as a classroom experiment (Martin, Williams, and Gray, 1974; Hare, 1976). The essential finding is that when a situation is ambiguous, and there is no external reality for determining the "right" answer, people are especially influenced by the group for a reality. They look at the light and have no objective way of determining how much light moves; they make a judgment as best they can. When they are in a group, they hear one another's judgments, a clarity develops for them, and they adjust their answers to within the range of the others. Generally, the greater the ambiguity of the object, the greater will be the influence of other group members in determining the judgment of the subject (Mills and Kimble, 1973; Keating and Brock, 1974; Luchins, 1963).

In real life, this translates to mean that membership in a group determines for individuals many of the things they will see, think about, learn, and do. Given a change in the price of gold or in unemployment statistics, union members will hear a different set of "facts" to clarify that situation than will Chamber of Commerce members. College students at Berkeley or San Francisco State College will understand student involvement quite differently from students at a small midwestern college.

How an event or situation becomes less ambiguous is to a large extent determined by the memberships of the person. Because of the limited range of events in a group, there evolves a common set of perceptions and convictions among members. Discussion groups, bull sessions, and rap groups all serve this function of helping an individual develop clarity in an ambiguous situation. The process of each member of a group giving his or her own opinion, even without attempting to influence an individual, can be highly influential in developing what that person consequently thinks.

A second classic and ingenious series of experiments was designed by Asch (Asch, 1951, 1955, 1956). Asch was interested in understanding when individuals would be independent of the group and when they would conform.

In his experiments, the stimulus materials were two sets of cards. On one set, each card had a single black line (the standard). Each card on the other set had three labeled lines, one of which was the same length as the standard, with the other two easily recognizable as different from the standard. Individuals from psychology classes who volunteered for the experiment were arranged in groups of seven to nine. They were seated at a table, and asked to state in turn, starting at the left, which line was closest to the standard.

READER ACTIVITY

Pretend you are one of the students in the Asch experiment. You are seated in a position to give your opinion sixth in a seven-person group.

Trial 6

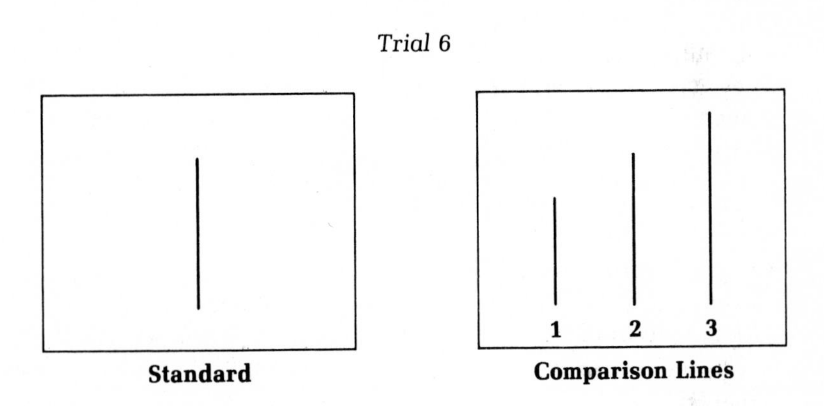

Which of the comparison lines is closest in size to the standard line? Person 1 says line 2, next person says line 2, next person says line 2, next person says line 2, next person says line 2. It is now your turn, what do you say? ______ . Person 7 says line 2.

Trial 7

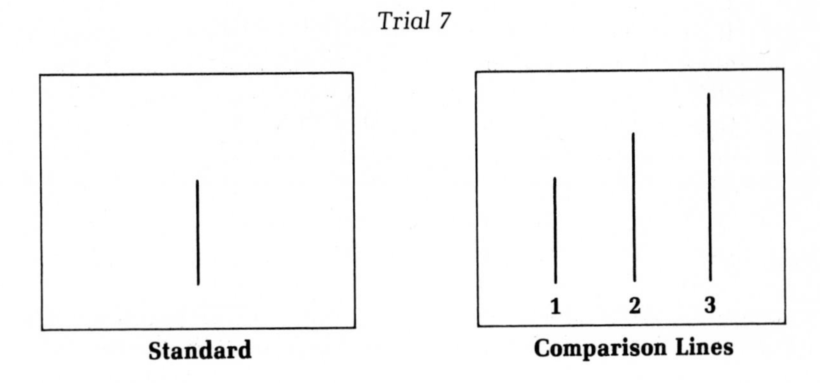

Which of the comparison lines is closest to the standard line? Person 1 says line 2, next person says line 2, next person says line 2, next person says line

2, next person says line 2. It is now your turn to answer, what do you say? ______ . Person 7 says line 2.

What would you say if you were actually at a table with six others? You look at the lines at the same time they do; they see what you see. How is it that sometimes all of you see the lines the same way and agree as in trial 6? At other times they all agree, but you see it differently. And this is not an issue of opinion, but perception. It is simple to see which line is closest to the standard. What is happening to you? What would you do?

____

____

____

What Asch did was to coach six of the seven to give the same incorrect reply on twelve of eighteen trials. In each group, there was one naive subject. The experiment was conducted with 123 naive subjects. The findings were overwhelming; nearly 37 percent of the subjects' responses were in error, as compared to almost no errors in the control trials. On one trial, less than 50 percent gave the correct answer. Remember, the subjects did not know each other, there was no overt group pressure to conform, each had the solid information of his or her senses to rely on and the situation was not ambiguous. There was no promise of future favor or advancement, nor was there threat of ostracism or punishment. How did it happen that so many, over a third, gave the wrong response, and they saw with their eyes that it was the wrong response?

Asch explains the situation as follows: Individuals come to experience a world they share with others: events occur that they see and others see simultaneously. They understand that an environment includes them as well as others, and that they are in the same relation to the surroundings as others. If it rains, it rains on them as well as those standing beside them. If an automobile in which they are passengers stops suddenly, they as well as other passengers will be shaken. They know (in the basic internal sense of knowing) that they, as well as others, are in a similar experience and that each responds to certain identical properties (they all are wet due to the rain, the smoothness of the ride is interrupted for all because of the sudden stop). Because individuals are aware of similarity of experience, and they are aware of similarity of response—which seems to be the inevitable direct response to an identical experience—an intrapersonal conflict arises for them.

In these circumstances, the response similar to what others have said becomes understandable. Since they know they experience a shared reality with others, there is an internal tension to hear what they are saying, even if it is not what they think they saw. The Asch findings continue to be powerful data of how a group influences its members, and how members conform (Ross Bierbrauer, Hoffman, 1976).

Although a person was likely to give an incorrect answer when responding in a large, unanimous group, especially over time, it should be noted that if even one other person gave the correct answer, the person trusted his or her own sensory information and gave the correct answer. No matter what the size of the majority, even one supporter encouraged the subject to trust his or her own senses (an important understanding in attempting to lead or to change norms).

This pull toward the group occurs only in making a judgment based on internal conflict; it does not occur in a judgment based on personal preference. Here, individuals do not perceive themselves as being in a situation similarly experienced by others; they perceive themselves as having idiosyncratic preferences, which are theirs alone. If, for example, they were sitting in a group with others, and a researcher asked for each person's favorite flavor of ice cream, the internal conflict would not occur. The first person might answer "chocolate," the next might also say "chocolate," and even if the next three reply "chocolate," the sixth will answer "coffee" or "burnt almond" or whatever his or her favorite flavor is. Personal preference replies are in a different category from judgment replies, in which each person is aware of being in a common situation.

Tendencies to Create a Social Reality

In addition to members conforming to norms in situations of ambiguity and conforming under conditions of overwhelming group agreement, they also conform as they seek a social reality.

Festinger (1954) proposed that there is a basic drive within each of us to evaluate our own opinions and abilities. It is potentially dangerous, or at the very least embarrassing, to be incorrect or to misperceive how well we can do various things. To avoid that, we constantly seek relevant evidence. For some opinions and abilities, the evidence is directly available to us in our contacts with the physical world. If we are not sure of the time, we can check our watch. If we are not sure whether we can jog two miles, we can go out and find out. However, for most of our opinions and abilities, there is no objective, nonsocial way to evaluate ourselves. All we can do is turn to other people. If we are not sure about how well we speak, we find out by how others respond to our talk. If we are not sure about how we should relate to the opposite sex (to open doors, light cigarettes, pay for meals), we listen and look for others like us to help us develop a *social reality*. This social reality is as important as, and sometimes more important than, the world of physical reality. In evaluating ourselves, there is a tendency to seek fairly similar others as a comparison group. We seek a similar, attractive reference group and use these individuals as a basis for comparison.

When there are vast changes in social norms, styles, morality, taste, criteria of beauty, child-rearing practices, divorce, and myriad other aspects of our lives, our reactions to the changes are primarily based on the evaluations of those around us. For that reason, friends or members of our peer group have greater influence than others.

Tendencies to create a social reality also occur whenever people confront an unfamiliar social situation that arouses the emotions (Schachter, 1959).

Just as there is pressure to establish the correctness of an opinion, there are pressures to establish the appropriateness of an emotional or bodily state. And since emotion-producing situations are often novel and outside the realm of our past experience, it could be expected that the emotions would be particularly vulnerable to social influence. Consider an example:

Each weekday morning, year in and year out, there were, with few changes, the same people standing on the suburban station platform waiting for the 7:40 train to take them into the city and to work. As each arrived, they nodded "good morning," or perfunctorily commented on the cold or the rain. For the most part, they got their papers or their coffee and waited in small clusters at where they thought the doors would be when the train stopped. There were a few women, and they boarded first; then the older men, and then the others.

On the train, they read, worked, or slept. There was almost no conversation. It was just another morning, except on October 15th, it wasn't.

In an incredible accident, the 7:40 crashed into a stopped train in front of it with such force that over three hundred people were knocked unconscious and, thought to be dead, were strewn about the cars. Bloody noses and bleeding heads were commonplace as the crash stop hurled passengers into the seats in front of them. Broken glass from windows spattered all within range. Smoke created a darkness in the cars.

It could have been a stampede for safety through the broken glass windows. It could have been a trampling horde concerned only with reaching a door and getting out. It could have been a caged mob regressed to the basic instinct of personal survival. It was the first commuter accident in fifty-two years. No one could have been prepared for it or known how to act, given the incredible nature of the catastrophy, the early hour of the morning, and the fact that most were half asleep as well.

The norms of the suburban station platform prevailed. The first concern was for the women, as the persons next to them helped them with their injuries, while they themselves were bleeding. Then, passengers shouted for the older men and rushed to their aid; one who was knocked unconscious was picked up by three men who moved him to a bench, covered him with three overcoats, and attempted to revive him. Not one person left; those who were not visibly injured were helping others.

Even when the police arrived, people who had been helping others stayed with them on the trip to the hospital, so that they would not be frightened by being injured and alone. People on the train emerged helping others who were unable to walk, and remained at their sides, helping them to phone booths to call their relatives and waiting with them until a family member arrived.

Police, newspaper reporters, and hospital workers alike commented that they had never seen anything like it. No looting or robbing, no pushing or crushing, no abandoning others for the police to care for in due time. Over and over, those injured reported, "I can't believe strangers can be so caring." A woman whose nose had been crushed and whose head was badly lacerated reported, "The man who had been seated next to me, and to whom I had never said more than good morning, held my arm and kept telling me that I would be all right. He would stay with me in the hospital. He would call my husband and daughter. I was so frightened and bloody and hurting, I don't know what I would have done without him. And it was like that throughout the car."

READER ACTIVITY

As you read this incident, how do you explain the norms that developed in the train crash? How do you explain people responding in a caring, altruistic manner, which is totally contrary to how people might have been expected to act?

____

____

____

Imagine yourself on that train. How do you think it happened that train group members influenced each other, and were able to get such conformity?

____

____

____

External Forces Based on Direct Influence of Others

Coming into a group means interacting, which means influence attempts. From the simplest, "Why don't you sit here with us?" to "John is by far the best candidate. How could you consider anyone else?" we influence each other. When we want a person to act in a certain way under certain circumstances, we are using direct influence.

There are a number of reasons why people attempt to influence others to comply to certain norms. One is that it will help the group accomplish its goals; another is that it will help the group maintain itself. Both these functions must be developed with strong supporting forces if the group is to succeed.

To Achieve Group Goals Pressures toward uniformity among members of a group may occur because uniformity is necessary for the group to achieve its goal (Festinger, 1950).

If a group is attempting to raise funds, they will develop norms for standardizing pledge cards, assigning members to districts to be solicited, turning in money, and reporting results. All of these procedures will develop because they will help the group achieve its objective of successfully raising the money needed. Consider, how will members respond to the member who turns in pledges on the corner of a menu or the back of a shopping list? How will they respond to the member who solicits people in someone else's territory? How will they respond if the campaign ends and a member doesn't turn in the money? People exert pressures on others to follow approved procedures that are directed toward achieving the group goal. These are sources of uniformity that are seen as legitimate.

Another example of norms developing to help achieve a goal can be seen in a library. The procedures for owning library cards, determining which books may or may not be taken out, setting the length of time a book may be held—these are all procedures developed to

them to like us (ingratiating behavior), the more we search for cues of what they want and give it to them (Jones, 1964). In general, the more we like the group, the greater the conformity (Kiesler, 1963; Savell, 1971).

Compliance

Although in the conformity situation there is no overt pressure on a person to behave as the others are, he or she responds as if there were pressures to comply. In the Asch line experiment, each person responded as he or she saw the line, but when there was unanimous group agreement, it created an internal pressure on the individual to conform (to say what they said). In compliance situations, the request is direct. A person asks you to do them a favor, to vote for a particular candidate, to contribute to a fund, or to do an unpleasant job.

In conformity, the pressure is invisible; in compliance, it is obvious. And that compliance seems to arouse resistance. We are bombarded by requests, from being asked to do a colleague a favor by covering a class for him or her, to buying tickets to help a cause; from taking minutes at a meeting, to being asked to chair a committee.

The request comes, and there is a pause as we think, "Now how can I get out of it, and still have them like me," or not seem difficult or ungrateful.

For some, a request almost automatically induces compliance; it is hard to say no. For them, there is now a whole world of literature on assertiveness training (Fensterheim and Baer, 1975) and learning how to say no. It is a measure of how ubiquitous the pressure to comply is, that there are scores of books and courses teaching people how to say no and how not to comply.

For others, compliance requests are met with resistance. We would otherwise be overwhelmed with constantly doing others' bidding.

Compliance research has focused on how, despite resistance, people can be induced to comply.

One way to get someone to comply is to do them a favor so that they are indebted to you. If someone asks to borrow your class notes, there is now an unequal relationship in terms of costs and rewards. His or her costs in the relationship have increased and your rewards have increased (Homans, 1961). The relationship is no longer equitable. You will feel more comfortable and equity will be restored, if you do something nice for him or her (something which increases your costs and his or her rewards).

Regan (1971) designed a simple experiment to test the hypothesis that doing someone a favor later induces compliance. In the experiment, presumably on art appreciation, subjects waited in groups of two (however, one of the two was a confederate). In half the situations, the confederate was pleasant and amiable as they chatted waiting for the "experiment" to begin; in half the groups, the confederate was purposefully unpleasant. They viewed art material, and then there was a pause in the experiment. In one condition, the confederate bought cokes for the two of them; in another, the two just sat and waited. They looked at more art; then again there was a pause. This time, the confederate asked the subject to buy raffle tickets to help the confederates' home town high school build a gym, since the person who could sell the most tickets would win a $50 prize. The results overwhelmingly indicate that whether the confederate was pleasant or unpleasant, the subject was most likely to buy the raffle tickets if the confederate had performed a favor for him or her. The business world knows this well: prospective clients are wined and dined, provided with gifts, from notebooks from insurance companies to free samples at food markets—all under no obligation. Thus compliance is strongly increased when a person feels obligated to the one making the request.

The foot-in-the-door technique is another method to get a person to comply. First, ask the person to comply with a small, innocuous request, then ask for a big one (Freedman and Fraser, 1966). In a fascinating field experiment, they asked subjects to place a large sign on their front lawns that said "drive carefully"; less than 17 percent of the subjects were willing to do so. Another group of subjects were first asked to place a small sign saying "drive carefully," and then at a later time were asked to put the large sign on their lawns. Now, over 76 percent were willing. And interestingly enough, a small initial request seems to be a better technique for inducing compliance than a moderate one (Baron, 1973).

Margaret Singer, whose interest in brainwashing and cults involves interviewing every prisoner of war who returns from North Korea and Southeast Asia, has developed an understanding of how cults operate. Singer (Freeman, 1979, p. 6) believes that members of religious cults are victims of this foot-in-the-door technique.

> In many cults, when people join they don't realize what it is they are joining, and they are lured a step at a time by very persuasive and deceptive practices. Middle-class Caucasian kids, in particular, are not very street smart. They are trusting and naive and they believe people who approach them with offers (of a dinner, a place to stay, a weekend away, instant

> companionship) much more than lower-class kids, who are wise to the fact that no one gives you anything without expecting something in return.

What starts as a response to friendship, or an invitation to dinner, insidiously moves with each step of compliance to a greater act of compliance, until after years of being in a cult, leaving is extremely difficult.

> On the whole, when people who have been in these very high-structured groups come out, they have a lot of difficulty making decisions on their own, because their lives were decided and planned, and someone else told them what to do for so long, sometimes getting going again in decision-making is a big problem. Even if these people were twenty-six years old and had a master's degree when they joined, when they come out they need a period of time to get rested. Often they have been on a very low-protein diet and have been working anywhere from eighteen to twenty-one hours a day, and they need a period of time to just get some good food and rest. That alone usually takes about a month. Then, many of them report, it takes anywhere from twelve to eighteen months to get back into functioning professionally, or as students or home members. (Freeman, 1979, p. 7)

The gradual, foot-in-the-door technique escalates to the point where compliance becomes automatic, and decision making as an individual extremely difficult.

Obedience

In compliance, people can say yes or no; there are options and alternatives. They may feel compelled to say yes, but they can be resistant. They don't have to accept the invitation to the group meeting, or the dinner, or acquiesce to the first request. There will not be severe consequences. Obedience is different.

If one person has power over another person, obedience can be demanded. If the second person fails to obey, power is exercised in the form of negative sanctions such as demerits, demotions, fines, imprisonment, even death. Those who refuse to follow orders are labeled bad, rebellious, uppity, even psychopathic.

Here, the social influence is very direct and very explicit. There is a sense of "Do it, or else . . . ."

Obedience is instilled in us as children, first when we are expected to obey our parents and later when we hear "Listen to the teacher. Do what she wants." Enlisted men are expected to obey their officers, no matter what the personal consequences or moral concerns, as in the incident of the Vietnam veteran reported earlier.

Whether Vietnam or Nazi Germany, how is it that otherwise moral, ethical, sympathetic people can be induced to be obedient to commands that inflict great harm to others?

Stanley Milgram (1963, 1964, 1965), at Yale, designed a series of experiments that are among the most important in social psychology. He sought to understand how it could be that people would actually harm another. (This was behavior, not words, called *action conformity.)*

The supposed intent of his research design is "to test the effects of punishment on memory." The question Milgram really raises is whether individuals who received a command from a legitimate authority would obey, even though the authority figure had no real power to make them obey.

In the experiment, subjects were ordered to give a confederate an electric shock whenever he or she got a word wrong. The more words wrong, the greater the intensity of the shock. The confederate was trained to writhe in pain and to pound on the wall. Despite the apparent pain, the intensity of the shocks was to increase whenever he or she was wrong.

When a group of college students was asked to predict how the subjects would respond, the majority guessed that they would refuse to administer extremely strong shocks to an innocent victim. However, most subjects did, in fact, obey the command to continue shocking the victim up to the maximum level. After watching the confederate pound on the wall and refuse to answer the next task, only 13 percent of the subjects defied the experimenter and stopped. Even when the shock level reached the danger level and beyond, well over half the subjects were still administering shocks to the victim.

How did the subjects react as they were carrying out an order to deliver painful shocks to an innocent victim? They were not indifferent or cold-blooded torturers. Rather, they seemed to be very nervous, and the tension appeared to be extreme. Subjects perspired, trembled, stuttered, bit their lips, groaned, and dug their fingernails into their palms. Over a third engaged in nervous laughter. Subjects afterwards were embarrassed about the laughter and explained that it was not under their control; it did not indicate that they enjoyed the task.

This experiment puts subjects in a conflict situation, a conflict between their moral values about harming others and their tendency to obey an authoritative command. What is somewhat frightening about the results is that, in this case, the authority figure had no real power over the subjects. They would never see the authority figure again. Presumably, obedience would be even higher if the commands came from someone who had some control over the subject's life.

Milgram, in a modification of the original experiment, wondered whether subjects could be induced to perform acts that they would not perform individually in response to group pressure (Milgram, 1964). The same general experiment was used, but now the subject administering the shock was joined by two associates (confederates). The level of shock to be administered was suggested on each trial by the subject and his or her two associates; the learner received the lowest of the three suggestions. If he or she wished, the subject could administer the mildest shock simply by naming the lowest (15 volt) level each time. The assistants proposed increased shock levels each time.

As a control condition, to determine how individuals respond without group pressure, some subjects had no assistants and each was told that he or she could select any shock level he or she wished on the various trials.

The group pressure led to much more intense shocks being given when compared with the subject-alone condition. When subjects were by themselves, only 5 percent went past the 150-volt condition in which the learner asked to be released because of a heart condition. Better than two-thirds of the subjects went past this shock level when the confederates urged them on. Once again, it was shown that subjects can rather easily be influenced to inflict pain on an innocent victim.

Although persons said they would not harm someone and knew it was morally wrong, they did so on command of the experimenter or the suggestion of the associates in the group condition.

If group pressure can induce the subject to continue to give shocks and intensely harm an innocent person, what will encourage the subject to defy the experimenter?

In yet another revision of the Milgram experiments (Milgram, 1965), in the group condition, the two confederates went along with the experimenter until the 150-volt shock level. At this point, one of them said he or she did not wish to participate any further because of the learner's complaints. Despite the experimenter's insistence, the associate got up and went to another part of the room. The experiment was then continued, but, after 210 volts, the second confederate also decided not to continue. He or she said, "I'm willing to answer your questions, but I'm not willing to shock that man against his will. I'll have no part of it."

With defiant peers for support, 90 percent of the subjects were able to defy the experimenter and stop shocking the victims. Without such support, only 35 percent did so.

Milgram suggests that in the regular experimental condition, moral pressures probably lead many individuals to come near to defiance, but the moral pressures are not quite enough for them to take the final step. Additional pressure from fellow group members who are defiant is sufficient to push them over the threshold of disobedience. As one subject said, "Well, I was already thinking about quitting when the guy broke off." For others, it was as if the defiant peers had suggested something new. "The thought of stopping didn't enter my mind until it was put there by the other two."

Not surprisingly, most subjects denied that the confederates' behavior had anything to do with their defiance; the data, however, clearly indicate the confederates exercised powerful influence.

In real-life situations, if only a few individuals speak out and refuse to engage in acts they consider wrong, their influence can be surprisingly powerful. People have greater freedom to defy immoral standards than they realize, and the defiant behavior of someone else can make that fact obvious.

ACCEPTING GROUP NORMS: UNDER WHAT CONDITIONS?

Under what conditions does a member act like other members and conform to the norms of the group? He or she is most likely to conform when one or more of the following circumstances are a factor.

Continued Membership Is Desired If membership is desired, people are more likely to be influenced by other members of the group. (For a further discussion on attractiveness of membership, see Chapter Two.) If the group is a reference group for a person, he or she will be especially influenced by the norms of the group (Gould, 1969; Buchwald, 1966). It is the fact that we want to "fit in" that we look for dress cues on a campus, listen to the others and let their opinions mold ours, and withhold our views when they disagree with the majority (Singleton, 1979).

Salience of Membership Is Heightened Imagine a research experiment designed to acquire information on attitudes toward birth control. Let us assume the population to be studied is divided into two groups. One group will be asked their opinion on birth control. For the second group, the design will be changed slightly. One additional question will be asked prior to the central one; each person will be asked his or her religion. How will the responses differ? The question on religion heightens the salience or cues of membership, and the respondent is more likely to remember "how one of us is expected to feel." When a member receives cues that he or she is an American Federation of Teachers' member, a professional, a Christian, or a Boy or Girl Scout leader, he or she is more likely to accept the pressures toward uniformity for that group.

The Group Is Cohesive A cohesive group is one that members find meets their needs, or one in which they desire to remain for some other reason. The more cohesive the group, the greater the likelihood that members will conform to the group norms, and the greater the pressure the members exert on others to conform (Festinger and Thibaut, 1951; Janis, 1971). It is often said that management or administration should help their lower staffs or employees become cohesive units. The dilemma here is that inducing cohesiveness may or may not be compatible with the goals of management. A highly cohesive work group might develop group pride and produce at a high level. A highly cohesive group might also decide to restrict production by bringing "eager beavers" and "rate busters" into line.

Sanctions Are Expected Norms exist for specific purposes, and if these purposes are valued by members, those who are deviant can expect to be sanctioned. The sanctions may be fines (for lateness, swearing, and so forth), a negative comment, sarcasm or ridicule, or even exclusion from the group. In some situations the sanctions may be even more severe, such as being dishonorably discharged, being fired, being banished or imprisoned, or even being put to death.

In the movie and play "One Flew Over the Cuckoo's Nest," the cocky hero, McMurphy arrives at a state mental institution, which is his lazy man's alternative to the chain gang at the state prison. He becomes interested in the patients and involves them in basketball, watching the world series, even reviving their interest in sex and women—all the while defying the rules of the head nurse, which are expressed as the "community norms."

The patients know the system and the obedience expected; group therapy is a farce. Briefly, there is renewed hope that McMurphy will defy the system. As they marvel at his questioning and noncompliance when it doesn't make sense, they are exhilarated.

It becomes increasingly clear, however, that in the end the system will prevail. It will take away McMurphy's identity (he has defied the rules), make him and the others dependent on the authorities (who stand futilely by, or must punish infractions no matter what).

The tragedy unfolds, from electric shock treatments to the ultimate sanction for defying the rules and the authority, a lobotomy.

Sanctions are powerful mechanisms to force members into compliance. When patients, clients, or students are the involuntary members, sanctions can enforce conformity beyond belief. Although the term sanction is usually used with a negative connotation, it is important to note that sanctions can be used for positive reinforcements. Sanctions may also be one's name on the honor roll, the designation "good citizen of the week," a raise, a bonus (to the person with the best safety record in the plant), or a promotion.

Another aspect of sanctions is that they carry two messages, one about present behavior and another about future behavior. For example: "That was quite a mistake (present); that had better not happen again (future)"; or, "You're doing a great job (present), at the rate you're going, you'll be a vice-president in no time (future)". Sanctions can be expected to influence both present and future conformity.

DEVIANCE FROM NORMS

What about those who do not conform to norms? An act that violates a shared idea about what should or should not be done at a particular time (or by a particular person) is called a deviant act. It is deviant to disagree with Nurse Rachett in "One Flew Over the Cuckoo's Nest," to sit on the table during a board meeting, or to arrive at class in a bathing suit. However, every behavior that departs from the expected is not deviant. The quiet member may become more active; the active member may take an unpopular position in the course of discussion; one member may change his or her feeling toward another member. (None of these are deviant acts unless the group has norms pressuring that quiet members remain quiet, that no opinions other than the traditional may be discussed, that members maintain the same relations with others.) What happens when a person is deviant?

Interaction with the deviant increases when the group first recognizes a member's deviancy; there will be efforts directed toward amplification and clarification of his or her position, opposing information will be directed toward him or her in an effort to lessen the deviancy. If the pressure brings the deviant to return to the norms of the group, the discussion will continue with the deviant in his or her usual role. If, however, the deviant continues the behavior, the group may redefine its boundaries to exclude him or her by ignoring him or her in discussions. This is most likely to occur in a cohesive group, and if the area of deviance is relevant to the purpose of the group (Schachter, 1951).

However, if the member is a respected and valued member, the situation may be different. There is a theory (Hollander, 1960) that prestigious members have "idiosyncrasy credits," that is, a kind of credit for helpful behavior in the past that allows them leeway in following the norms of the present. It is as if their previous work for the group entitles them to some rewards in the form of increased flexibility of behavior and immunity from punishment for deviance. A prestigious member who deviates from the norm may be dealt with in a number of ways. One is that he or she may be purposely misinterpreted or "not heard." Another is that his or her behavior may be perceived as idiosyncratic ("You know he has these bad days; disregard anything he says; he doesn't mean it"). Or nonconforming acts are now approved rather than disapproved. (A successful, grant-getting physician and corporation president continues to wear green scrub suits to staff meetings when other staff members are stylishly dressed in three-piece suits and current women's fashions. No one comments on the green scrub suit, however.)

Until recently, deviance from norms was regarded only as a negative function, a behavior requiring reinterpretation and individual or group pressures to return to accepted norms. However, deviance can also be viewed as serving a positive function (Dentler and Erikson, 1959). Deviance helps members to master norms (it is a demonstration of what should not be done) and helps them to be more articulate about these norms. In addition, it helps them to comprehend what their group is and what their group is not (to feel offended by an act and to see others similarly offended provides information on oneself and the group that could be gained in no other way).

Deviance or nonconformity has become an interesting target of research from yet another viewpoint (McDavid and Harari, 1968, Grinder, 1969). A number of studies call attention to the importance of distinguishing true independence (indifference to the normative expectations of other people and groups) from rebellion (direct rejection of the normative expectations of other people and groups). Although both represent nonconforming behavior, it is important to recognize that they are very different in both the attitudes they represent and the consequential behaviors. Conforming behavior (to norms) is a consequence of a person's awareness of expectations held for him or her by others or as a member of a group, coupled with a decision to adhere to these expectations. Rebellion against norms is a consequence of awareness of the norms coupled with a decision *not* to adhere to them. True independence represents indifference to the norms and expectations of others—personal criteria for behavior are seen as motivators of behavior rather than expectations of others. In some cases, norms may be adhered to, in others, disregarded. In some ways, rebellion and indifference are alike in that both represent nonconforming behaviors. However, it should be noted that conformity and rebellion are also alike in that both represent behavior dependent on norms and expectations of others. This is important to understand with regard to some of the radical movements on many college campuses. Many of the dissidents were rebelling against some institutionalized issues—against the "military-industrial complex," against "computerization," against the "establishment"—rather than being for some alternative. The adolescent who sees himself or herself as independent is often more likely to be rebelling than to be truly independent (Willis and Hollander, 1964).

NORMS FOR THE FUTURE

John Kenneth Galbraith (1970), in a *Time* magazine article, refers to regulatory bodies, but the phenomenon he describes is familiar: "the fat cat syndrome."

> Regulatory bodies, like the people who comprise them, have a marked life cycle. In youth they are vigorous, aggressive, evangelistic and even intolerant. Later they mellow, and in old age—after a matter of ten or fifteen years—they become, with some exceptions, either an arm of the industry they are regulating or senile (p. 88).

The analogy to youth and vigor in young people and organizations is apt.

Group norms are basically conservative mechanisms: they tend to preserve the status quo. Procedures that at one time might have been appropriate and helped the group achieve its purposes may still be in existence long after their appropriateness has diminished. Procedures developed at a time when there was only one organizational structure (the formal hierarchical model) may still be in effect for lack of consideration of possible alternatives. Other rules and procedures may be appropriate for some tasks but not for others. The overwhelming tendency is to consider what is as what should be. Of course, there may be problems; progress is "part of the times," but it should be considered only within "our policies, our framework, our image."

Yet these very resistances to change that promulgate increased difficulties and problems in a differentiated society comprise some of the group's prime assets. Norms and pressures toward uniformity create a security and order in interaction. Members know the rules and procedures for working in that group; they know what is expected; they hope to be rewarded for conforming to group norms.

In working with groups, the question with regard to norms is not whether they are good or bad. Rather, the questions are: Which norms help the group achieve its purposes, and which are harmful or inhibiting? Which norms are compatible with the goals and values of the group and under what conditions? How can the norms be changed or reconsidered to permit the group to achieve its purposes under conditions of maximizing its resources? (McGregor, 1967; Ford, Nemiroff, and Pasmore, 1977; Bonacich, 1972).

CHANGING GROUP NORMS

How many people are giving up smoking, or going on a diet? Why is it that every year there is a new fad diet book that enough people buy to produce a best seller for the authors? How many people make New Year's resolutions of how they are going to change—this year, definitely?

How many resolve, and do it? There are people who can systematically review the number of times they have stopped smoking and present reasons why they had to return. Dieters are notorious about how many pounds they have painfully shed, and how they have regained every pound, and more. (An oversized friend cheerfully comments that he has lost over one thousand pounds in a lifetime of dieting.) Even the words *New Year's resolution* evoke a smile in the hearer; we know that New Year's resolutions, by definition, are momentary fantasies about change, having almost nothing to do with actual change.

If an individual, convinced of a need to change (he or she reads the diets and stocks the foods or throws away the pack of cigarettes) has such difficulty, imagine how hard it is for a group to change—at least ten times as hard. Once norms, which are the group's procedures or expectations for its members, are developed and agreed upon, they are exceedingly difficult to change. What happens, according to Lewinian force-field theory (Lewin, 1947) or the Homans equilibrium model (Homans, 1947), is that when forces are increased for change in a given direction, the equilibrium (status quo) is interrupted. There then arise counter forces in the other direction, resulting in no change.

For example, a university may decide that there needs to be a reexamination of its grading procedures. There may have been complaints that students are essentially interested in grades for impressive graduate school transcripts rather than learning; that grades produce competition rather than cooperation among students; and that students are reluctant to take courses that may be difficult or represent an outside interest for fear of the grades they may receive.

Over two years, the system of grades may be reviewed. There may be hearings and committee meetings, and even articles in the college paper pressing for the adoption of another system. Finally, the empowered committee may make recommendations for a system of credit-no credit rather than traditional letter grades for a certain percentage of students' courses. The administration may accept the recommendation and approve it. (An example of a changed norm for the organization.) However, respecting a degree of departmental autonomy, the administration may say that each department will implement the new grading system in ways appropriate to that department.

Those departments who prefer the previous system may stall on implementing the change. ("Our long-range planning committee is currently reviewing practices in the department, and we cannot make changes until their report is complete. Then we'll see how a change in grading might fit.") Departments may never get around to using the new system or they may change the grading for an obscure course which is not conducted because of insufficient registration. The resistant forces have effectively maintained the old norms; the result is no change in norms except at the written level.

Of course, the classic illustration of resistance to group norm change is the Brown-Topeka 1954 Supreme Court decision, which stated that segregation was, by definition, unequal education. Twenty-five years later, in a reexamination of integration in the schools, it was evident that the counter forces had prevailed. The primary method for maintaining the status quo (nonintegration of schools) in the South was for whites to send their children to a newly organized system of private schools rather than to the public schools; and in the North, white families moved to suburban areas, where there were almost exclusively other white families. Twenty-five years later, there was limited integration, especially in the North. (It had been assumed that the South would be opposed to integration, but that there would be no such problems in the "liberal" North.)

Change involves not only increasing the forces in the direction of the desired change, but also holding the resistant forces constant or reducing them. How can that be done?

Lewin describes three stages in the change process: first, there must be a *disequilibrium*. People need to reexamine the present system, or feel a tension or dissatisfaction with it, or experience themselves in new ways (for example, to shop for clothes and be shocked at the size that fits). There need to be "ripples," or incidents, or a crisis in their lives. As a result of that disequilibrium, people experience a sense of urgency, feel off balance, and acquire a different perspective—all leading to the beginning of a process for change. The women's movements of the late sixties stirred organizations to reexamine their norms toward women. Declining enrollments induced universities to reexamine their admission and recruitment procedures. A deficit city budget induces concern about RIFs (reductions in force) and employee morale.

One of the most powerful effects of T-group training was that individuals experienced themselves in a different culture and learned in such a different way from their normal method of learning. They experienced personal disequilibrium and, from that vantage point, began a reexamination of themselves and their organizations.

Sometimes, heads of organizations say, "We can't change now. Wait until the crisis is over, then we'll look at what needs to be changed; but not now." However, the greatest inducement for change occurs when there *is* a crisis, an urgency for change. And, when there is not that urgency for change, when there is no perceived need for training, a team development program produces no change, because the members of the groups being trained see no need. They go through the training motions, but there is no effect on the organization. In fact, consultants, through the reporting of their diagnostic surveys, will sometimes deliberately emphasize discrepancies and problems to create a sense of urgency and crisis thereby producing that disequilibrium that can lead to changing norms. People in equilibrium (complacent with their present situation) are most resistant to changing norms, no matter how antiquated they seem to others. They know what is expected, and are comfortable as is.

At the second stage in the change process, behavior changes—people act in a way different from the previous norm, or they come from the disequilibrium (being off balance) to a new position (behavior) in what is called *freezing*. Some examples of the famous Lewin "action research" during World War II may illustrate these concepts.

Americans are used to eating meat (especially in the forties); men especially enjoy red meat. However, there was a war, and meat was in short supply. The question became, how could families be induced to try low-priority meats like sweetbreads, kidneys, and liver (Lewin, 1947)? How could their food habits be changed so that they would buy the low-priority meats?

The war effort and desire to be loyal Americans were the factors that created the disequilibrium; the next step was to produce behavior change. Lewin determined that women were the "gatekeepers," they were the major influence with regard to which foods were brought into the house, and he set about changing the behavior of their families through them. Two methods were used to convince them to buy and serve the unfamiliar, unpopular foods. Some women were given lectures on the values of eating the new foods (information); others participated in discussions (which involved information and social comparison).

After discussion, the women reached consensus about what they would do. That consensus was stated and agreed to publicly. Once this public standard existed (for example, "We believe loyal Americans will agree to substitute low-priority meats for high-priority meats once a week"), it was easier to change individual behavior in accord with this new standard. Lewin found there was greater change among the women who had participated in the group discussion, and that more of these women continued in the new behavior.

There is extensive research to support these findings over a wide range of behaviors; in community problems (Lipsitt and Steinbruner, 1969); in raising productivity (Jenkins, 1949; Hall, 1971); and in improvement of group skills (Hall and Williams, 1970).

In attempting to understand why group discussion seems to be an important element in change, researchers have found evidence pointing to a number of factors. One area of research indicates that breaking down the old value system prior to adopting a new one is a crucial element in changing norms (creating the disequilibrium stage first) (Rokeach, 1971). Another factor is that group discussion is an emotional as well as an intellectual process, and involving the whole person becomes more effective in producing change (Watson, 1967). Finally, lecture involves a direct bid for compliance, which is more likely to induce resistance, but discussion is more likely to induce internal pressures and conformity.

Others feel that what produces the changed behavior is not so much the public decision as the consensus (Bennett, 1955). The pressure for conformity is greater when there is consensus, as in the Asch experiments, in which a person was most likely to conform when there was not a single dissenter.

In the third stage of change, *refreezing* or stabilization of the behavior change occurs. This is often the most difficult aspect of change. Participants can go through a laboratory experience, be in a T-group, and learn about themselves in different ways. Under these special circumstances, they can try a variety of new behaviors and be quite proud of how they have changed. Often, they are convinced that when they return they will make major changes in their organizations (Luke, 1972). What a disappointment! They return ready to make organizational change, and find that the behaviors that were so appropriate in the laboratory community are discrepant with the norms of the organization. They learn that individual change is very different from organizational change. Although trust and open communication are valued in the laboratory, there are norms of competitiveness, secret information, and distrust in the organization. It is only a matter of weeks before they revert back to the prelaboratory behaviors—often with a renewed frustration that although they were able to make some personal changes, the newly recognized restrictiveness of the organizational norms is even more confining. Some of the major disillusionment with sensitivity training is that participants return high and excited on Sunday, and that "it lasts all the way 'til Wednesday."

Pilot projects are often developed with high hopes, behavior is changed, and then at the end of the project, the system reverts back to where it had been. Initial norms continue. Educators in public schools become wary of innovations; they have been enthused, worked hard, seen projects come and go, and watched the system revert back to where it was before the project. Businesses have hired experts to create more effective interpersonal relationships, to reduce competitiveness with their staffs, and to encourage team development. There were transient changes; a couple of months later, the ripple created by the training, like a stone thrown in the water, was nowhere in evidence.

Behavioral scientists, only too familiar with the phenomenon, are now changing their emphasis. Persons going for training are encouraged to attend in teams. An individual seems to need the support of at least one other member to disagree or begin a shift. Previous reinforcement in the group appears to further encourage being deviant to the established norm or taking a risk for promoting change (Levine, Sroka, and Snyder, 1977).

Currently, there is a reduced emphasis on the length of the initial training and a greater emphasis on one-month, two-month, and six-month follow-ups. These follow-ups are built into the training proposal. There is no assumption that any changes that occur during the training will be sustained; rather, there is a recognition that the prevailing norms will continue unless there is monitoring at intervals to halt the almost inevitable reversion.

Also, being the person or persons pressing for change can be exceedingly difficult. In the short run, they will be bombarded with questions and information in an effort to have them conform; in the long run, they may be isolated and rejected.

When faced with a puzzling situation such as nonconformity, group members attempt to interpret and negotiate the meaning of that behavior, rather than let the nonconformity speak for itself. They make use of interpretations of other members, whether expressed in words, gestures, or facial expressions. Ultimately, the group reaches one effective interpretation and produces one effective response to the behavior, and that response is not predictable from the individuals' isolated responses (Wahrman, 1977). Persisting for change can be devastating; knowing this, there is an increased effort in creating support groups, self-help groups, professional networks, and other ongoing methods to provide a belonging and a reference group.

For example, the Federal Women's Program Coordinator in an agency finds herself in an unenviable role. Each agency is required to have such a position, but usually is complying with a federal regulation rather than demonstrating a concern for hiring or promoting women.

The coordinator is supposed to be an advocate for increased women's positions, for training, and for equality in pay and job descriptions. If she takes her position seriously and makes an issue of the discrimination she sees or hears reported, she is quickly branded a troublemaker and a strident "women's libber." Other women are fearful that being seen with her will reduce their chances of promotion or acceptance by the overwhelmingly male administration. Frequently, the loneliness and rejection are too much, especially after previous respect.

She started with high hopes for effecting change, but resigns in despair. The frustration of being viewed as blatant, when in fact her style has been very conciliatory, and the rejection of even the women, over time become too much. She either gives up actually or burns out emotionally. A support group, a council, a group of other FWPCs or a women's group are methods that would help her persist in pressing for change.

Changing norms is not easy. However, there are a number of ways:

Through contagion, as in dress style or patterns of speech.

Through influence on the group from the external environment. Issues here relate to the reexamination of values about the family: sexuality, relationships to children, and the effects of divorce. Some attitudes that reflect changing values include: a mother asking for time off to attend a conference with her child's teacher would once have been viewed as not committed to a career; it would have been unimaginable that an executive father would ask for such time; divorce would have seriously impaired a man's career; and a working wife would have been viewed suspiciously.

Through high-status members, who have earned their *idiosyncracy credits*. Hollander (1958, 1960, 1964) offers the hypothesis that those who have reputations for demonstrating competence or living up to the expectations of the group have built up idiosyncracy credits. These credits are like a bank account of favorable impressions built up over time. They may be exchanged for freedom from criticism or rejection following nonconforming or individualistic acts. A high-status member can attempt to change norms by playing the devil's advocate, or offering a "wild idea" or a suggestion of "let's try this for size." However, even high-status members may be viewed as deviant if the changes they are suggesting are too drastic. This hypothesis would indicate that a new member of the group who has not had time to build up idiosyncracy credits would be accorded less freedom for nonconformity than a senior member who has had ample time to build up good will. In a sense, senior members have "paid their dues" and are rewarded with increased freedom from censure.

By groups diagnosing their own norms and modifying them, so that they are compatible with the group's goals and resources.

By an outside consultant. Outside consultants, who are behavioral scientists and experts in organizations, bring an objective perspective to the organization; and because they are not part of the hierarchy, they can influence changes in norms without the same fear of sanctions. They may offer a variety of alternatives and methods for examining change that may be acceptable to the organization.

By trained internal consultants. In this situation, members of the organization are specifically designated to review procedures in the light of organizational goals.

Through group discussion. Norms formed through interaction can be changed by interaction. Group discussion is generally found to result in more change than other forms of persuasion, such as lectures or directives (Lippitt, Watson, and Westley, 1958).

By those with high self-esteem (Constanzo, 1970; Stang, 1972) and who are willing to risk.

By those previously reinforced by the group (Endler, 1965, 1966).

By generations of members who over time have acquired influence in creating change.

In many areas, we have witnessed enormous societal changes, with an impact on norms that would have been difficult to even fantasize.

An order of nuns was founded in the middle of the last century and remained virtually unchanged for one hundred and thirty years. There were one thousand women dedicated to treating the poor and the sick.

Women who entered the order knew that they could be sent home for even the simplest disobedience. They wore long black habits made from fifteen yards of material, held together with seventy-two tiny buttons that had to be closed with a button hook. They wore immaculate, heavily starched bonnets with attached veils.

Convents housed fifty or more nuns. At the convent, there were no visitors, not even family, and conversation was frowned upon as frivolous or gossip. At dinner, each had her seat at a large table; seating was by longevity in the order. Those who were novices sat at the foot of the table, served, and washed the dishes. No one spoke. Each evening was spent in work or prayer in the chapel. They were not permitted to attend family functions—not even a sister's wedding.

Contrast that description with the same order today. Members are still committed to the poor and sick, but now may live in a small house in an urban neighborhood where they can be close to those they serve and where they share the lifestyle and community of their people. In small groups of four to six, they become a family for each other. All work, cook, serve on community committees, and can go to the movies or lectures or classes. They are dressed in clothes typical of women their ages, and enjoy being feminine (but are still embarrassed at being admired for their appearance). Discussions on how the community should be conducted are essential elements to their planning for the future of the order and they vote on many issues. They are consulted rather than ordered. Visitors are welcomed, and nuns are regularly in touch with their families.

Who would have believed all of this could have occurred in the last ten years?

New members will enter a group confused, the norms will be unseen or feel ambiguous. They will feel restrained in their behavior until they can determine what is appropriate behavior and what is expected of them. As they interact with others, the rules become less absolute and more flexible. They appear not only as constraints, but also as guides. Newcomers are then better able to see the norms as outer limits within which they are free to operate. When they learn what is allowed, they gain both a sense of confidence and a greater latitude in their behavior—they have options to conform, to deviate, to change, or even to leave.

EXERCISE 1 Awareness of Norms

Objectives

To develop an awareness of norms by changing the norms

To develop the understanding that norms function to create an order in the group and regulate the interaction among members

To note the changes in the group when the normative structure is violated

Setting

The following exercise is illustrated within a classroom setting, but the procedures would be equally applicable in a variety of other situations.

Action

The exercise should not be used until after the group has met for some time; then the teacher (facilitator) examines a number of norms of the class (group) and for one session purposely changes them. This need not continue for more than 15 minutes of the classroom period.

An example, as the authors have used it, may be helpful. A course in group dynamics is taught to college students. Typically, the chairs are arranged around tables; smoking and drinking coffee in class are accepted; attendance is not taken; observer's reports are returned at the beginning of each class; examinations are not given unless scheduled. The teacher is available for questions or problems that have occurred since the last class; sometimes there are a number of questions and class does not begin promptly. All these norms are purposely broken.

The students enter the class one day and encounter the following:

Chairs are set in rows.

A pile of examination blue books are on the desk (no examination had been announced).

The teacher (could be a substitute) sits at the desk and appears to be busy reading; he or she answers no questions, nor has discussion.

The class starts promptly.

The teacher announces the observer's reports are not ready to be returned.

The teacher calls the roll.

The teacher makes a sarcastic comment to someone smoking and to someone drinking coffee.

The teacher answers no questions on her or his strange behavior and appears not to notice the confusion in the class.

Blue books are distributed.

Members of the class are asked to write one or two sentences on how they feel at this moment.

The teacher explains that the norms of the class were purposely violated.

EXERCISE 2 Survey of Dress Norms

Objectives

To understand that norms develop uniquely to a particular group

To develop the understanding that the "character" of a group can be determined from observation of its norms

To collect data in order to determine whether there are differences in the norms of groups, and to explore the reasons that might be behind these differences

Situation

It is sometimes thought that norms are not developed by a group itself. Rather, they may exist for a variety of reasons, that is, they may be imposed by exterior forces, they may be common to the environment, they may be typical of all people and so forth. A simple research project can be another method for seeing that groups do develop their own norms, and that these transcend generalized norms for an age group or college population. For example: among the colleges in the Philadelphia area are University of Pennsylvania, Temple University, LaSalle College, Drexel Institute of Technology, Beaver College, Philadelphia Community College, and Philadelphia College of Textiles. One hypothesis is that college students in the Philadelphia area will dress reasonably alike, because all the students will conform to the dress codes of large eastern cities. They will wear what college-age students wear at an urban eastern college, and there will be no difference in dress among the colleges. (A student at Penn will be dressed similarly to a student at Drexel or one at Temple.) Another hypothesis is that each school will have its own norm for dress and that typical dress will be different and recognizable among the schools.

Action

The class is divided into work teams of four to six on a team. Teams stand at places on the various school campuses where they can observe the dress style of the students. Each team views a sample of fifty students. Each team visits the various schools at their convenience in no special order over a common two-week period. Team members tally their observations on specific items (see categories of dress, below) by a simple categorizing system. Each team tallies its results and describes the attire of a typical male and female at each of the colleges. On a given day, all the teams distribute their reports to the other teams. Data and typical attire for each school are examined for replication in other team reports. The data will indicate the norms that each school develops.

Discussion

How do norms develop? What function do they serve? Are they the result of outside influences? How are they changed? Interviews may also be conducted with some of those being observed and used as a stimulus to further discussion.

Variations

Any number of variations are possible in gathering data on contrasting populations; for example, dress codes of students or teachers in schools in different parts of a city or the city and suburban areas, the dress of ministers or priests in different orders or denominations, the dress of members of various civic organizations. Data relating to other norms could also be observed after hypotheses are developed for various groups. Dating behavior, conversation patterns in different offices, or between classes, hall behavior in local high schools are a few examples.

CATEGORIES OF DRESS

(Basis for tallies; separate sheets for each institution)

1. Hair length
   a. Females: Short—to nape of neck
      Average—to shoulders
      Long—beyond the shoulders
   b. Males: Short—no sideburns, well-trimmed at back and sides
      Average—sideburns, length not passing the collar in back, sides full but not over the ears

Long—that which exceeds the previous description
Clean-shaven
Facial hair—moustache, beard

2. Slacks
   a. Females: Jeans
   Designer jeans
   Other pants
   b. Males: Dungarees
   Levis
   Designer jeans
   Pressed slacks
3. Other Attire
   a. Females: Skirts
   Dresses
   Suits (skirt and jacket)
   b. Males: Suit pants (tweeds, stripes)
   Suits (pants and jacket)
4. Shirts
   a. Males: Button-down collar
   Ivy league
   White
   Colored (yellow, blue, and so forth)
   Patterned
   Tee-shirt
   Turtleneck or other casual shirt
5. Blouses/sweaters
   a. Females: Shirts
   Tee-shirts
   Sweaters
   Jerseys (turtleneck, jewel neck)
   Shirt with sweater
6. Shoes
   a. Males: Penny loafers
   Laced dress shoes
   Sneakers
   Sandals
   Boots
   None
   b. Females: High heels
   Flats or small heel
   Loafers
   Sandals
   Moccasins
   Boots
   Sneakers
   None
7. Stockings/Socks
   a. Males: Crew
   Dark

Light
None
b. Females: Stockings
Knee socks
Colored stockings (yellow, green, and so forth)
Patterned stockings
None
8. Pocketbooks
a. Females: Shoulder
Handbag
Bookbag
None

EXERCISE 3 Toward Changing Group Norms

Objectives

To understand what is meant by group norms

To recognize the difficulty in changing norms

To offer an opportunity to examine and change norms at various levels

To develop insight into how norms can be changed in organizations

When appropriate, this exercise can be used after a group has been working together for some time. It can be a work team, a task group, an organization, a segment of a class group, or a seminar.

Action

Phase 1 The facilitator presents a small lecture on norms—what they are, their influence on a group, and so forth. The material in the first part of the chapter might be the basis for development of such a talk. (Approximately 10 minutes)

Phase 2 The groups are asked to examine their norms. They list as many as possible (dates, times, seating arrangements, order of meeting or work, typical behaviors, and so on). These should be listed on paper. (Approximately 20 minutes)

Phase 3 The group is then asked to change some of its norms. The facilitator says, "Which norms can be changed?" He or she has the group change them and holds a brief session under the new conditions.

Typically, groups change superficial norms: they will sit on the table instead of sitting in chairs; or will shout rather than talk to one

another; or attempt to conduct the session nonverbally. This encourages laughter and a reduction in inhibitions—perhaps even a party atmosphere. (Approximately 20 minutes)
(Usually a break is indicated here.)

Phase 4 The groups are asked to examine which norms they changed, and how relevant the changes were in helping accomplish their goals. What impeded changing norms? What norms need to be changed? The group then goes into its work session on this basis. (The ensuing discussion is very different from the party atmosphere of the first change; it raises difficult issues for the group and involves members in high-risk behaviors.) An observer might be assigned to watch for behavior that changes norms.

Phase 5 Later, perhaps the next day, the next week, or the next group meeting, a discussion of what norms were changed, which were difficult to change and why, should be conducted. Who was most influential in changing or proposing changes in norms? Who has the highest status and role in group? What are the problems involved in changing norms to increase movement toward the group's stated goals?

EXERCISE 4 Group Influence

Objectives

To understand the pressure of a group on the individual

To understand the internal conflict that arises when the group opinion is different from the individuals

To understand how powerful group influence can be

Action

Invite a cult member (for example, Hari Krishna or Moonie) to come to class to describe how he or she became involved in the cult. Have the cult member talk about what he or she was like before and after joining. How did he or she come to modify his or her beliefs on family, clothes, education, sex, friends, future? Then, in a question and answer period, discuss why he or she left, how it felt to be out of the group, what difficulties he or she continues to have, and how he or she now views the experience.

Variation

Another way to achieve this understanding is to stage the event. The teacher/facilitator contacts a person to enact the role of a former cult member. There are some vivid first-person accounts of what it was like to be a cult member, as well as books that describe the process of inducing members to accept the cult group norms. The person (who is unknown to the class) can prepare by reading the materials.

The teacher/facilitator announces that there will be a guest who was a former cult member who will discuss how cults work and how they pressure members to adhere to group norms above all. This exercise is even more effective if several persons come, and one is assigned to each small group. After the talk and discussion, the teacher/facilitator announces the staging, and the cultists explain who they really are.

Still another method is to have group members design a day (class time) in which they role-play being members of a cult, creating a mood so that they can experience the fear and urgency of remaining compliant to the group.[4]

EXERCISE 5 Differences between Group Norms (Parsons Model)

Objectives

To examine the basis for norms within a group

To examine the differences in norms between two groups

To learn more about norms within a group

To understand more about the effect of norms on group productivity

Setting

The facilitator explains that each group develops its own norms that affect both the relations among members and the group's productivity. An understanding of a group's norms will help it examine the appropriateness of its norms and the areas of stress.

Questionnaires such as the one following the exercise should be distributed.

[4]The Anti-Defamation League of B'nai Brith has materials that will be helpful. Their national headquarters are in Washington, D.C. and they have branches in most large cities.

Action

Members fill in the questionnaires. Members, by groups, analyze the results and each group reports back to the entire group. The data may be tabulated as the mean average reply on the rating scale or it may be on the median reply of persons responding.

Discussion

What kind of norms does this group have? On what are the norms based (explicit, implicit, formal, informal)?

What are the areas of stress?

How appropriate are these norms in terms of the purposes of this group?

Could there be more effective norms? What would they be? How could these norms be introduced and effectuated?

Variation

This exercise can be a means to contrast two groups with which members are involved. It is effective if one is a small group and the other a secondary organization. The discussion questions can be the same as above, but there will be an increased understanding that norms for one group may be inappropriate for another. Yet another basis from which to examine norms is to use a questionnaire similar to the one below, but having respondents answer the question: "In what groups do I derive the most satisfaction?" Then the four questions are asked. Ensuing discussion develops understanding that there is not one desirable model; that varying norms are appropriate in different situations; and that many norms in existing organizations are outmoded.

Reference

Parsons, T. and E. A. Shils, Eds. *Toward a General Theory of Action.* Cambridge: Harvard University Press, 1951, pp. 80–88.

QUESTIONNAIRE ON NORMS

1. What are some of the formal rules held by this group that are not followed and for which informal norms have replaced the established rules? (For example, consider time, absenteeism, the use of Robert's Rules, criteria for membership.)

2. Suggest norms that exist around the following areas:
   a. Who speaks and who is listened to (consider the implications of status, age, achievement)?
   b. What kind of language is permissible (slang, swearing, tone of voice, punitive, supportive)?
   c. How are decisions actually made and what unstated norms govern this process (vote, few power people, assuming agreement, rule of discussion)?
   d. How are feelings shown in the group, both in terms of the positive and negative attitudes that may exist?
3. How would you rate this group in the following areas?

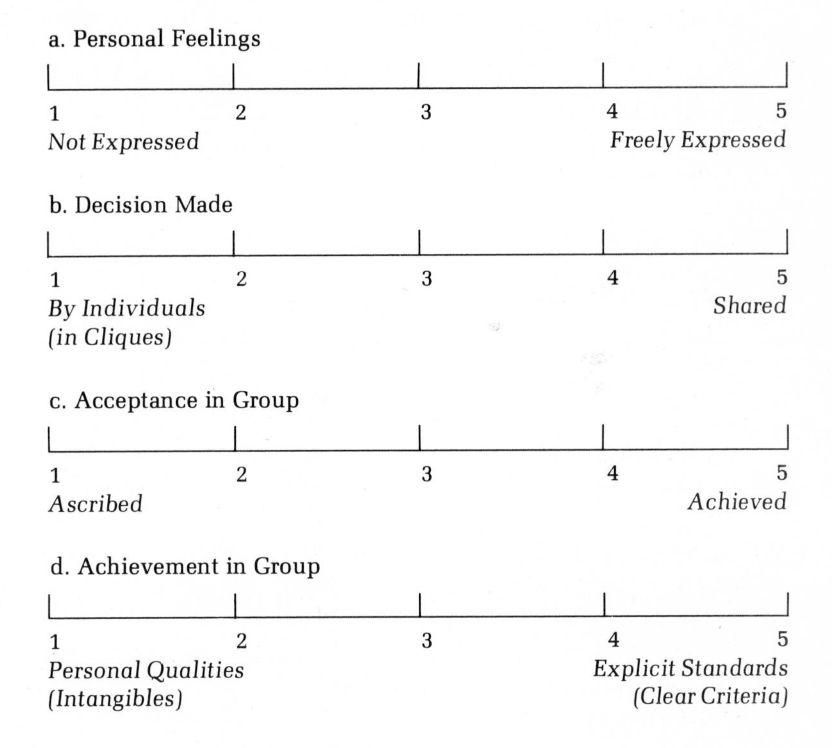

EXERCISE 6 Group Pressures on Issues

Objectives

To understand how group pressure affects an individual

To illustrate the influence of group decision making

To learn the distinction between individual responses and individual responses after consensus or discussion

Design

The facilitator distributes the following questionnaire to each of the participants. He or she requests that the participants fill in the answers as honestly as possible; the questionnaires may be answered anonymously. After the questionnaires are collected, the participants are divided into small groups of six to ten. The facilitator distributes additional copies of the questionnaire to each participant as well as an additional one to record the group answer. He or she asks the participants to discuss each question by groups and to arrive at a *single decision* on each of the questions, a decision that represents their group's feelings on the subject. A discussion follows, and the summary sheets are collected. Each group should be named at the top of the sheet. Each person answers his or her questionnaire as an individual. (If possible, this might be done after a break or at a later session.) The questionnaire may be responded to anonymously, but the group name appears at the top.

The facilitator tabulates the replies to the first questionnaire (the individual replies) and records (on a blackboard or newsprint) the group decisions on each of the questions. He or she tabulates and records the replies to the individual questionnaires following the group decisions and indicates the individual replies under the group answer.

Discussion

1. How are replies the first time different from the second individual replies?
2. How do groups influence individuals? What are the implications?
3. Do people act differently in a group than they act as individuals? How? What difference does it make?
4. How does group discussion influence individual judgments? Are there implications for social change?

Was there a difference in degree of group replies on the first two questions, as opposed to third and fourth questions? (The first two were concerned with personal choices, the second two with social issues). How were the final individual replies influenced by the group on the first two questions, on the last two questions?

QUESTIONNAIRE

The following issues generate a great deal of discussion. How do you feel about each of these issues? Circle the answer you consider most representative of how you feel.

1. Would you marry someone of a different religion?
   Definitely
   Probably
   Undecided
   Probably not
   Definitely not
2. Would you marry someone of a different race?
   Definitely
   Probably
   Undecided
   Probably not
   Definitely not
3. Should abortion be legalized?
   Definitely
   Probably
   Undecided
   Probably not
   Definitely not
4. Should marijuana be legalized?
   Definitely
   Probably
   Undecided
   Probably not
   Definitely not

EXERCISE 7 Effects of Group Pressure Toward Uniformity

Objectives

To better understand the process by which groups function to obtain conformity

To provide the basis for examining what group members do to exert influence on one another

To better understand the deviant's position

To better understand the dynamics of intervention

To understand the effects of a latecomer

Method[5]

The facilitator groups the participants into clusters of twelve to eighteen. He or she tells each cluster to designate half their group as

[5] This design was developed from an experiment by Stanley Schachter in "Deviation, rejection, and communication," *Journal of Abnormal and Social Psychology*, 46 (1951), 190–207.

group *A* and the remainder as group *B*. Group *A* is given sheets 1, 2, and 3 (included with this exercise) and instructed to scan the materials. Group *B* is given sheet 4. The facilitator asks that they study the observation instructions, be prepared to observe, and report on their observations. He or she then asks group *A* to select two of their members to be latecomers to the meeting and instructs the two latecomers, briefing them separately. He or she gives each a copy of the latecomer instructions.

Action

The participants sit around a table; the meeting begins and continues until a unanimous decision has been reached. Then the facilitator sends latecomer 1 in; eight minutes later he or she sends latecomer 2 into the meeting. (When the facilitator sends the latecomers in, he or she reminds the observers to begin their observations.) The meeting is continued until a high point has been reached, or a decision is reaffirmed or changed. The facilitator then pairs members of group *A* (participants) and members of group *B* (observers) and instructs them to do the following task:

1. Identify how the group's behavior toward each latecomer affected the group's attitude toward its decision.
2. Determine what the group did to influence each latecomer.

SHEET 1–THE SITUATION

You are citizens of a community in which a new superintendent of schools has just been hired. The superintendent is eager to obtain some knowledge of your attitudes concerning the proper treatment of children who get into trouble in the schools and in the community.

The superintendent has asked you to discuss what you would want her to do with a typical case. Her interest is not in the specific method of treatment, but rather in your attitudes toward children who are in trouble. You have been invited to meet in the living room of a member of the group.

The superintendent has decided it would be wise for her not to be present at the meeting. She has furnished you with two documents: one document is a brief summary of the case on which she wants your counsel, the other is a scale she calls a "love-punishment scale." She has asked you to arrive at a unanimous decision concerning the point on the scale that best expresses your opinion.

You are informed upon arrival that two persons are late and will be coming in soon. You are not to wait for them but are to proceed with making your decision.

SHEET 2–REGIONAL CITY PUBLIC SCHOOLS

Summary of Case No. 217 Name: Johnny Rocco

Johnny is the third child in an Italian family of seven children. He says that he has not seen his father for several years. His only recollection of his father is that he used to come home drunk and would beat every member of the family. Everyone ran when Father came staggering home. Mother, according to Johnny, has not been much better. She is constantly irritable and unhappy. She has always told Johnny that he would come to no good end. She has had to work, when her health allowed her to do so, and has been so busy keeping the family supplied with food and clothing that she has had little time to be the kind of mother she would like to be.

Johnny began to skip school when in the seventh grade. He is now in the ninth grade and is having great difficulty in conforming to the school routine. He seldom has lessons prepared, often misbehaves in class, is frequently a truant, and has been in a number of fights with schoolmates in the past year.

Two years ago he was caught stealing from a local variety store. Since that time, he has been picked up by police for stealing, for destroying property, and for being on the streets at a very late hour. Police have spotted him as a "bad one."

The court dealt with the matter by appointing a "big brother" to care for Johnny. The man, Mr. O'Brien, has brought the first semblance of discipline into Johnny's life. Through Mr. O'Brien, Johnny got a job running errands in a grocery store. Thus far, he has worked well on the job, although he complains that his boss is too strict.

One teacher has great appeal for Johnny. She teaches English. He says that she is the only kind and thoughtful person he has known and that he would do anything for her. Despite this statement, Johnny has not shown good work in her classes. He apparently spends most of his time in English class in some sort of daydream. The teacher has had very little contact with Johnny outside of her class.

Next year Johnny will be in senior high school. This will enable the school system to make counseling services available for him. The schools do not provide professional help for students in junior high school or below. The school principal has been attempting to deal with Johnny for the past two years.

In the senior high school, a number of things may be done, or arranged, for Johnny. A well-organized program of study fitted to Johnny's abilities and interests can be developed. It is also possible to have Johnny put into a foster home, through the help of the State Children's Institute, or to have him committed to the State Vocational School for Boys.

What plan the school system will follow next year depends, of course, on how Johnny behaves in the next few months. In general, the schools want to follow policies that are acceptable to the citizens of the community.

SHEET 3–LOVE-PUNISHMENT SCALE

It is important to note that Johnny is not an attractive child, and he is weak, sickly, and shows signs of malnutrition. What kind of attention should the public schools try to arrange for Johnny?

1. Give Johnny very much love, warmth, and affection so that he learns that he can depend on others and that they will protect him and overlook his misbehavior.
2. Give Johnny understanding treatment of both his personal and his family difficulties, based on careful diagnosis, so that Johnny can learn to handle his problems with the help of others when he needs it.
3. Help Johnny's mother set up a more wholesome family life.
4. Give Johnny impersonal attention in an orderly routine so that he can learn to stand on his own feet.
5. Give Johnny a well-structured schedule of daily activities with immediate and unpleasant consequences for breaking rules.
6. Provide strict control over Johnny's activities and immediate attention to misbehavior, so that he will learn adult standards for behavior.
7. Create very strict and very strong controls over every event in Johnny's daily life together with immediate and strong punishment for misbehavior.

SHEET 4–OBSERVATION INSTRUCTIONS

1. Keep a tally of the number of times comments are made to each latecomer.
2. Jot down the nature of the comments as much as possible.
3. Make notes on what members of the group *do* to the latecomers.

LATECOMER ONE INSTRUCTIONS

You believe that point 6 on the scale (sheet 3) best expresses your opinion. When you arrive at the meeting, please stick to this opinion.

You like this group very much; you like the topic under discussion. You are proud to have been asked to attend the meeting and think you will get a lot out of it. Many of your friends admire your activity in this advisory group.

LATECOMER TWO INSTRUCTIONS

You believe that point 6 on the scale (sheet 3) best expresses your opinion. When you arrive at the meeting, please stick to this opinion.

You dislike this group very much; you don't like the topic. You are sure that this discussion will be a waste of time and that your friends would laugh at your spending time discussing this topic.

Discussion

How does pressure to conform affect interpersonal relations?

How does pressure to conform affect the group task?

What are the relationships?

Other Possible Areas for Discussion Can Include the Following:

1. A group that has formed a mutually agreeable decision tries to protect its decision (in this case it was their own property reached by joint discussion).
2. A person whose beliefs deviate from this opinion is put under pressure to conform.
3. Most of the communications are directed toward the deviant.
4. If members perceive that they are making the latecomer uncomfortable, they may try to adjust communications so that no difference of opinion is seen to exist; thus there is no deviant.
5. The deviant under pressure is often quite uncomfortable.
6. It is easier to resist the group pressures when one is not attracted to a group than when one is highly attracted to the group.

BIBLIOGRAPHY

Asch, S. E. "Effects of group pressure upon the modification and distortion of judgments." In H. Guetzkow (ed.), *Groups, Leadership, and Men.* Pittsburgh: Carnegie Press, 1951, pp. 177–190.

Asch, S. E. "Opinions and social pressure." *Scientific American* 193, No. 5 (1955), 31–35.

Asch, S. E. "Studies of independence and conformity: I. A minority of one against a unanimous majority." *Psychological Monographs* 70, No. 9 (1956).

Bandura, A. *Social Learning Theory.* Englewood Cliffs, N.J.: Prentice-Hall, 1977.

Baron, R. A. "The foot-in-the-door phenomenon: Mediating effects of size of the first request and sex of requester." *Bulletin of the Psychonomic Society,* 29 (1973), 113–114.

Baron, R. S., G. Roper, and P. H. Baron, "Group discussion and the stingy shift." *Journal of Personality and Social Psychology,* 30 (1974), 538–545.

Bennett, E. "Discussion, decision, commitment and consensus in 'group decision.'" *Human Relations,* 21 (1955), 251–273.

Bonacich, P. "Norms and cohesion as adaptive responses to potential conflict: An experimental study." *Sociometry,* 35, No. 3 (1972), 357–375.

Boszormenyi-Nagy, I. and G. M. Spark. *Invisible Loyalties: Reciprocity in Intergenerational Family Therapy.* New York: Harper & Row, 1973.

Buchwald, A. "The grown-up problem." In *Son of the Great Society.* New York: G. P. Putnam and Sons, 1966.

Caplan, G. and M. Killilea, eds. *Support Systems and Mutual Help Multidisciplinary Explorations.* New York: Grune and Stratton, 1976.

Codol, Jean-Paul. DU PROVENCE, LAB DE PSYCHOLOGIE SOCIALE, AIX-EN-PROVENCE, FRANCE. "The phenomenon of superior conformity of one's self to group norms in a situation requiring perceptual estimation of physical stimuli." *Cahiers de Psychologie,* 73, No. 16(1) (1973), 11–23.

Codol, Jean-Paul. DU PROVENCE, LAB DE PSYCHOLOGIE SOCIALE, AIX-EN-PROVENCE, FRANCE. "Concept of superior conformity of one's own group to accepted norms—does such a phenomenon exist?" *Cahiers de Psychologie,* 73, No. 16(1) (1973), 25–30.

Constanzo, P. R. "Conformity development as a function of self-blame." *Journal of personality and Social Psychology,* 14 (1970), 366–374.

Dentler, R. A. and K. T. Erikson. "The functions of deviance in groups." *Social Problems,* 7 (1959), 98–107.

Dustin, D. S. and H. P. Davis. "Evaluative bias in group and individual competition." *Journal of Social Psychology,* 80 (1970), 103–108.

Endler, N. S. "The effects of verbal reinforcement on conformity and social pressure." *Journal of Social Psychology,* 66 (1965), 147–154.

Endler, N. S. "Conformity as a function of different reinforcement schedules." *Journal of Personality and Social Psychology,* 4 (1966), 175–180.

Fensterheim, H. and J. Baer. *Don't Say Yes When You Want To Say No.* New York: Dell Publishing, 1975.

Ferreira, A. J. "Family myth and homeostasis." *Archives of General Psychiatry,* 9 (November 1963), 457–463.

Festinger, L. "A theory of social comparison processes." *Human Relations,* 7 (1954), 117–140.

Festinger, L. "Informal Social Communication." *Psychological Review,* 57 (1950), 271–282.

Festinger, L. and J. Thibaut. "Interpersonal Communication in Small Groups." *Journal of Abnormal and Social Psychology,* 16 (1951), 92–99.

Ford, D. L., P. M. Nemiroff, W. A. Pasmore. "Group decision-making performance as influenced by group tradition." *Small Group Behavior,* 8, No. 2 (May 1977).

Freedman, J. L. and S. C. Fraser. "Compliance without pressure: The foot in the door technique. *Journal of Personality and Social Psychology,* 4 (1966), 95–102.

Freeman, M. "A conversation with Margaret Singer." *APA Monitor,* (July/August 1979), 6–7.

Galbraith, J. K. *Time,* March 30, 1970, p 88.

Gardner, J. *Self-renewal: The Individual and the Innovative Society.* New York: Harper and Row, 1963.

Gerard, H. B., R. A. Wilhelmy, and E. S. Conolley. "Conformity and group size." *Journal of Personality and Social Psychology,* 8 (1968), 79–82.

Goffman, E. *Behavior in Public Places.* Glencoe, Ill.: Free Press, 1963.

Goffman, E. *The Presentation of Self in Everyday Life.* Garden City, N.Y.: Doubleday Anchor, 1959.

Goffman, E. *Interaction Ritual: Essays in Face-to-Face Behavior.* Chicago: Aldine Publishing Co., 1967.

Gould, L. J. "The two faces of alienation." *Journal of Social Issues,* 25, No. 2 (1969), 39–63.

Grinder, R. E. "Distinctiveness and thrust in the American youth culture." *Journal of Social Issues,* 25, No. 2 (1969), 7–19.

Hall, J. "Decisions, decisions, decisions." *Psychology Today,* 5, No. 6 (1971), 51–54, 86–88.

Hall, J. and M. Williams, "Group dynamics training and improved decisionmaking." *Journal of Applied Behavioral Science,* 6, No. 1 (1970), 39–68.

Hare, A. P. *Handbook of Small Group Research,* 2nd Edition. New York: The Free Press, 1976, 19–59.

Heckel, R. V. and H. C. Salzberg. *Group Psychotherapy: A Behavioral Approach.* Columbia, S.C.: University of South Carolina Press, 1976.

Hollander, E. P. "Competence and Conformity in the Acceptance of Influence." *Journal of Abnormal and Social Psychology,* 61 (1960), 365–370.

Hollander, E. P. "Conformity, status and idiosyncracy credit." *Psychological Review,* 65 (1958), 117–127.

Hollander, E. P. *Leaders, Groups, and Influence.* New York: Oxford University Press, 1964.

Homans, G. C. "A conceptual-scheme for the study of social organization." *American Sociological Review,* 12 (1947), 13–26.

Homans, G. C. *The Human Group.* New York: Harcourt, Brace, 1950.

Homans, G. C. *Social Behavior, Its Elementary Forms.* New York: Harcourt, Brace, 1961.

Jahoda, Marie. "Psychological issues in civil liberties." *American Psychologist,* 11 (1956), 234–240.

Janis, I. L. "Group Think." *Psychology Today,* 5, No. 6 (1971), 43–46, 74–76.

Jenkins, D. H. "Feedback and group self-evaluation." *Journal of Social Issues,* 2 (1949), 50–60.

Jones, E. E. *Ingratiation: A Social Psychological Analysis.* New York: Appleton-Century-Crofts. 1964.

Keating, J. P. and T. C. Brock. "Acceptance of persuasion and the inhibition of counter-argumention under various distraction tasks." *Journal of Experimental Social Psychology,* 10, No. 4 (July 1974), 301–309.

Kelman, H. C. "Compliance, identification, and internalization: three processes of attitude change." *Journal of Conflict Resolution,* 2 (1958), 51–60.

Kiesler, C. A. "Attraction to the group and conformity to group norms." *Journal of Personality,* 31 (1963), 559–569.

Laing, R. D. *The Politics of the Family and Other Essays.* New York: Vintage Books, 1972.

Levine, J. M., K. R. Sroka, and H. N. Snyder. "Group support and reaction to stable and shifting agreement/disagreement." *Sociometry,* 40, No. 3 (1977), 214–224.

Lewin, K. "Frontiers in group dynamics." *Human Relations,* 1 (1947), 5–42.

Liberman, R. "A behavioral approach to group dynamics, I. Reinforcement and prompting of cohesiveness in group therapy." *Behavior Therapy,* 1 (1970), 141–175.
"II. Reinforcing and prompting hostility-to-the-therapist in group therapy." *Behavior Therapy,* 1 (1970), 312–327.

Lippitt, R., J. Watson, and B. Westley. *The Dynamics of Planned Change: A Comparative Study of Principles and Techniques.* New York: Harcourt, Brace, 1958.

Lipsitt, P. D. and M. Steinbruner. "An experiment in police-community relations: a small group approach." *Community Mental Health Journal,* 5, No. 2 (1969), 172–179.

Luchins, A. S. "Focusing on the object of judgment in the social situation." *Journal of Social Psychology,* 60 (August 1963), 231–249.

Luke, R. A. "The internal normative structure of sensitivity training groups." *Journal of Applied Behavioral Science,* 8, No. 4 (1972), 421–427.

Martin, J. D., J. S. Williams and L. N. Gray. "Norm formation and subsequent divergence: prediction and variation." *Journal of Social Psychology 93, No. 2 (1974), 261*–269.

Martin, R. and D. Lauridsen. *Developing Student Discipline and Motivation: A Series for Teacher In-service Training.* Champaign, Ill.: Research Press, 1974.

McDavid, J W. and H. Harari, *Social Psychology.* New York: Harper and Row, 1968.

McGregor, D. *The Professional Manager.* New York: McGraw-Hill, 1967.

Milgram, S. "Behavioral study of obedience." *Journal of Abnormal and Social Psychology,* 67 (1963), 371–378.

Milgram, S. "Group pressure and action against a person." *Journal of Abnormal and Social Psychology,* 69 (1964), 137–143.

Milgram, S. "Liberating effects of group pressure." *Journal of Personality and Social Psychology,* 1 (1965), 127–134.

Mills, J. and C. E. Kimble, "Opinion change as a function of perceived similarity of the communicator and subjectivity of the issue." *Bulletin of the Psychosonomic Society* 2, No. 1 (July 1973), 35–36.

Mills, T. M. *The sociology of small groups.* Englewood Cliffs, N.J.: Prentice-Hall, 1967.

Moore, J. C., and E. Drupat. "Relationships between source status, authoritarianism, and conformity in a social influence setting." 34, No. 1 (1971), 122–134.

Parsons, T. and E. A. Shils, Eds. *Toward a General Theory of Action.* Cambridge, Mass.: Harvard University Press, 1951.

Regan, D. T. "Effects of a favor and liking on compliance." *Journal of Experimental Social Psychology,* 7 (1971), 627–639.

Rokeach, M. "Long-range experimental modification of values, attitudes, and behavior." *American Psychologist,* 26, No. 5 (1971), 453–459.

Rose, S. D. *Group Therapy: A Behavioral Approach.* Englewood Cliffs, N.J.: Prentice-Hall, 1977.

Ross, L., G. Bierbrauer, and S. Hoffman. "The role of attribution processes in conformity and dissent, revisiting the Asch situation." *American Psychologist,* (February 1976), 148–157.

Savell, J. M. "Prior agreement and conformity: An extension of the generalization phenomenon." *Psychonomic Science,* 25 (1971), 327–328.

Schachter, S. "Deviation, rejection and communication." *Journal of Abnormal and Social Psychology,* 46 (1951), 190.

Schachter, S. *The Psychology of Affiliation.* Palo Alto, Calif.: Stanford University Press, 1959.

Schmuck, R. "Helping teachers improve classroom group processes." *Journal of Applied Behavioral Science,* 4, No. 4 (1968), 401–436.

Schmuck, R., P. Runkel, and D. Langmeyer. "Improving organizational problem-solving in a school faculty." *Journal of Applied Behavioral Science,* 5, No. 4 (1969), 455–482.

Sherif, M. "A study of some social factors in perception." *Archives of Psychology,* 27, No. 187 91935).

Sherif, M. *The Psychology of Social Norms.* New York: Harper, 1936.

Sherif, M. "Conformity-deviation, norms, and group relations." In I. A. Berg and B. M. Bass, eds. *Conformity and Deviation.* New York: Harper, 1961, 159–1981.

Singleton, R. "Another look at the conformity explanation of group-induced shifts in choice." *Human Relations,* 32, No. 1 (1979), 37–56.

Stang, D. J. "Conformity, ability, and self-esteem." *Representative Research in Social Psychology,* 3 (1972), 97–103.

Wahrman, R. "Status Deviance, Sanctions, and Group Discussion." *Small Group Behavior,* 8 (May 1977), 147–168.

Watson, G., ed. *Change in school systems.* Washington, D.C.: National Training Laboratories, 1967.

Watzlawick, P., Beavin, J. H., and D. D. Jackson. *Pragmatics of Human Communication: A Study of Interactional Patterns, Pathologies, and Paradoxes.* New York: W. W. Norton, 1967.

Watson, G. ed. *Change in school systems.* Washington, D.C.: National Training Laboratories, 1967.

Willis, R. H. and E. P. Hollander. "An experimental study of three response modes in social influence situations." *Journal of Abnormal and Social Psychology*, 69 (1964), 150–156.

Four

Goals

Almost since birth, we have set goals and accomplished them. Infants learn quickly to find mother's breast, and later to reach for the brightly colored mobile. They learn to cry when they are wet or uncomfortable with the goal of getting out of that situation. Later, they learn how to get out of the prison of their cribs so that they can wander around the house, free to explore. As individuals, we set goals, such as being in a desired location or out of a difficult or undesired location. We see an attractive person and arrange to sit next to him or her in class. We are harangued by a bore, and we get up and leave, making some excuse about having an appointment.

We set goals for positive accomplishments, such as getting a good stereo, or losing ten pounds, or working up to jogging eight miles, or becoming a lawyer, or getting married, or living in Arizona. We set goals routinely (identifying a desired situation in which to be at a future time). We can also set goals for getting out of a present undesired place, such as not flunking algebra, or making a conflicted marriage work so that it doesn't end in divorce, or not going into debt.

As individuals, we are constantly concerned with our needs and how to fulfill them (achieving our goals). We are always in touch with what we want; walking down the street, we may unexpectedly see ourselves in a shiny, plate glass window, and seeing the reflection, button our jacket, hold our heads up a bit higher, and pull in our stomachs—in keeping with our goal of maintaining a certain image.

We understand goals for ourselves. But accomplishing goals with groups is something else again.

The *Reader's Digest* has a special way of looking at our society. It does so with humor, assembling quotes that express what we feel through a well-turned phrase. Under the heading, "Meeting of the Minds," the following comments aptly described the functioning of groups in meetings:

> In any given meeting, when all is said and done, 90 percent will be said—and 10 percent will be done.
>
> A committee meeting provides a great chance for some people who like to hear their own voices to talk and talk, while others draw crocodiles or a lady's legs. It also prevents the men who can think and make quick decisions from doing so.
>
> Having served on various committees, I have drawn up a list of rules: Never arrive on time, or you will be stamped a beginner. Don't say anything until the meeting is half over; this stamps you as being wise. Be as vague as possible; this prevents irritating the others. When in doubt, suggest that a subcommittee be appointed. Be the first to move for adjournment; this will make you popular—it's what everyone is waiting for.[1]

Then, of course, there are the stock clichés, not original enough to warrant magazine space, such as "committees take minutes and waste hours," or "a camel is a horse designed by a committee."

[1] *Reader's Digest*, September 1977, p. 158.

What each of these comments expresses is our pervasive reluctance to work with a group, especially if we want to get something done. What emerges from the humor is the sense that it will be time consuming, frustrating, meandering, and conflicted—that in the end, the product (goal) will be less important than getting out of the endless wrangling, and what is accomplished will have so many compromises that, like the camel, it will be a different animal from what was initially intended. Furthermore, with more people, there will be more conflict, bickering, and confusion, as we see in the following example.

An international conference is a special event. It is an assemblage that demonstrates that there are certain common concerns that transcend differences in nationality, race, religion, or economic status. As part of the celebration of the twenty-fifth anniversary of the UN, it was agreed to sponsor a World Youth Assembly. It was decided to bring together the youth leaders of the world, the leaders of the future, to demonstrate their ability to work together (the next generation will not be crippled by our mistakes) and to make their recommendations based on discussions of international problems. It was understood that there would be many differences of opinion in the substantive areas, and that it might be difficult to arrive at recommendations that would be acceptable to all. However, it was assumed that the organizing would be simple and create a cooperative atmosphere for the more difficult later discussion. The following account appeared in the newspapers:

> The World Youth Assembly, in a disorderly session at times approaching bedlam, completed its organization yesterday. . . . A Puerto Rican participant withdrew from the youth parley . . . charging that leftists were converting it into a propaganda festival. An Israeli, seeking to raise the issue of treatment of Jews in the Soviet Union, was shouted down from the rostrum. Communist demands that participants from South Vietnam, South Korea, and Nationalist China be expelled from the Assembly were voted down.
>
> The Assembly overrode its eighteen-member steering committee and voted to throw open the session of its four commissions which will discuss problems of peace, development, education, and environment. The steering committee had recommended that these be closed sessions so that the participants could speak freely.
>
> Lars Thalen of Sweden, elected permanent chairman of the steering committee, had difficulty controlling the boisterous

> meeting as delegates shouted points of order, objection to statements, and demands that new courses of action be followed. "I am ashamed that this World Youth Assembly behaves in such a manner," Thalen shouted at one point as he pounded the gavel."[2]

It would seem that creating and achieving goals for an individual is possible, although, in some situations, difficult. When two people are involved, getting them both to agree to a common goal is more difficult. Attempting to get even more people to agree raises the level of difficulty exponentially. The task becomes an endless round of compromises, negotiations, and frustrations.

DISTINGUISHING BETWEEN INDIVIDUAL GOALS AND GROUP GOALS

Once there was a serious controversy over whether there was such a construct as a group. F. H. Allport (1924), an early social psychologist, would have said there is no such thing as a group mind and there could not be a group goal. He argues that:

> Alike in crowd excitements, collective uniformities and organized groups, the only psychological elements discoverable are in the behavior and consciousness of the specific persons involved. All theories which partake of the group fallacy have the unfortunate consequence of diverting attention from the true locus of cause and effect, namely the behavioral mechanism of the individual . . . If we take care of the individuals, psychologically speaking, the groups will be found to take care of themselves.

Kurt Lewin (1939) argued to the contrary. He noted that groups were different from individuals and that groups as systems could not be explained solely through knowledge of their constituent parts. And second, there is a difference between an individual and an individual goal, and a group and a group goal. Although the raging controversy is over, there continues to be confusion in this area (Quey, 1971).

We can examine individual goals and hope to better understand the behavior of individuals. In the World Youth Assembly example, we can focus on individual goals by exploring the reasons the Puerto Rican delegate resigned or the Israeli delegate took the rostrum to

[2] The Philadelphia *Evening Bulletin*, July 11, 1970, p. 4. Reprinted by permission of The Bulletin Co., Philadelphia, Pa.

question treatment of Russian Jews during a period of organization of the conference. We can also focus on the group goals by asking such questions as what was the delegate body's goal in reversing its executive committee and opening hearings to the public? In addition, we can look at the interplay of individual and group goals by examining why delegates were constantly raising points of order, since this was a method to play out individual goals and at the same time created norms of constant arguing. Were they trying to forestall the possibility of discussion and recommendations by the Assembly as an outcome? Were they playing out their seniors' organization, but now, instead of carefully articulating formal statements, expressing feelings of power and powerlessness directly? Our perspective in viewing produces different interpretations, but that is not to say that one is more important than the other.

An example will clarify the distinction between individual and group goals and suggest the necessity for understanding the group as a system that is different from individual goals (Mills, 1967).

Consider competitive games in general and tennis in particular. Each tennis player plays to win—to win is the personal goal of each—and so the goals of the two players are similar. However, although they have identical individual goals, we have no information about their collective (group) goal. The group goal is not "to win"; similar personal goals do not make a group goal.

To arrive at a group goal, we must first conceive of the dyad as a unit or group, and then ask what present or future state of this unit is thought to be desirable to the pair. Clearly, the state cannot be "to win," because the dyad as a unit is not a contestant and has no opponent; the dyad itself can neither win nor lose. Consequently, to refer to the goal of the unit as "to win" is meaningless. What is the goal of the unit?

READER ACTIVITY

Consider the illustration of the two tennis players. Each has an individual goal—to win, or perhaps, not to lose badly. They meet the criteria of a group, which is usually stated as two or more people interacting with a common goal. They will be interacting.

Are they a unit (two or more individuals)? ______ Yes ______ No

Why? ____

Is there a *common* goal? ______ Yes ______ No

Explain. ____

____

If yes, what is their common goal? ______________________________

____

There are a number of common goals possible, but in connection with tennis, the goal of the players is to have a good quality game, where they can land and receive some good shots, and where superior play wins. The group goal, as distinct from individual goals, is to have a good match. That last word helps to clarify the group goal, which is to have well-matched players for a good game. Why else are there handicaps and rankings?

How much satisfaction is there in winning over a player who hardly knows how to play, and in trouncing him or her easily? How does it feel to play with someone who is tired or who plays a lackluster game? And conversely, how does it feel to have lost after exhilarating play and a tight score?

One needs to understand two levels of goals: an individual goal, which is to win; and a group goal, which is to have a high-quality contest. Of course, the two are interrelated. The individual needs to be motivated to win to help produce a good contest; the rules of play and the process of selecting contestants are designed to stimulate good play.

Two points need emphasis. First, the group goal is not the simple sum of individual goals, nor can it be directly inferred from them. It refers to the desirable state of the group, not just the individuals. Second, the concept of a group goal is not some mental construct that exists in some mythical group mind. What sets the group goal apart is that, in content and substance, it refers to the group as a unit—specifically, to a desirable state of that unit. The concept resides in the minds of individuals, as they think of themselves as a group or unit.

HOW ARE INDIVIDUAL GOALS FORMED?

We recognize that we are motivated, and we move with direction, but how is this motivation formulated into a goal? How is it that we keep working in a chosen direction? In some of the most imaginative research, Zeigarnik (1927) and later, Morrow (1938) explain how. According to Zeigarnik's theory, when individuals set goals for themselves an internal tension system is aroused that is correlated to each goal. That tension system continues to motivate an individual until the goal is actually achieved, or there is a psychological closure so that a person feels that it has been achieved.

In the experiments on which the theory is based, subjects were given a large series of tasks to do (typical experimental tasks such as putting pegs in holes or crossing out a given letter on a page). On some of the tasks, the subject continued until completion; on others he or she was interrupted prior to completion. A significant finding was that subjects remembered the incompleted tasks more frequently than those they had completed. This experimental finding was validation that there is a tension system connected to a goal, which continues until the goal is met. This theory has been tested and verified with a wide variety of subjects.

For example, students have a goal to do well in a course, and a subgoal to pass a mid-term examination. They review the texts, study their notes, and apprehensively submit themselves to the ordeal of taking the examination. At the end of the test period, is the tension system reduced? Certainly not.

They wait anxiously in the hall and query others on their responses to difficult or ambiguous questions; they enter the next session of class eager to know if the papers are marked. Someone can be depended upon to ask almost routinely at the beginning or end of each session, "When will the papers be back?"

The tension system subsides when the students have their papers returned and they know more about their progress toward their goal. They then can determine what their next goal will be—with its coordinated new tension system.

A tension system coordinated to a goal and its recall as an incompleted task weighs heavily on all of us. The authors can attest to the recognition of a tension system coordinated to an incomplete task. The tension system was in action when friends invited us to dinner, or to see a play, or to take a couple of hours off while we were writing this book. Immediately, the tension surfaced in thoughts of "the book will never get done that way"; "at the rate we're going this won't be a book but a lifetime project; better stay home and work on it."

Zeigarnik found that a tension system was connected to an incomplete task, and Horwitz (1954) wondered if that tension system would also apply to a group. In his variation, he had teams of two go through the series of tasks. In some, he interrupted one person, but the other member was allowed to complete the task. In other tasks, the individual who started completed the task. A group tension system did emerge. If the task was completed by one person, the other individual did feel that the goal had been achieved (by the team), and the task was recalled with the same frequency as the task he or she had actually completed. Members feel a closure when a task is completed whether they perform the final stages of a task or another member of their group actually does.

HOW ARE GROUP GOALS FORMED?

How do goals change from individual goals for the person to goals for the group? In the transmittal, what are the problems, the implications?

Individuals Have Goals for the Group

Basically, individuals participate in a group because they believe they will derive more satisfaction than if they did not participate or belong.

Individual motives may be characterized as "person-oriented" or "group-oriented." Although the reasons are roughly classified in one category or the other, these motives should be viewed as a mixed bag—a percentage of each will motivate an individual's behavior. Person-oriented motives exist mainly to satisfy the personal goals of an individual. Recreational groups (tennis clubs, golf clubs, bowling teams); educational groups (classes in marine navigation; Great Books discussions; stock market clubs); therapeutic groups (group therapy; marriage counseling); "growth centers" (personal awareness groups, yoga-meditation groups, "games" groups)—all exist to serve primarily person-oriented motives. The individual seeks to meet his or her goals through the group activity.

Others belong to a group for what is termed a group-oriented motive, that is, they accept and conform to the group objective even though accomplishment promises no special benefit to them individually. The person, here, is satisfied by results favorable to the group as a unit. For example, people may be active in their political party although they do not personally know the candidates, are not anticipating a job in government, and do not expect any direct rewards. They are motivated to act because they believe their party represents the better choices in the forthcoming elections; they will be satisfied if their party wins and especially pleased if their party "wins big."

Another illustration is the community center board members who recognize that facilities at the agency are inadequate and antiquated. They recommend the establishment of a committee to raise funds for a new center. They are fully aware that their children are grown and will not utilize the new center, that a building campaign will mean they have less time to spend on their businesses, and that it will cut into their already limited free time. It will entail the onerous task of asking people for money, and it will be a thankless job. Yet they vote for the establishment of the committee, knowing full well that if it is approved, they will become a member.

As previously stated, this classification is not to be viewed as a dichotomy. However, the distinction helps to understand individual motivations in conjunction with a group goal. An individual whose prime concern is the person-oriented motive is likely to consider a suggested group goal in terms of alternatives for himself or herself. Which of the alternatives will be most satisfying? Which permits the greatest benefits at the least personal costs? The individual whose prime concern is the group-oriented motive will consider which goal, if attained, will be most beneficial to the group, even when the consequences may not be of benefit to him or her. In the person-oriented situation, the individual thinks, "If we agree on this goal, how will I look? How can I look best?" In the group-oriented situation, the individual thinks, "What chance do we have of making it, of accomplishing it? What would block us, how can we get around it and be successful?"

Although the terms we use are *person-oriented* and *group-oriented*, the reader is aware of the similarity of these terms and the age-old distinctions between "selfish" and "altruistic" motivations, or "ego orientations" as opposed to "task orientations"—which still interest researchers (Cartwright and Zander, 1968; Lewis, 1944; Lewis and Franklin, 1944). Does it really make a difference for which reasons a person helps a group achieve its goals? Isn't the real issue that he or she be willing to accept the group's goal and move in that direction? It seems that it does make a difference (Fouriezos et al., 1950). Research has indicated that groups with more self-motivated behavior had longer meetings, yet covered fewer items on their agendas. Also, they reported being less satisfied with both the decision making and leadership in their meetings than those groups having more group-oriented members (Ricken and Homans, 1954).

Individual Goals Are Converted to Group Goals

There are certain limitations set on the determination of goals. First, there are the limits set in the purposes of the organization. A United Fund Committee does not determine what its goal will be, for example, whether to raise money or not, but, rather, how much money. The organizational purposes determine the goal in this case, and it is the subgoal, how much money this year, which is discussed and agreed upon. A second limitation concerns changes in the group or its environment that may necessitate a reevaluation of its goals. Within these limitations, groups develop goals based on the criteria of fairness or effectiveness, or some combination of the two.

There seems to be a sustaining myth that groups arrive at goals in a manner reminiscent of a New England town meeting. Each person speaks and makes his or her point of view known. The others listen, consider the information, and arrive at a decision that represents the most effective method for dealing with the situation, or at least, the best decision possible. The group decision arrived at is compatible with the individual interests of the majority. In reality, the picture is quite different. Although ideally each member should have an equal say in the determination of goals, it rarely happens. Some, by their personalities, are more verbal and forceful than others; some speak with interest on many subjects, but others only speak in an area in which they disagree; still others simply do not participate. On another level, some are excluded from even an opportunity to speak as decisions are made by the executive board or the planning committee. The criterion of fairness, that is, full participation, in setting of goals is not met. Frequently the decision making or setting of goals by a select few or even one person is justified in terms of effectiveness. It is assumed that the head of a company knows best what a group can achieve; the expert is most knowledgeable in setting a goal for the whole group. The argument frequently goes, "If we had unlimited time, we could allow all members to participate, but it becomes such a long and frustrating procedure that it is a time-saver and more effective to have goals set by one person or a few people."

For some, an attempt at fairness means reduced efficiency; they believe time expended in arriving at goals could be better expended in progress on the goal. However, it is possible that the preceding two criteria may be compatible, that is, it may be possible to widen member participation in setting goals and direction for the group (increasing the fairness criterion), and through this increased participation, there may be goals arrived at which are also most effective (based on discussion of alternatives, resources, interests of members). A problem-solving method of arriving at group goals involves discussing alternative choices, examining the resources of members for developing each of the alternatives, considering the time factor for accomplishment, and questioning the probability of success.

In terms of the previous discussion of person-oriented and task-oriented motives of participants in setting goals, some differences in behavior may be discernible (Kelley and Thibaut, 1954). If task-oriented motives are most dominant, members are more likely to arrive at group goals through problem-solving approaches, that is, through exchange of information, opinions, and evaluations. If person-oriented motives are more dominant (Fouriezos, et al., 1950), goals are apt to be determined only after arguments, negotiations, bargaining, and forming of coalitions.

Group goal-setting behavior can be illustrated by Robert Kennedy's report on the 1962 Cuban missile crisis. Typically in meetings there were the usual fights, coalitions, tensions, and decisions by a special few. The late Senator Kennedy described the tension and disagreement among the men meeting to recommend an action that might "affect the future of all mankind." After considerable talking, they divided into groups and wrote their recommendations. They submitted their recommendations to other groups for criticism, then reworked their original ideas. Gradually there developed the outline for definite plans.

> During all these deliberations, we all spoke as equals. There was no rank and, in fact, we did not even have a chairman. . . . As a result, the conversations were completely uninhibited and unrestricted. It was a tremendously advantageous procedure that does not frequently occur within the Executive Branch of the Government, where rank is so often important.[3]

Goals were set with optimal opportunities for all to participate, with development of a plan based on consideration of alternatives, resources, time factors, and the probability of successful accomplishment.

READER ACTIVITY

Our membership in groups is not random. We join groups for a number of reasons, and are committed to the group's goals in different ways. Consider two situations:
Think of the group you most want to be associated with.

What do you want from this group? ____________________

____

What are you prepared to give to it? ____________________

____

What does the group want from you? ____________________

____

[3]Kennedy, R. F. "The 13 days of crisis." *Washington Post*, November 3, 1968, B–1.

What does it give to you? ______________________________

____

Think of a group you recently left or are considering leaving.

What do you want from this group ______________________________

____

What are you prepared to give to it? ______________________________

____

What does the group want from you? ______________________________

____

What does it give to you? ______________________________

____

How do these different situations influence you in working for the group's goals? ______________________________

____

CLASSIFICATION OF GOALS

In addition to understanding goals both from an individual and a group perspective, we can also understand goals in terms of a number of classifications. We can classify goals as formal or informal, and as operational or nonoperational. Further, we can describe the movement on goals as action on a *surface* agenda or a *hidden* agenda (Bradford, 1961).

These concepts can be explained through a case study of a highly conflicted national peace organization. This is the oldest peace organization in the United States. It has consistently spoken out against military expenditures, governmental coups, dictatorships, and oppression of the impoverished. Even before World War I, it was active in strongly campaigning for enfranchising women. It is consistently and vehemently opposed to war and has special representative delegate status to the UN, encouraging peace at every level. It steadfastly maintains that differences can be resolved through peaceful means and has served as the coordinator of a number of international conferences seeking to resolve complex, difficult problems through face-to-face talk and negotiation. It is a highly respected organization.

The organization is governed by an elected, volunteer board of directors. They are well-educated, overwhelmingly liberal, idealistic, and have been committed to the organization for at least a decade. They reside throughout the United States, with large blocks on the East Coast and a rival block residing on the West Coast.

The affairs of the organization are conducted by a young, nominally paid, individualistic staff working out of national headquarters in a large Eastern metropolitan center. The organization survives financially through dues from members belonging to chapters in each state, through sale of a newsletter, and through sales of literature and buttons.

Peak times, in recent years, occurred during the late sixties and early seventies, when membership boomed and activities mushroomed in protest of the Vietnam War. Some of the most successful and charismatic community organizers were hired to be members of the national staff. They were young and brilliant, excellent writers and moving speakers. They were less concerned with salary and more concerned with being opposed to the military-industrial establishment. They lived in communes, ate and dressed very simply, and hoped through protests and speech to change the course of U.S. policy, and at least, the actions of young people on college campuses. Their thrust was for a better, fairer, more peaceful world.

The executive director had peace credentials that went back forty years. She had written brilliant legislation; was an informed, eloquent speaker; was invited regularly by hostile governments to visit; and served on a number of international commissions. She also had ideas about supervising staff, assigning duties based on board policies and priorities, and an expectation that a staff would perform duties as assigned.

The organization's goals were well-known for over sixty years; they were to encourage peace and freedom. This was to be accomplished through increased memberships, conferences, literature sales, and legislative influence.

But, by the mid-1970s, there were serious difficulties on all fronts. The war had ended, and membership had declined by some 70 percent. The financial situation was so serious as to be precarious. Lower staff members were laid off, and remaining staff members were expected to assume additional responsibilities covering what previously had been two jobs. It was a time for creating a community within the staff—in cooperatively getting out the newsletter, or mailing out bills, or rising to each current emergency.

In this situation, the three most charismatic, individualistic young staff members, who headed essential departments, stopped coming into the office except for a few minutes once or twice a week. They preferred to work independently. They said they found the office atmosphere "bureaucratic" and "repressive." They rejected the "authoritarianism" of the executive director, with her requirements of filling out time sheets, working at the office, attending staff meetings, and her expectations that her priorities would determine their work assignments. The executive director was frustrated and irritable at their rejecting her requests, memos, and even orders, to carry out the work she assigned.

They were talented. They were knowledgeable. They wrote excellent articles and carefully researched policy statements. On the other hand, they adamantly refused to be involved in membership recruitment or retention, in chapter programs, or travel into "the hinterlands."

While the executive director was out of the country for two weeks at an international conference, the three staff leaders conducted a series of meetings with the remainder of the staff, encouraging them to express their grievances with the executive director, even goading them on. By the end of the first week, a letter was drafted by the three, and all of the staff signed. It was a letter to the board of trustees demanding that the executive director be fired.

The staff listed, with their interpretations, all of the staff grievances against her. Further, they said, since the organization was committed to new forms of structural relationships, it did not make sense to have a traditional, hierarchical organization.

It was their suggestion that there be a small increase to each of the top three staff members, and that they would develop a communal relationship among the staff that would be a model organization. There would be a further advantage of reducing financial difficulties by eliminating the executive director's salary.

As a final inducement, the staff stated that unless their recommendations were met, they would resign as a staff and call a newspaper conference to express their dissatisfaction with the "traditionalism and hypocrisy" of the organization. In closing, they added that they looked forward to working with the organization, increasing membership, and moving on to new heights.

Each member of the board received a copy of the letter. The dissident West Coast board members also received telephone calls asking that they, especially, be supportive of the staff. Certainly the West Coast members knew how it was to have the "Eastern Establishment" run everything from national headquarters to the board presidency. The West Coast board members responded that they would be "delighted" to speak at the board meeting on behalf of the staff.

At the next board of trustees meeting, the prime items on the agenda for discussion were "Restructuralization of National Headquarters" and "The Functions of the Executive Director."

Formal–Informal Goals

At a formal level, the president announced at the board meeting, "The staff has recommended that we reexamine the structure of our national staff toward it becoming more innovative and less traditional." It sounds as if the goal, at a group level, was to examine the current organizational structure with the aim of finding one that was a paradigm of a different view of the world. It would seem there was no special impetus for the discussion, but, rather, a periodic updating.

At an informal, or sometimes called *implicit,* level, the group goal was to decide whether to back the executive director or the staff. If they back the executive director, how will the organization function without a staff, or will the staff really resign? If they back the executive director and the staff calls a news conference, what will be the reputation of a peace organization that is so conflicted that its entire staff resigns in protest? And, how do they feel about being blackmailed?

It will be difficult to understand the discussion that ensues if we focus only on the formal goal. It will seem as if the board is discussing all kinds of irrelevancies. However, the real discussion will be based on the informal goals, and it is at that level that decisions will be made.

Movement on goals is understood in terms of the dynamics at the informal level.

Operational–Nonoperational Goals

Suppose someone comes in to the board of trustees meeting incensed that there has been a Russian takeover of Afghanistan. Suppose that she or he insists that the first goal of the organization is to speak out in behalf of peace and freedom, and that "petty internal squabbles" should not occupy valuable board time.

Suppose she or he insists that the board clarify its interpretation of peace, and that there be a discussion on their moral responsibility to draft a resolution condemning the aggressive takeover of another government. Suppose a long discussion then evolves on whether the Russians "were invited" to come to the aid of the government in Afghanistan to help keep the peace, or whether that story was a blatant fabrication and the takeover was an act of aggression and war.

A heated discussion then develops of the meaning of peace. Members get involved in emphasizing how important it is that organizations speak up in behalf of peace, and how "military expenditures always take on a high priority, but peace takes on a low priority." They discuss the issue, "Do increased military budgets deter war, or do they increase the opportunities for two armed nations to create a spark that ignites a war?"

Note that as the discussion goes further and further afield, and becomes more and more removed from solving the problem on the agenda, the goal has become increasingly nonoperational. What purpose does such a discussion serve? What are the benefits derived from such a discussion?

For one, it is a safe discussion. The board can rally around peace and carefully avoid the conflict that would emerge from discussing the real issues. Instead of dealing with the tension and divisiveness that will undoubtedly occur, it is safe to "get off the subject" and deal with a higher one, "peace." It also builds cohesiveness at a time when a frightened board is overwhelmed by the difficulty of the task it must face. It is much easier to know "rights" from "wrongs" in the outside world than in the interpersonal world of this organization, where board members know all of the central staff members.

Nonoperational goals are broad and vague (for example, favoring motherhood—although there is doubt about whether that is a universal any longer—or favoring full employment, or peace). There is fundamental agreement among members, and the subject is safe. There are words, but few actions other than vague recommendations and general resolutions.

Operational goals, on the other hand, are specific, with well-defined targets, action plans, and evaluations set for specific times. An operational goal would be a six-month membership campaign to gain additional members for the organization, or strategies to increase the organization's income.

An operational goal would be to decide how the situation between the staff and the executive director will be resolved. It would invite discussion from the participants through the personnel committee; it would involve discussion within the board about the issues raised and their feelings about the situation; it would involve information on previous policy in termination or resignation—and on and on. It would involve taking the steps necessary to solve the crisis, from assembly of facts, through discussion, toward resolution.

Operational goals are specific and carry with them clear direction for movement and recognition of what solution or goal attainment will be. Operational goals can also involve conflicts, as members have different ideas about what steps should be taken or have different priorities. Operational goals also carry with them ideas of success and failure, of time limits and evaluations, and of responsibility and achievement.

SURFACE AND HIDDEN AGENDAS

In addition to classifying goals, we can describe movement on goals on two levels; the surface level or *surface agenda*, and the below-surface or *hidden agenda*.

In the above illustration, the statement of the surface agenda is that there will be a reexamination of the structure of the national staff. It would seem that if the task is clear and the leader competent, there should be no reason why the group should not proceed logically and calmly to an intelligent conclusion. The president of the board is hard-working, intelligent, and competent; she has conducted many meetings. The task is clear. However, the discussion will not go logically nor will people react from intelligence. They will wrangle over inconsequential points. The West Coast group will align in favor of the staff. East Coast members, who work closely with the executive director, will favor her. The arguments will go, "It is time we reorganized our structure into a team/commune model. We have had this traditional, big-business model for too long. Social psychologists and researchers say that a new model is more effective."

Others will argue that "in a volunteer organization there must be a staff where the executive director is accountable to carry out board policy. It is too difficult to maintain accountability now, without further ambiguity as to whose responsibility a given task is." The discussion may proceed to, "While we are reconsidering structure, we might as well consider location. Our offices have been on the East Coast since we were established; isn't it time to move to the West Coast, where there are more chapters and a lifestyle compatible with our interests?"

It will go on. The West Coast people will push for changing the national headquarters location. Some may back the staff as a way to increase their power and make an alliance with youth rather than the establishment. Someone who has had difficulties with the executive director will be sure to recount the incident, for personal sympathy or as a legitimate way to "get even."

As the discussion goes on it may seem that all patterns of logical thinking have been forgotten; emotionality will be high; there will be calls for votes and all kinds of shenanigans, such as introducing amendments, recalling votes, voting and then rescinding, calling everyone in to vote (even using proxies); decisions may be made that would be an embarrassment to a ten-year-old.

Before we become too judgmental and exasperated, we need to realize that groups are working simultaneously and continuously on two levels. One level is formally labeled. Whether confused or clear, simple or difficult, this is the obvious, advertised purpose for which the group meets. Unlabeled, private and covered, but deeply felt and very much the concern of the group, is another level (Bradford, 1961). On this level are all of the conflicting motives, desires, aspirations, and emotional reactions held by the group members, subgroups, or the group as a whole that cannot be fitted legitimately into the accepted group task. Here are all of the problems that, for a variety of reasons, cannot be laid on top of the table. These are called *hidden agendas*.

In the example given, what are the hidden agendas of the three staff people who waited for the executive director to be out of the country before meeting and probing with the staff to create a list of grievances against her? What is the hidden agenda in advocating a new staff form?

At the board meeting, what is the hidden agenda of the West Coast group? Or of members who have had previous differences with the executive director?

Hidden agendas are neither better nor worse than surface agendas. Rather, they represent all of the individual and group problems that differ from the group's surface task, and therefore get in the way of the orderly solving of the surface agenda. They may be conscious or unconscious for the member or for the group.

Hidden agendas can be held by group members (as individuals or as subgroups/cliques), by the leader, and by the group itself. Each of these, in turn, can be divided in terms of the cause of the hidden agenda held and the person or group unit to which its action is directed.

Each agenda level affects the other. When a group is proceeding successfully on its surface agenda with a sense of accomplishment and group cohesiveness, it is evident that the major hidden agendas have either been settled, are being handled concurrently with the surface agenda, or have been temporarily shelved. Let the group reach a crisis on its surface agenda (the situation with the peace organization) and the repressed hidden agendas rage forth.

Groups can work diligently on either or both agendas. A group often spends endless time getting nowhere on its surface agenda, seemingly running away from its task, and yet, at the end, gives the impression of a successful, hard-working group. Often, group members leave a meeting saying, "Finally, we're getting somewhere." When asked what they had accomplished, they might mention some trivial aspect of the surface agenda. What they were really saying was that an important issue on the hidden agenda had been solved.

A group may have been working vigorously without visible

movement on its assigned task. Suddenly, it starts to move efficiently on its task, and in a short time completes it. The group had to clear its hidden agendas out of the way before it could work on its assignment.

Hidden agendas need to be understood, or a great deal of energy at meetings will go into frustration and feelings of powerlessness (the mood reflected by the comments at the beginning of the chapter).

Hidden agendas can be dealt with in the followng ways: Consider that hidden agendas can be present. Recognition of hidden agendas at individual and group levels is the first step in diagnosis of a group difficulty.

The leader, or a member, can make it easier for members to bring the hidden agenda to the surface with a statement like, "Perhaps we could go around the room, and each of us comment on how we feel about the project before we vote."

Remember that the group is continuously working on two levels at once. Recognize that the group may not move as quickly on the surface agenda as the more impatient might wish.

All hidden agendas cannot be brought to the surface; they may hurt the group seriously if discussed openly. Other hidden agendas can be talked about and do become easier to handle. It is an important judgment to know which can and which cannot be faced by the group.

Don't scold or pressure the group because there are hidden agendas. They are legitimate; each of us is constantly working out individual needs in the group as well as group needs. It is a legitimate part of group life that we see things differently and want different things accomplished.

Help the group work out methods of solving their hidden agendas just as they develop methods for handling their surface agendas. Methods may vary, but basically they call for opening up the problem, gathering data on it, generating alternatives for a solution, and deciding on one. Data from people, and feelings, are important.

Help the group evaluate its progress on handling hidden agendas. Each experience should indicate better ways of more openly handling future hidden agendas. There may be short evaluation sessions (ten minutes) at the end to review progress. Were they able to talk more freely in areas that were previously difficult? Is there a greater feeling of comfort?

READER ACTIVITY

Think of a group of which you are a member. What blocks the group's movement toward its objectives; why do things seem to "go off on a tangent"? List as many things as you can think of.

____

____

____

Who has the hidden agenda? (A member? A subgroup? The leader?)

____

____

____

Think about each of these hidden agendas. Given your knowledge of your group, what are some things that could be done with one of the hidden agendas?

____

____

____

As the group increasingly deals with hidden agendas, it becomes able to see with greater clarity what the group's goal really is. From the viewpoint of members, goals are sometimes classified as "clear" or "unclear." When clear goals exist, each member, if polled, could respond with a statement of the goal and the steps for attainment of that goal. Clear goals are more likely to be operational goals, they are more likely to be stated formally, and they are more apt to be on the surface agenda. When unclear goals exist, as in the peace group illustration, members, if polled, would give a variety of replies depending upon their personal interests, and they would have a variety of ideas about how these goals should be attained.

Generally, a successful group has clear objectives, not vague ones, and members of a group have personal objectives that are compatible with the group's objectives. The more time a group spends developing agreement on clear objectives, the less time it needs in achieving them, and the more likely the members' contributions will converge toward a common solution.

RELATIONSHIP BETWEEN GROUP GOALS AND GROUP ACTIVITIES

Goals themselves are powerful inducers of action. What the goal is, what kind of goal it is, influences relations among members.[4]

Content of the Goal Affects the Group

Let us assume that the goal of a correctional institution is the rehabilitation of prisoners (Zald, 1962). Staff members might be given a great deal of autonomy and be encouraged to be creative. There would be a large emphasis on programs. There might be a reading program or even a college program on the premises. There might be lectures, plays, and special speakers brought in to keep members informed. There might even be group activities like an orchestra, chorus, or dramatic group that performed for community groups outside the prison. The goals would be to encourage prisoners to develop their vocational and artistic skills and to become more competent in the straight world while serving time.

Contrast these activities with another correctional institution where the goal is custody of prisoners—holding them so that they cannot inflict damage on citizens of the community. In this situation, there would be fewer professional staff and more custodial staff (guards). Authority would be centralized, and rules plus penalties would be the basis for relationships.

The difference in content of goals will result in a difference in relationships among staff and prisoners, as well as in a difference in activities.

Difficulty of the Goal Affects the Group

Suppose that out of a twelve-game season, a football team won eight games. What will be the team's goal for next season? Will it set a goal at winning two games? To win only two games would be regarded as a disaster; it would be too easy. Will it set a goal of winning all twelve games? This seems too difficult, and would seem to doom the team to failure.

Assuming none of the key players graduate or have sustained serious injuries, the group would set an aspiration level of probably nine or ten games. To win this number would be regarded as a successful season, a fine performance by the team.

The example is meant to illustrate the aspiration level of a group, that is, when a group confronts a set of alternatives ranging from

[4]Korten, 1962.

easy to difficult and selects one, this is referred to as the group's aspiration level. Performance above this level will be considered successful; performance below this level as failure. The level of aspiration will influence members' self-evaluations, group activities, attractiveness of membership, and subsequent group cohesiveness. Groups that are successful tend to be realistic about their aspirations (Atkinson and Feather, 1966).

In the fifties, a generally held belief was that groups would make more conservative decisions than individuals. In contrast, Ziller (1957) found that decisions made by group-centered decision-making groups were more risky than decisions made by leader-centered groups. Stoner (1961) in a master's thesis, made a more direct comparison of individual and group decision making. He found that decisions made by groups were riskier than prediscussion decisons made by individual members of the group. After that finding, there was a rash of experiments in the sixties and early seventies dealing with *risky-shift phenomenon.*

Risky-shift has dropped from being a popular topic as the plethora of research indicated there were special circumstances when the group would make a riskier decision, and others when it would make a more conservative decision than an individual. The risky-shift phenomenon is more likely to occur when the group discussion provides pertinent and persuasive information (Bateson, 1966; Ebbeson and Bowers, 1974; when responsibility for the decision spreads over the group (Meyers and Arenson 1972; Zajonc, Wolosin, and Wolosin, 1972); when other group members approve of risk taking (Teger and Pruitt, 1967; Wallach and Wing, 1968); and when the experimenter indicates he or she approves of a risky decision (Meyers and Arenson, 1972).

A shift toward conservatism will occur if the same factors create a conservative environment. People are more comfortable making a conservative decision when there is social approval for making a conservative decision, or the experimenter favors a conservative decision (Rabow et al., 1966; Stoner, 1968).

It seems that the same factors that affect the members and the group in making other decisions also influence decisions toward risk or conservatism (Hare, 1976).

Type of Goal Affects the Group

Whether the goals are competitive or cooperative greatly influences the activities toward the goal and the relations among members. The earlier illustration of the tennis match was an example of a competitive goal; one player can attain his or her goal only if the other does not. Baseball is another competitive game. If one team wins, the other loses; there is never a tie. However, baseball also has cooperative goals. Each member of the team can only attain his or her goal of winning if the entire group also attains its goal—the team must win as a unit. Members therefore attempt to cooperate with one another, coordinate their efforts, and use their resources jointly. Observers report significant differences between groups working under competitive conditions or cooperative conditions. Where there were competitive goals, members would seek to "one up" each other, withholding information, and displaying hostile feelings and criticism (Klein, 1956). For example, in schools where grade-point competition is keen and colleges will accept only a limited number from one high school, competition is unbelievable. High-ranking students push to become presidents of obscure clubs, thus gaining one more degree of status to edge out competition. It comes as no surprise that a student will remove the notice of a prospective visit from a prestigious college from the bulletin board, or remove information on scholarships to reduce competition. And the competitiveness induces hostility and criticism toward those who figured out the best "angles."

Cooperative groups, in a task requiring collaborative activity, show more positive responses to each other, are more favorable in their perceptions, are more involved in the task, and have greater satisfaction with the task (Church, 1962; Julian and Perry, 1967; Wheeler and Ryan, 1973). As a result of their cooperative efforts, members are less likely to work at cross-purposes (Gross et al., 1972), are more efficient, and have a better quality product (Deutsch, 1960; Zander and Wolfe, 1964; Workie, 1974).

For example, some high schools have organized tours for students to visit a number of colleges in a given area. All students who are considering attending colleges, let us say, in the Massachusetts-Connecticut area, are asked to list them. An itinerary is then developed and posted, and students interested sign up for the weekend tour. The school makes arrangements with the admissions offices for those desiring interviews; students write to graduates of the high school for appointments to learn about how they see the schools. Following the trip, students meet to share their impressions of the schools and discuss who will be applying to which schools.

The two different methods of handling student needs for information about colleges have very different outcomes.

Group Goals Themselves Are Inducing Agents

Previously, we discussed the tension systems coordinated to a goal. It seems that when a group accepts a goal, those who most strongly accept the goal display a strong need to have the group achieve its goals. Acceptance of the goal is for them an inducing agent (Horwitz, 1954). However, if the group goal is not accepted by a significant section of the group, there is likely to be a high incidence of self-oriented behavior rather than group-oriented behavior, with activities being coordinated to personal rather than group goals.

GROUP PRODUCTIVITY

Group goals are meant to be a guide for action. One of the methods by which a group measures its success is whether it accomplished its goals. Were they clear enough and operational enough to be measured? At what costs? Are members disillusioned and are relationships strained? Are they glad the project is over so that they can terminate their associations?

Some might question the validity of raising the question, "At what costs?" For them, the productivity question is the only one to be asked. Did they accomplish the goal—raise the money, develop the recommendations, increase membership, or resolve the situation between the executive director and the staff in the best way?

Nonetheless, it has become standard (Barnard, 1938) to describe the adequacy of group performance in terms of both concepts: *effectiveness*, the extent to which the group is successful in attaining its task-related objectives; and *efficiency*, the extent to which a group satisfied the needs of its members.

Each factor can be examined independently of the other. It is possible to examine only task accomplishment, and frequently that is the only factor considered. It is possible to examine only relations among members and the degree of satisfaction each feels as a member of the group independent of task accomplishment, although this is much less frequently considered. Yet it is important to remember that a group expends energy on both aspects of performance, and the effectiveness and efficiency of a group set upper limits on each other. Some illustrations will clarify this relationship.

Cohesiveness of the Group Affects Productivity

If members spend their time strictly on business, the surface agenda, and ignore interpersonal relationships and hidden agendas, misunderstandings can increase. Communication may be severely limited, subjects to be discussed are highly controlled, and there becomes an emphasis on the "grapevine" and informal systems to meet members' personal needs. In this situation, each person does his or her attain the goal of having books available to the community in a manner that allows maximum use.

If a football team is to win, members are expected to learn the plays and execute them. The maverick who disregards the rules (pressures toward uniformity) will be ridiculed—"So you thought you were the whole team, or better than the whole team?"

Members are expected to adhere to the norms viewed as necessary to help the group achieve its goal; any member who does not adhere will be seen as a threat to that end, and efforts will be made to induce him or her to return to the group procedures.

For Group Maintenance Some group standards are sustained to help the group maintain itself. Procedures for paying dues, pressures for attending meetings regularly, norms of delaying the start of a meeting until sufficient members are present are all norms that are conducive to maintaining a group or an organization.

Frequently, other norms develop in an effort to sustain the group. For example, members may avoid areas of conflict in discussion for fear that such conflict may evoke anger or loud voices and, consequently, cause some members to leave as a result of the unpleasant experience. (It is a common norm that outward display of harmony is a necessary condition for group survival.)

Areas of tensions may be ignored. A discussion of reduced attendance at meetings may be studiously avoided, because to talk of why there is reduced attendance might create a trend of others dropping out or an image of a declining group. Certainly information that has brought disgrace on a member will not be brought up (a fraternity member dropped for academic failure; a civic leader indicted for tax evasion).

On the other hand, norms develop for announcing honors to the group. Whether a member is named in the newspaper, or makes the varsity team, or is included in a prestigious representative council, groups brag because it conjures up an image of being with the winners and enhances the desirability of membership. These norms will be enforced as members are solicited to report honors such as a speech, a publication, or a promotion—and even reprimanded for not letting others know. These pressures toward uniformity are also deemed legitimate as a means of sustaining the group.

THE POWER OF GROUPS

What is the power of groups to make people conform? How does it happen that people can be induced under the pressure of a group to commit acts that are beyond belief?

Social psychologists, incredulous about what they see about them, have sought to understand and theorize how it happens that people conform, comply, and obey. Kelman (1958) sought to understand how officers had been brainwashed as prisoners during the Korean War. Jahoda (1956) was aroused by the civil liberties issues and loyalty oaths of the fifties. More recently, the power of the group to induce people to conform is a continuous source of questions, speculation, and incredulity. Consider the following:

Patty Hearst, the abducted heiress daughter of the San Francisco publishing family was photographed robbing a bank with an ultra-radical group, after only two months of being with the group. She testified that she was forced to conform to the group's demands out of fear for her life.

The charismatic James Jones brought his followers, the People's Temple, to the jungles of Guyana. In response to complaints, a congressman and a committee were dispatched to investigate the movement in behalf of United States citizens. The congressman was killed, and in a desperate move to avoid predictable resultant actions, over nine hundred men, women, and children committed mass suicide by drinking a fruit punch laced with cyanide.

Cults like the "Moonies," Scientology, Hari Krishna, and others are able to convince typically white, middle-class, college youth to forego their families, education, career goals, and material possessions.

The most incredible, and horrifying example, is the experience of a Vietnam veteran[3] who called a Boston radio station on Labor Day, 1972:

> I am a Vietnam veteran, and I don't think the American people really, really understand war and what's going on. We went into villages after they dropped napalm, and the human beings were fused together like pieces of metal that had been soldered. Sometimes you couldn't tell if they were people or animals. We have jets that drop rockets, and in the shells they have penny nails. These nails—one nail per square inch (over an area) the size of a football field—you can't believe what they do to a human being. I was there a year, and I never had the courage to say that was wrong. I condoned that. I watched it go on. Now I'm home. Sometimes my heart bothers me, because I remember all that, and I didn't have the courage then to say it was wrong. . . . When you come home, you can't believe that you didn't have the courage to open your mouth against that

[3]*Time* Magazine, October 23, 1972 (p. 36). Reprinted by permission from TIME. The Weekly Newsmagazine; Copyright Time Inc. 1972.

> kind of murder, that kind of devastation over people, over animals.
>
> . . . A lot of guys had the guts. They got sectioned out, and on the discharge, it was put that they were unfit for military duty— unfit because they had the courage. Guys like me were fit because we condoned it, we rationalized it.

Each of the incidents is incredible and almost impossible to understand. How do groups induce people to conform to this level?

In thinking of how people respond to group pressures, three concepts have been helpful. They are conformity, compliance, and obedience (Baron et al., 1974).

Conformity

It is hard to resist explaining conformity through an old, old joke. It goes like this: a person is standing on a corner intently looking at the sky. Another person comes by, sees the first person looking at the sky, and also looks up. After a while there is a small crowd clustered around the original person, all looking at the sky. After looking for a while, and waiting, someone asks the first person what he is looking for. "Oh," he replies, "I'm not looking for anything. I have a stiff neck and I'm waiting for my friend to pick me up and take me to the doctor." That's conformity! It involves the way we are influenced by others simply on the basis of what they do. Our implicit assumption is that similar behavior will elicit approval, and dissimilar behavior will elicit disapproval.

At a rock concert, almost with the first bar of music, people scream and mill around; and in the forties, the sound of big band music was a signal for the audience to get up and dance in the aisles and on the stage. On a college campus, one sees masses of look-alikes and it becomes evident what the perceived norms of acceptance are. And high school! Is there anyone who will ever forget the narrowness of the boundaries of what was acceptable appearance, and how important it was, to our very being, to conform to those standards?

READER ACTIVITY

When you were in high school, how were you supposed to look?

Hair ______________________________

Clothes ______________________________

Shoes/sneakers ______________________________

Cosmetics ______________________________

Jewelry ______________________________

Do you remember someone who didn't look that way? What happened? How were they treated? ______________________________

____

____

Although clothes are readily evident illustrations of conformity, of course they were not the only example. Which activities were "in"? ______________

____

Which activities were "out"? ______________________________

____

Why? ______________________________

What did you want to do, but didn't because it would be "uncool" (or whatever the derisive expression was in your day)? ______________

____

____

Conformity is illustrated in the previous sections in discussions of intrapersonal conflict: in the findings of Sherif and the autokinetic phenomenon, the Asch group pressure experiments, and the Festinger social comparison theory. In each, there is an intrapersonal force to conform to the group.

Conformity occurs because we internally decide to go along with the group. It may be because we are ignorant of the subject, they are more knowledgeable, and we listen to the experts.

Conformity increases as the size of the group grows; the larger the number, the larger the group with whom we compare ourselves, and the more likely we are to conform (Gerard, Wilhelmy, and Conolley, 1968).

The more conforming behavior is reinforced, the more likely we are to continue (Endler, 1966). And of course, the more we want job, but steadfastly remains uninvolved with other members as people.

On the other hand, if members spend a great deal of their work time getting acquainted, building personal relationships, developing increased listening skills and influence on each other, there may be a high personal satisfaction, at least for some, but no time or energy to work on the task. High personal involvement may mean high morale, but little effort on task activity, and, consequently, low productivity. The dilemma of whether to sacrifice productivity or member relations is ever present.

However, there is evidence (Berkowitz, 1954; Thelen, 1954) that if the group spends more time initially in the interpersonal relationships, there will be greater long-run efficiency. If, during the initial phases of the group, members talk to each other, discuss their personal goals, and get to know each other, they build a common frame of reference, a set toward problem solving. Consider the following example.

One small, newly-formed task force was having a difficult time deciding on how it would proceed with its charge. It was not clear what the first steps in the process would be, and there were quite a number of possibilities.

Two of the seven members knew each other; the others had, at best, a "nodding" acquaintance. Although resumes of members had been distributed, they provided members with a knowledge of the others in terms of affiliations and education, but not as persons. The situation was further complicated by the fact that each was busy, and could only allot two hours per month to meeting time with the task force. It had been a mammoth feat even to find a time when all could meet.

Although members were willing and committed, they were frustrated by the first two meetings and questioned why they had agreed to serve. Very, very little was happening.

Since the meetings occurred from 8:00 to 10:00 A.M. (to reduce the amount of time taken from work), the group decided to meet at 7:00, have breakfast together and then proceed. That should have produced horror at getting up and to a central location at such an early hour. Instead, the group, in its first enthusiastic decision, agreed.

The effectiveness of the committee markedly changed after that first breakfast meeting. Members met, talked while eating their eggs about all kinds of personal trivia, and developed a knowledge of one another that greatly enhanced their ability to make decisions on how they would proceed with their task. Members came to like each other in the process and became interested in becoming a "terrific" task force. One member jokingly said that he enjoys the breakfast meetings so much that he eats breakfast at the restaurant every Thursday morning so that he won't miss the meetings (the group met the second Thursday of the month).

Members learn over time that some issues are to be avoided, and others can be readily discussed. They learn which subjects are special favorites of particular members, and in what areas members agree. Members develop a clearer view of their roles and where they fit into the group.

Frequently, more cohesive groups are more productive than less cohesive groups. The more the group is attractive to members, the more membership is valued, and the more members can influence each other. There seems to be a general circular relationship between group solidarity and efficiency. As members work together and see one another as competent, they are drawn even closer together, and this relationship increases the likelihood of successful performance.

However, increased cohesiveness does not necessarily mean increased productivity. Increased cohesiveness means that members are able to influence one another more. If they decide to use this influence for increased productivity, they could be, potentially at least, very effective. However, increased cohesiveness could also mean that workers might band together to restrict production, and thereby restrict productivity.

When groups seem bogged down in movement on their goals, one way of understanding the difficulty is to examine the relationships between time spent on task and time spent on interpersonal relations (task and maintenance behaviors). Or, to put it differently, study the relationship between effectiveness and efficiency.

Personalities in the Group Affect Productivity

How is productivity related to personality style? Which is better, people who are all task oriented, want to get the job done, and leave or members who want warm, intimate personal relationships? Which will be more productive? Or is it better to have a mix, so that they can complement each other?

The prime consideration is the nature of the task. If the group has a purpose that emphasizes problems of affection (an alcoholics recovery program), members should not have difficulty being close or expressing emotions. On the other hand, a project that emphasizes problems of control (procedures for conducting a census) should not be composed of members who have difficulty following directions or working on an ordered task.

If the task involves the major steps in problem solving, persons who have high individual scores on intelligence tests or problem-solving ability reflected in higher levels of education, usually form a more productive group than those who are less able (Tuckman, 1967; Turney, 1970).

Findings in general indicate that a group composed of those who prefer more formal relationships when working on a task were most productive (Berkowitz, 1954), but groups composed of members who prefer closer and more intimate personal relationships were also productive (Schutz, 1958). That is, groups made up of similar types were more productive than groups made up of some who preferred closer relations and others who wanted more distant interpersonal relations. This last group was characterized by recurring personality clashes, with resultant lower productivity (Roethlisberger and Dickson, 1939; Reddy and Byrnes, 1972).

The compatible groups were found to be more productive, presumably because they were able to agree in the social-emotional areas, and in so doing, freed themselves to work on the task.

For the same reasons, groups tend to be more productive when they are made up of members of the same sex (Gurnee, 1962), and groups of friends are usually more productive than groups of strangers (Weinstein and Holzbach, 1972).

These are only a few of the personality factors affecting productivity. Stress, disinterestedness, and self-oriented behaviors are also illustrations of personality factors that might influence productivity. It is important to recognize that productivity is influenced by, and needs to be understood by, examining both the nature of the task and social-emotional levels.

Productivity Affects the Group

What happens to a group in the process of achieving a goal? How are members influenced as they work together to determine their goals, to integrate personal goals into a group goal, to synchronize a series of activities with specialized roles for members, to evaluate the outcomes and perhaps modify or change their goals? How is the group different after this process?

They are different. First, groups have real, practical knowledge of their resources. After a group has worked together they know who attends meetings and who rarely attends; who talks a "good show" but cannot be depended upon to carry out a responsibility, and who can be depended upon; who really knows how to coordinate activities, and who is readily overwhelmed by a minor disturbance; who has contacts that work and who doesn't; who rarely speaks but is interested and hard working; who causes less friction if he or she works alone; who is highly committed to the group and group goals. Members know one another and what can be expected of each.

Second, they have increased experience in working together as a group. They now know how to go about determining a goal; they know how to get information they may need; they know what is involved, what skills are required, to complete a task.

Especially if they were successful at the first task, members have increased confidence in the resources of the group. The first time is the hardest and the most uncomfortable. The initial floundering to begin lessens, the fear or reticence to make a decision diminishes, and the early inexperience is replaced by growing experience and confidence. Members may begin to feel more comfortable and express themselves more freely and clearly.

Third, there may be an emergence of new group procedures and norms. Working together, members may realize that they could be more effective if they revised some of the procedures. They may even have changed some of the patterns of working relationships. Based on experience, they may decide to modify their methods of arriving at goals; they may modify their aspiration level based on their previous experience; they may even develop different criteria for success and for failure.

Also, the emotional level of the group changes. The initial surface politeness is gone. Members know one another at least at an acknowledgement level, and friendships develop. There may be greater flexibility in role behaviors; people who were too shy or fearful to volunteer for certain roles may now feel free to volunteer. It may mean that members feel more comfortable in the group, and so may say what they are thinking more spontaneously.

Then again, it might mean just the opposite. There may be a small status group, and others may feel less acceptance. The high-status members may speak only to each other, and most may feel outside the ruling clique. There can be increased hostility among members (as each may blame the others for failure), and a reduced willingness to accept group goals or to work together.

Most successful groups, when compared with less successful ones, seem to make a fuller commitment to the group goal, communicate more openly, coordinate activities aimed more specifically toward the goal, and achieve better personal rapport. And the effect is a spiral one. Work on one task influences the next, and there is a greater likelihood of success in pursuing goals in each subsequent task.

CHANGING GROUP GOALS

Goals in the broadest sense include a forward time perspective—a future different from the present (Bennis et al., 1969). Sometimes, group goals are inappropriate and should be reexamined and changed, but goals, like norms, are difficult to change. However, there is a greater likelihood that new goals will be supported, with concomitant implementation, if there is active discussion in creation of the new goals. If those involved at a later date in carrying out the new policies or procedures are also involved in setting them, it is more likely that they will integrate the new goals into their personal goal structures. Where behavior change is desired, setting goals through group discussion is more effective than separate instruction of individual members, external requests, or imposition of new practices by a superior authority.

The paradox, however, is that although it is easier to change people as group members than as isolated individuals, it is difficult to change the group's goals because of the very support members receive from one another. A number of steps have been suggested (Lippitt, 1961) to help a group be more productive:

1. The group must have a clear understanding of its purposes. It is important at the beginning of any group's life that it have a clear understanding of the goals it wants to reach.
2. The group should become conscious of its own process. By improving the process, the group can improve its problem-solving ability. The group should be encouraged to continuously diagnose its process and act accordingly. Time invested in building interpersonal effectiveness may lead to quicker and more effective task performance.
3. The group should become aware of the skills, talents, and other resources within its membership. Over time there is a tendency for groups to lock persons into certain roles; a continuous flexibility to other roles based on resources should be encouraged.
4. The group should develop group methods of evaluation, so that the group can have methods of improving its process. Data from evaluations help members become aware of the opinions and feelings of others.
5. The group should create new jobs and committees as needed and terminate others as outdated when no longer compatible with the goals.

GROUP GOALS AND THE INDIVIDUAL MEMBER

Initially members act in a new group as they acted in others (Mills, 1967); there is for them an undifferentiated membership role. They scan others for guides of norms and expected behaviors. As they become familiar with the processes of the new group, they learn which behaviors are rewarded and which are deviant. They widen their boundaries of what is acceptable behavior in this group. Their personal goals are no longer the only considerations.

They begin to operate at a higher level. They come to understand the group's goals, they accept them, they commit their personal resources to accomplishing the goal, and give it higher priority than their own goals. They eventually come to evaluate their performances and the performances of others in terms of accomplishment of the group's goals. They even modify their behavior to help the group become more effective.

Goals are such a central concept of groups that the most common definition of a group is two or more people interacting with a common goal. Some theorists define a group as a goal-seeking system (Parsons and Shils, 1951; Deutsch, 1963). All agree that group goals influence all aspects of group behavior.

EXERCISE 1 A Series of Skill Exercises

Objective

To increase skills in goal areas

Rationale

Goal setting seems obvious; participants often feel they have no difficulty setting goals, or agreeing on them. Frequently, any difficulties that arise are seen as "personality conflicts," which is another way of saying that nothing can be done. These exercises give participants an opportunity to check out their perceptions on goals and movement toward goals. There are also exercises that allow members to build skills in goal setting or statement of the problem. These focus on the group problem rather than inducing individual defensive behaviors. The following exercises should be used individually as appropriate.

1. Setting up the Problem

Usually when defining a problem, we do so in a way that implicitly suggests a solution. This may cause some people to become defensive and work on their private or personal goals rather than the group goals. This exercise attempts to help participants overcome that difficulty.

Action

Participants are divided into groups of six to ten. The facilitator introduces the exercise by saying, "Although all of us publicly state we want the group to make a decision, we behaviorally don't mean it even when we think we do. For example, we say the office secretaries use the phone too much and ask what we can do about it. But this question does not allow the group to make a decision based on determination of the situation. Rather, it puts the secretaries on the defensive. We do this all the time. How can we state the problem in such a way that some people do not begin to feel guilty and in which there is no implied solution? This will be an excuse to practice these skills."

The facilitator may state one or several problems that have occurred in the life of the group (one is preferable). In each case, the facilitator asks each participant to assume the appropriate position for asking the question (in the illustration cited, he or she could be the office manager). Each person writes the problem so as not to make anyone feel guilty and so as not to imply a solution. Then each member reads his or her statement and the others critique it for meeting the criteria. The group suggests improvements, and the next person is heard. As the analysis goes on, some general principles of stating the problem emerge. Each group reports its best statement of the problem and the general principles. As additional skill building, groups make up a problem and submit it to the next group. The same procedure as above is used, and there is a testing of the general principles. This exercise is cognitive, but members usually find the experience interesting in that they come to appreciate the difficulties of avoiding predetermined solutions as they refine their skills. Some problems for restatement, if the group does not create its own, would be as follows:

1. A bus driver reports that children in the buses are destroying property, using abusive language, and picking on younger children. The high-school buses especially have this problem on the morning run at 7:00 A.M.
2. Shortly, we will be electing class officers. I believe that they should be truly representative of the class. In the past, this has not been done.
3. Mr. Brown, from the American Federation of Teachers, came to visit me yesterday, and he urges us to affiliate. Last year Bill and John led the opposition, and we did not join.

2. Clarity of Goal Setting

Here the objective is to increase observer skills in goal setting and to increase awareness in various aspects of goal setting. The facilitator introduces a role-playing situation (the hidden agenda example in this chapter is appropriate, or another that involves a current group issue). Depending upon the facilitator's objectives, he or she may have one group role play and all others observe, or he or she may divide the total membership into a number of role plays with two observers for each role-playing group. The first method builds common skills in observation and goal setting; the second develops an understanding of personal and group goals with observations.

Observers are instructed to note whether behaviors are person-oriented or group-oriented, which behaviors helped to clarify the problem, and which impeded movement on the problem. The facilitator cuts the role playing when a decision is reached, or if it becomes evident that a decision will not be reached. Role players report back how they feel, especially with regard to movement toward a goal. How did their private agendas help them? Or did they impede them? What would have helped them become involved in the group goal? Was there a group goal? The observers report back. In the hidden agenda role play, there are usually so many more individual behaviors it becomes obvious that a decision cannot be reached until these factors are in some way dealt with. Some might be brought out into the open, some consciously ignored.

After the exercise, members begin to understand both the problems and pervasiveness of work at several levels on goals. They also develop increased awareness of the behaviors needed to help the group center on group issues rather than personal goals.

3. Diagnosis of Goal Clarity and Goal Movement

If groups are to work efficiently at goals, at both task and maintenance levels, it is essential that they become aware of their own processes. It is important to gather data on the current state of the group and use this information for helping the group set its goals, clarify them, and learn the degree of involvement.

Simple Reporting One method is to stop each session 10 minutes before the end. The members of the group discuss their answers to these two questions: How much progress do you feel we made on our goals today? What would help us?

This can be done in a workshop at the end of each session; it can be used effectively in ongoing work groups. Initially there is resistance to the concept as well as the process. If it is begun at a routine session or becomes part of an ongoing process, it loses much of its threat and becomes a simple, effective device for helping the group get feedback on its goal movement.

Individual-Group Reaction, Reporting The design is similar to the one above. It is used about 20 minutes before the end of the session. Sheets with the following questions are distributed to each group member:

What did you think the explicit goals of the group were?

What do you think the group was really working on (implicit goals)?

What was helpful?

What hindered movement?

Each person replies to these questions privately and individually. Members then share their replies and consider actions based on data.

This method is also initially threatening to members, but if it becomes routine, it develops increased skills in diagnosing group problems on goals and allows for greater group productivity.

Feedback on Goals, Instrumented Another method for achieving clarity of goals as well as movement on the goal is to use a chart that is distributed, scored, and the results fed back to the members. Because it has a more objective, statistical format, it sometimes encourages members to be more open to the findings and less defensive. It takes more time, and perhaps a half hour should be allowed. The group is rated on three dimensions as illustrated below.

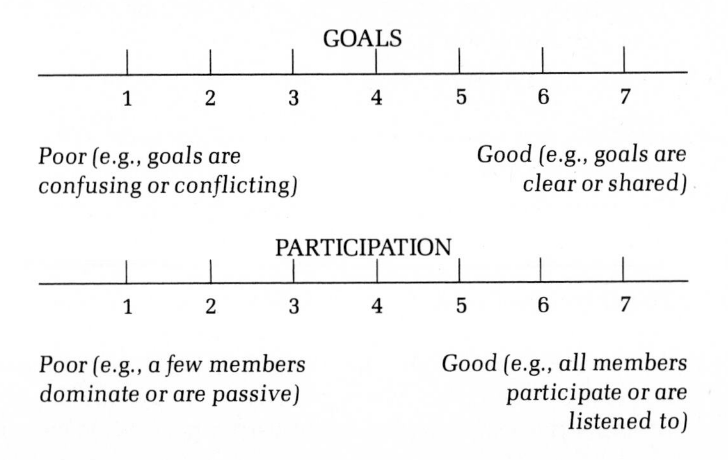

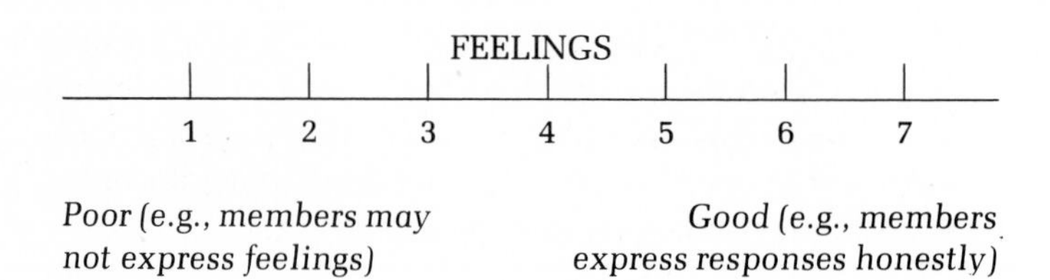

Poor (e.g., members may not express feelings) *Good (e.g., members express responses honestly)*

The sheets are collected. One member tabulates the data. A check is scored at its numerical value on the scale (a check between 3 and 4 is scored 3.5). The numbers are totaled for each question. The findings on each dimension are reported in terms of highest and lowest scores as well as average. The higher the score, the closer to the objectives of goal clarity, group participation, openness in response. Members then discuss the findings.

Although these problems may meet with some initial resistance (and therefore the simple open-ended discussion is the first recommended), each helps the group diagnose its own situation and hopefully modify its behaviors toward increased productivity.

EXERCISE 2 Setting Individual Goals and Reassessing Them

Objectives

To increase the understanding of what is meant by setting goals as an individual

To develop skill in stating goals clearly and specifically

To recognize that goal setting is an ongoing process

To periodically reevaluate goals and determine whether any changes or modifications are needed

To increase understanding and skill in giving and receiving help

Rationale

We participate in many group experiences with only vague ideas of what we expect to derive. This exercise is designed to help the participant formulate his or her goals specifically and realistically, and then reevaluate them at regular intervals. It is hoped that in the course of the experience he or she will not only develop skill in formulating goals, but that once this has been brought into consciousness, he or she will be motivated to pursue these goals in his or her activities. As we move toward a goal, we have new insights, new obstacles, new understandings. Thus, as a result of increased understanding at time A, we can make revisions of goals and move into time B. At this time we again reassess goals for time C, and so on.

Timing

Phase I of the design should take place early in the program. It might be used after an initial "micro-lab" or "getting to know each other" opening session. It is appropriate for a workshop or course that will continue over a period of time. Phase II should occur about the middle of the program, and Phase III at the end. Each phase takes approximately an hour.

Action

The facilitator announces the exercise, and may informally state some of the objectives. The group is divided into trios. These trios become a support system for each individual, helping to redefine his or her goals as well as analyze the forces that help or hinder him or her. Each helps and receives help from the others. The facilitator reads and explains Phase I, Steps 1 and 2. When completed, the facilitator suggests the group continue with Step 3. Discussion questions may be considered at this point.

Phase II is scheduled midway through the learning experience. The groups form into their original trios. The facilitator reads and explains Phase II. The trios work. If there is time, a discussion similar to the one after Phase I occurs; however, the exchange will be very different from the first one. Trio members will be much more comfortable with one another; there will be much less anxiety and more willingness to share and to help. More time should be allowed.

Phase III should be scheduled at the end of the learning experience. It will be conducted similarly to Phase II, but there will be a marked change in atmosphere. Trio members will feel closer and will honestly discuss their feelings and reactions to the workshop. The prospect of continuing in a home setting produces mixed responses. There are those who "can't wait" to continue on the goals, and others who are apprehensive about whether their situations will permit even tentative movements in the directions they would like to go. However, Phase III cannot be eliminated. It must be continued in "real life" if developing skills in individual goals is to have any permanent value. A discussion following Phase III might allow participants to discuss the aids and hindrances they expect to encounter, and to show how they may find alternatives for dealing with these factors.

Goal-Setting Procedure

PHASE I INITIAL GOAL SETTING

Step 1 Take five minutes to write one to three responses to the following question: *What do I want to learn most from this workshop* (or course or laboratory experience)?

Step 2 Take turns going through the following procedure: One person starts by reading one of his or her responses from Step 1. All discuss the response (goal). The following guideline questions may be helpful to clarify and amplify the goal under discussion:

Is the goal specific enough to permit direct planning and action, or is it too general or abstract?

Does the goal involve you personally, that is, something you must change about yourself?

Is the goal realistic? Can it be accomplished (or at least progress made) during the period of this program?

Can others help you work on this goal?

Is this the real goal, or is it a "front" for a subtle or hidden goal?

During or following the discussion, the person whose goal is being discussed revises his or her goal so that it is specific and realistic. This procedure is used for each goal listed. Allow about 20 minutes per person.

Step 3 Discuss in turn the barriers you anticipate in reaching your goal(s). Write them down as specifically as you can. Take a few minutes to list ways you individually, or with the help of others, can overcome these barriers. Discuss your lists with your trio.

PHASE II REASSESSMENT

Earlier this week (day, month, session) you prepared your initial assessment of goals for this workshop. One purpose of this session is to reexamine your goals in light of your experience so far. Refer to your earlier responses. Look at the goals stated there. Also look at the helping and hindering factors you listed earlier.

Step 1—Goal Reassessment Within your trios, take turns reassessing and discussing your goals. As you discuss them, one of the other members should enter your modified or reconfirmed goal on the list of goals. These statements should be checked out to your satisfaction.

Step 2—Analysis of Helping and Hindering Factors After your goals have been reformulated or reconfirmed, discuss in turn your perception of the present helping and hindering forces—in yourself, in others, in the setting—and make a list.

PHASE III REASSESSMENT

We are now in the concluding period of this workshop. Within your trios, reexamine your goals as modified and the helping and hindering forces you listed. How much progress did you make? What still needs to be done?

Step 1—Evaluation of Progress on Goal Each person in turn discuss how much you feel you accomplished on your goals. What helped you most? What hindered you?

Step 2—Goal Setting for the Home Discuss how you could continue on your set goals at home. What forces will help? What will hinder? As you discuss these questions, another member of the trio should enter the new goals on the lists.

Discussion

For some people, revealing personal goals and inadequacies is extremely difficult. Attempting to verbalize expectations is a formidable task for others because they lack practice in doing it. Giving help or accepting help may also be new and difficult experiences. Some of the following questions give participants an opportunity to discuss their feelings in these areas, and thus reduce anxiety.

What problems did you encounter in first stating your goals? Why?

How did your goals change after discussion?

How do you feel about help on your goals?

Are these your real goals? How honest do you feel you were in stating them?

What is the relationship between your goals and the goals of this workshop?

Are they compatible? Where is there conflict?

EXERCISE 3 Group Goals

Objectives

To see how the characteristics of group goals affect behavior

Setting up the Situation

The facilitator (instructor) discusses the general nature of group goals ("a place the group members want to get in order to reduce some tension or difficulty they all feel") and how these goals are made explicit through a coordination of individual motives and needs.

The class is divided into subgroups of six or seven persons. Each group has an observer. The observers are briefed (by the facilitator, outside) to watch (1) for the number of times members attempt to clarify the goal, (2) for disruptive, ineffective behavior by members, and (3) for general productivity.

Action

Observers return and the trainer gives the groups a short time (7–8 minutes) to accomplish a vague, abstract, complex task. (Reach agreement on this statement: What are the most appropriate goals to govern the best development of group experiences in order to maximize social development in a democratic society?")

After the time is up, the facilitator then gives the groups a concrete, clear, simple task. (In 7–8 minutes, list all the names you can of clubs or organizations appearing in a typical community.)

Analysis

Observers report back to the total group what they have noticed (for example, "The first scene had long silences, considerable angry feeling, much asking for clarification, long, vague, intellectual comments. The second scene produced an initial burst of laughter, very rapid discussion, and nearly everyone took part, which was very different from the first scene").

The class members generalize about characteristics of effective and ineffective types of goals (for example, "attainable, clear, challenging"), and examine the reasons for the negative, disruptive behaviors appearing in the first scene.

Variation

A variation on the above exercise is the "pegboard" demonstration. In the first scene, the groups are told to apportion the work and determine how to solve a problem with pegs and a board in the minimum amount of time. In the second scene, they are given a pegboard, each member is given a different number of pegs, and they are asked to develop a procedure to put the pegs in the board in the shortest amount of time.

EXERCISE 4 Hidden Agenda Role Play

Objectives

To increase the understanding that groups work simultaneously at an explicit as well as an implicit or hidden level

To increase learnings about "hidden agendas"

Rationale

Each of us has all of the personal, private needs that motivate our behavior in addition to the publicly stated reasons for our being present in a group. We all operate from these hidden agendas to some extent; they are part of us. The issue is not to label them bad, or to ignore them, but to be aware of them in our efforts to understand what is happening in a group. The role play presented allows members to understand the hidden agendas and to become aware that in making sense out of the meeting, these implicit goals greatly influence movement on the group goals.

Design

The facilitator asks for four volunteers to participate in a role play. The volunteers are selected, and each is given a piece of paper with the role he or she is to play. The role players are given a few minutes to study their roles and be sure they understand them. The roles on the pieces of paper are as follows:

1. Not only are you very anxious to be elected to this committee but you seriously believe that unless you are the one elected, it might not be successful. You like your fellow faculty members, but you feel that they tend to be authoritarian, and you are afraid if anyone else is elected, the students will be allowed to make decisions in name, but they will not be allowed to really influence the decisions. You feel very strongly that young people should learn autonomy and responsibility.
2. You are anxious to be elected. You are taking courses for your doctorate. You want to become a principal. The activity will look good on your record. It might even develop into a vice-principalship in this school.
3. Not only are you very anxious to be elected to the committee, you seriously believe that unless you are the one elected, the committee might not be successful. You like and respect your fellow faculty members as teachers, but you seriously believe that they do not understand the sort of guidance that students (who are really still

children) need in order to learn how to behave responsibly on committees. You feel that if you are elected to this committee, you can guide these students to success, and this will then make a student-faculty disciplinary committee a permanent thing. You are afraid if the other faculty members get the job, the whole thing will result in chaos, and the whole movement will be dropped.

4. You are in a hurry to leave the meeting, but the principal has asked that you be on the committee, and you want to stay on his or her good side. You had a fight with someone you are considering marrying last night. You are anxious for the meeting to end so that you can meet her/him.

Aside from the four role players, the others are asked to be observers for the role play. They are asked to note:

How clear is the goal?

Is it understandable?

Is it attainable?

Are the steps to goal accomplishment known?

How productive is the group?

What impedes the members?

At this point the role players are asked to be seated at a table.

Action

The committee is told that they are members of a high school faculty. The students have asked for more responsibility in determining matters of discipline. The principal has agreed to the establishment of a committee to handle such matters. The membership will be made up of students with one faculty adviser also serving on the committee.

The principal feels that each of the members he/she invited would be a suitable faculty adviser. He/she hopes the faculty committee will determine among themselves whom they would like to represent them. The committee is meeting after school. The members will decide who the faculty adviser will be and submit his or her name to the principal so that he/she can announce it at the assembly in the morning.

The committee meets. The facilitator should break in after about 5–7 minutes if no decision is made, and the role play should be cut so that interest does not wane.

Discussion

Observers are asked to report what they saw.

If the group goal was clear, why was progress so slow? In many groups this role play is not resolved; there is no decision made. If all the members understand the goal, and it is attainable, and the members know the steps to attain it, why is the group not more productive? What impeded the group from arriving at a simple decision? Consider the roles. Who wanted the job? Why?

What were the members' hidden agendas? (What seemed to be motivating them?) How do private goals (hidden agendas) influence group goals and productivity? What are the implications for other groups?

EXERCISE 5 Cooperative versus Competitive Goals

Objectives

To observe differences between cooperative and competitive goals

To experience cooperative or competitive goals

To recognize that group goals influence individual behavior

Materials

A roll of wide, white shelf paper

Masking tape

A large number of felt-tipped pens (broad) or other similar pens

Preparation

Hang a large segment of shelf paper on the wall as the basis for creating a mural. Hang other segments, same size, on other walls. There should be a minimum of two, a maximum of any even number. There should be a batch of pens (and other materials, if desired) for each group. Divide groups into the same number of units as there are murals. (Groups might be created randomly by having members count off.)

Action

Give each group written instructions. For half the groups, the instructions should read, "This is a contest to create a mural. A prize will be given to the person who makes the greatest contribution to the mural."

The other half of the groups should receive instructions that read: "This is a contest. Your group is to create a mural; the group with the best mural will receive a prize."

There should be one observer for each group who observes the behavior of the group in creating the mural. The facilitator gives the signal to start, and announces how long the group has to complete the mural. He or she stops the groups at the end of the time.

Discussion

Ask members of each group, by groups, to discuss how they felt in the project and ask observers to report. Then lead a general discussion of how being in a cooperative group is different from being in a competitive group.

In some way, have groups decide which mural is best and who contributed the most—individual votes, observer's opinion, outsider's opinion—and have small prizes available. Then, discuss how it feels to receive or not receive a prize as an individual and as members of the group.

BIBLIOGRAPHY

Allport, F. *Social Psychology.* Boston: Houghton Mifflin, 1924.

Atkinson, J. W. and N. Feather. *A Theory of Achievement Motivation.* New York: John Wiley and Sons, 1966.

Barnard, C. I. *The Functions of the Executive.* Cambridge, Mass.: Harvard University Press, 1938.

Bateson, N. "Familiarization, group discussion, and risk-taking." *Journal of Experimental and Social Psychology,* 2 (1966), 119–129.

Bennis, W. G., K. D. Benne, and R. Chin. *The Planning of Change.* 2nd ed. New York: Holt, Rinehart, and Winston, 1969.

Berkowitz, L. "Group standards, cohesiveness and productivity." *Human Relations,* 7 (1954), 509–519.

Bradford, L. "Hidden agenda." In *Group Development.* Ed. L. Bradford. Washington, D.C.: National Training Laboratories, 1961, pp. 60–72.

Cartwright, D. and A. Zander. *Group Dynamics: Research and Theory.* 3rd ed. New York: Harper and Row, 1968, pp. 403–405.

Chapin, F. S. "Social institutions and voluntary associations." In *Review of Sociology: Analysis of a Decade.* Ed. J. B. Gittler. New York: John Wiley and Sons, 1957, p. 273.

Church, Russell M. "The effects of competition on reaction time and palmar skin conductance." *Journal of Abnormal and Social Psychology,* 65, No. 1 (1962), 32–40.

Deutsch, K. *The Nerves of Government.* Glencoe, Ill.: Free Press, 1963.

Deutsch, M. "An experimental study of the effects of cooperation and competition upon group process." *Human Relations,* 2, (1949), 199–231. Reprinted in *Group Dynamics.* 2nd ed. Ed. D. Cartwright and A. Zander. New York: Harper and Row, 1960, pp. 348–352.

Ebbesen, E. and R. Bowers. "Proportion of risky to conservative arguments in a group discussion and choice shift." *Journal of Personality and Social Psychology,* 29, No. 3 (1974), 316–327.

Fouriezos, N. T., M. L. Hutt, and H. Guetzkow. "Measurement of self-oriented needs in discussion groups." *Journal of Abnormal and Social Psychology,* 45 (1950), 682–690.

Gross, D. E., H. H. Kelley, A. W. Kruglanski, and M. E. Patch. "Contingency of consequences and type of incentive in interdependent escape." *Journal of Experimental Social Psychology,* 8, No. 4 (July 1972), 360–377.

Gurnee, H. "Group Learning." *Psychological Monographs,* 76, No. 13, 1962.

Hare, P. *Handbook of Small Group Research.* New York: Free Press, 1976.

Horwitz, M. "The recall of interrupted group tasks: an experimental study of individual motivation in relation to social groups." *Human Relations,* 7 (1954), 3–38.

Julian, James W., and Franklyn A. Perry. "Cooperation contrasted with intra-group competition." *Sociometry,* 30, No. 1 (March 1967), 79–90.

Kelley, H. and J. Thibaut. "Experimental studies of group problem-solving and process." In *Handbook of Social Psychology,* Vol. II. Ed. G. Lindzey. Reading, Mass.: Addison-Wesley, 1954, pp. 735–785.

Klein, J. *The Study of Groups,* London: Routledge, 1956.

Korten, D. C. "Situational determinants of leadership structure." *Journal of Conflict Resolution,* 6 (1962), 222–235.

Lewin, K. "Field theory and experiment in social psychology: concepts and methods." *American Journal of Sociology,* 44 (1939), 868–897.

Lewis, H. B. "An experimental study of the role of the ego in work. I. The role of the ego in cooperative work." *Journal of Experimental Psychology,* 34 (1944), 113–116.

Lewis, H. B. and M. Franklin. "An experimental study of the role of ego in work. II. The significance of task orientation in work." *Journal of Experimental Psychology,* 34 (1944), 195–215.

Lippitt, G. "How to get results from a group." In *Group Development.* Ed. L. Bradford. Washington, D.C.: National Training Laboratories, 1961, p. 34.

Mills, T. M. *The Sociology of Small Groups.* Englewood Cliffs, N.J.: Prentice-Hall, 1967, pp. 81–82.

Meyers, D. G. and S. J. Arenson. "Enhancement of dominant risk tendencies in group discussion." *Psychological Reports,* 30 (April 1972), 615–623.

Morrow, A. J. "Goal tension and recall. I and II." *Journal of General Psychology,* 19 (1938), 3–35, 37–64.

Parsons, T. and E. A. Shils, eds. *Working Papers in the Theory of Action.* Cambridge, Mass.: Harvard University Press, 1951.

Quey, R. L. "Functions and dynamics of work groups." *American Psychologist,* 26, No. 10 (1971), 1077–1082.

Rabow, J., F. J. Fowler, D. L. Braadford, M. A. Hoefeller, and Y. Shibuya. "The role of social norms and leadership in risk-taking." *Sociometry,* 29, No. 1 (1966), 16–27.

Reddy, W. B. and A. Byrnes. "Effects of interpersonal group composition on the problem-solving behavior of middle managers." *Journal of Applied Psychology,* 56, No. 6 (1972), 516–517.

Ricken, H. W. and G. C. Homans. "Psychological aspects of social structure." In *Handbook of Social Psychology,* Vol. II. Ed. G. Lindzey. Reading, Mass.: Addison-Wesley, 1954, p. 810.

Roethlisberger, F. J. and W. J. Dickson. *Management and the Worker: Technical vs. Social Organization in an Industrial Plan.* Cambridge, Mass.: Harvard University Press, 1939.

Schutz, W. C. *FIRO: A Three-Dimensional Theory of Interpersonal Behavior.* New York: Holt, Rinehart, and Winston, 1958.

Stoner, J. A. F. "Risky and cautious shifts in group decisions: the influence of widely held values." *Journal of Experimental Social Psychology,* 4, No. 4 (October 1968), 442–459.

Stoner, J. A. F. "A comparison of individual and group decisions including risk." Master's thesis, 1961, School of Industrial Management, Massachusetts Institute of Technology, Cambridge, Mass.

Teger, A. I. and D. G. Pruitt. "Components of group risk-taking." *Journal of Experimental Psychology,* 3 (April 1967), 189–205.

Thelen, H. *Dynamics of Groups at Work.* Chicago, Ill.: University of Chicago Press, 1954.

Tuckman, B. W. "Group composition and group performance of structured and unstructured tasks." *Journal of Experimental Social Psychology,* 3 (January 1967), 25–40.

Turney, J. R., "The cognitive complexity of group members, group structure, and group effectiveness." *Cornell Journal of Social Relations,* 5, No. 2 (Fall 1970), 152–165.

Wallach, M. and N. Kogan. "The roles of information, discussion, and consensus in group risk taking." *Journal of Experimental Social Psychology,* 1 (1965), 1–19.

Wallach, M. and C. W. Wing, Jr. "Is risk a value?" *Journal of Personality and Social Psychology,* 9, No. 1 (1968), 101–106.

Weinstein, A. G., and R. L. Holzbach. "Effects of financial inducement on performance under two task structures." Proceedings of the 80th Annual Convention of the American Psychological Association, Part 1, 7 (1972), 217–218.

Wheeler, R., and R. L. Ryan. "Effects of cooperative and competitive classroom environments on the attitudes and achievement of elementary school students engaged in social studies inquiry activities." *Journal of Educational Psychology,* 65, No. 3 (1973), 402–407.

Workie, A., "The relative productivity of cooperation and competition." *Journal of Social Psychology,* 92, No. 2 (1974), 225–230.

Zald, M. "Organization control structures in five correctional institutions." *American Journal of Sociology, 38* (1962), 305–345.

Zander, A. and D. Wolfe. "Administrative rewards and coordination among committee members." *Administrative Science Quarterly,* 9, No. 1 (June 1964), 50–69.

Zajonc, R., R. Wolosin, and M. Wolosin. "Group risk-taking under various group decision schemes." *Journal of Experimental and Social Psychology,* (January 1972), 16–30.

Zeigarnik, B. "Uber das behalten von erledigten undü unerledigten handlunger. *Psychologische Forschung,*" 9 (1927), 1–85.

Ziller, R. C. "Four techniques of group decision making under uncertainty." *Journal of Applied Psychology,* 41 (1957), 384–388.

Five

Leadership

In October, 1979, Pope John Paul II made an American pilgrimage, visiting five cities. In Philadelphia, the event far outshadowed the celebration of the nation's bicentennial at Independence Hall on July 4, 1976.

From the moment of the announcement that Philadelphia would be on his itinerary, planning went into high gear. For the retiring controversial Catholic mayor, the opportunity to play host to a pope, and Pope John Paul especially, was the "high point" of his life.

It was necessary to find a place where the Pope could celebrate Mass that was large enough and was an appropriate setting for the throngs of Philadelphians and residents of the whole metropolitan area who would want to experience and be blessed by the Pope. The usual sports stadiums were considered both inappropriate and inadequate for the crowds anticipated.

It was decided that Logan Circle, close to central Philadelphia and the terminus of a ten-lane boulevard, would be the best location. Usually, it is a fountain of spewing mermaids encircled by well-landscaped shrubs and flowers. For the Mass, a giant platform was to be erected, its base banked by thousands of yellow and white chrysanthemums. On the platform, there would be a huge (240 feet high) cross. Traffic was to be closed so that people would then be able to stand on all sides of the circle and in the entire area up to the Art Museum. Although the American Civil Liberties Union filed a complaint against the city for spending public money on a cross, a religious symbol (and was later upheld by the courts, so that eventually the Roman Catholic church paid the quarter of a million dollars), there were few, if any, other protests. The city was jubilant as it prepared for a once-in-a-lifetime experience with a great leader. Not only were Catholic schools and businesses closed, but even the public schools were closed because buses were being reassigned from their usual routes to convey people to see the Pope.

A giant human carpet of over one million people surrounded Logan Circle that day. Some people had come the day before with campers, sleeping bags, and provisions. People were everywhere, as far as you could see in any direction. While they waited, people sang, swapped anecdotes, and became instant friends. They listened to radio commentary on the Pope's whereabouts—his arrival at the airport, the tumultuous cheering as he was driven up Broad Street, and his arrival at the Cathedral a couple of hundred feet away.

Then, there he was. A chorus of high school choirs sang in welcome. He looked like radiance, in gold vestments, wearing a gold mitre, and holding a gold crozier (staff). He was beaming, and walking around the platform, waving to everyone. He was followed by a procession of about thirty cardinals and high-ranking priests also in gold, and then about fifty bishops in maroon. It was a spectacle!

From the vastness, everywhere, people were cheering and waving handkerchiefs. He ascended the platform (walked is the wrong word), and soon began the Mass. It was in English; everyone had become very quiet and reverent. When the hymns were being sung people felt enveloped in a wonderful humanity; hundreds of thousands of people joyously sang. Everyone was singing with tears streaming down their faces. All were deeply moved.

There were gifts, and communion, and a rousing final hymn as the Pope came down from the platform. Tough police reverently knelt and kissed his ring. Amid cameras, and flashbulbs, outreached arms, and TV cameras, he continued to enchant, embrace, and bless those near him. People had been touched, not physically, but by his presence.

Spectators hugged and kissed each other, and wished each other well as they said good-bye. They had not only become friends, but had shared the magic moment of being with the Pope. People dispersed to go back to the mundane world, but each of them was different; each face continued to be streaked with tears, and there was a very special sense of closeness. Can you imagine feeling close to one million people?

The experience was not a matter of being impressed with the leader's policies, or the service, or even of agreement with him. It was rather the compelling vision of this man as a leader, with the capacity to move and to influence so many. It was his extraordinary ability to move people by his presence. It was a very special experience, one that those who were there will never forget.

He possessed that special quality of being a "true" leader; more than being the Pope. For contrast, consider the following description of a leader:

His early life and times were the American dream come true, 1920s fashion. At twenty-three, he was washing dishes in a Brooklyn cabaret, and when he could get it, a part-time bartender there. At twenty-eight, half a dozen years removed from his dishwashing days, he was known throughout the nation. He was head of an organization making a hundred million dollars a year. He was glamourized, viewed as a hero. Capone didn't always have it easy, of course. He constantly faced the threat of machine gun and pistol fire—but what business does not have its problems? However, in a competitive nation, in a highly competitive business, where risks were daring and the stakes high, he was number one.

Al Capone was not only a leader, he led a charmed life. More than once he was saved by the steel armor and bulletproof glass of his limousine. The real sign of his "magic" however, occurred when eight full cars of enemy gunmen drove past his headquarters. The gunmen riddled the building with machine-gun fire but Capone emerged unharmed.[1]

The leader appears to be someone bigger than life, for whom the ordinary problems of life seem nonexistent. Whether heroes, or antiheroes, they are viewed as singular and unique. There is about them a special aura; they are not one of us.

[1]Based on *The American Heritage History of the 20's and 30's* (New York: American Heritage Publishing Co., Inc.), 1970, pp. 166–167.

Each of us has a remembrance of our moment of being briefly connected with them. We vividly recall the experience and the emotions that presence evoked in us—awe, respect, admiration, anger. We seem eager to recount the experience, to tell others our impressions—how the experience varied from other accounts, or how our personal judgment substantiated what others have written. Our lives are different somehow, for having been in the presence and seen with our own eyes a leader, a person who influences so many.

LEADERSHIP TRAITS

The concern with leadership has probably always existed, but it became a prime area for research by social scientists about the time of World War I. With our increased knowledge of testing and new statistical tools, there was a strong impetus to accumulate data and determine what traits were common to leaders. If these could be identified, perhaps those who had these traits could also be identified, and leaders could be selected quickly and efficiently. The usual procedure in studies on leadership has been to select certain personality attributes and relate them to success or lack of success in certain leaders.

READER ACTIVITY

Think of people you know who are leaders of groups. What traits do they possess?

____

____

____

____

____

Summarize these traits. What would you say are the traits that leaders have?

____

____

Do you think those traits apply to leaders generally?

____

Implicit in most of the research on personality traits and leadership is the belief that the qualitative components that make for effective leadership are consistent. In other words, you have it, or you don't. The leader might have been born with these traits (one theory) or he or she might have acquired them (another theory), but in either case, the person possesses the traits of leadership. The only problem, it would seem, is that personality traits are still poorly conceived and unreliably measured. It is thought that as we refine our methods of measuring personality traits, we will be able to determine what the traits are that we need to find or train. In this theory, the ability to create leadership effectiveness is just around the corner.

Results, however, of this approach have been disappointing. The sorting out of leaders with various leadership traits from those without them has been notoriously ineffective. One study (Bird, 1940) extensively reviewed the relevant research and compiled a list of traits that seemed to differentiate leaders from nonleaders in one or more studies. However, only 5 percent of the traits listed appeared in four or more studies; many of the other traits listed only appeared in a single study. Mann (1959) reviewed 125 leadership studies searching for a relationship between personality and performance in small groups. His search yielded 750 findings about personality traits, but no traits as conclusions. He found a lack of consistency among traits described as significant for leaders, and further, found that some traits listed as significant were diametrically opposed to significant traits listed in other studies. Researchers continue to search for the behavioral scientists' (if not the alchemists') gold, with similar results. In studying discussion leaders, Guyer (1978) found no statistically significant relationships between traits of discussion leaders, student evaluations of them, or the grades received by students of discussion leaders. His conclusion: "Attention to personality traits . . . would have been of limited value in the selection of discussion leaders" (p. 697).

There are those who say you can tell a leader by how he or she looks, that physical attractiveness is a qualifier for leadership. One researcher (Archer, 1974) decided to examine that idea. He studied physical attractiveness as a predictor of power and dominance in a small group setting. He found that physical attractiveness was a significant predictor of power and dominance—but the relationship was negative. Group members who rose to positions of power in their groups were those who had been rated as relatively unattractive.

There is some evidence that leaders tend to be a bit taller, more intelligent, more enthusiastic, have greater self-confidence, and have greater social participation than nonleaders (Berelson and Steiner, 1964; Smith and Cook, 1973; Sorrentino, 1973; Zigon and Cannon, 1974). However, it is impossible to predict and to use this information in selecting and training leaders. For example, it has been repeatedly demonstrated that the person who does most of the talking (greater social participation) becomes the leader, *unless* he or she talks so much that he or she antagonizes other group members (Stang, 1973). A student leader may be more intelligent, unless he or she gets all A's and is viewed as a "curve wrecker" (scores so high that other students get lower grades by comparison); then he or she becomes an outcast (Davie and Hare, 1956).

After extensive surveys of the literature searching to identify leadership traits, researchers are increasingly coming to the weary conclusion that leadership is not the possession of some combination of traits (Stogdill, 1948; Stogdill, 1974). Rather, "in every instance, the relation of the trait to the leadership role is more meaningful if consideration is given to the detailed nature of the role" (Gibb, 1954, p. 878). Since traits of an effective leader are so closely related to the functions that persons will perform, the most general rule would be to focus on what task needs to be performed, and to select those who are willing and have the skills to perform that task. Even more is involved. Leadership seems to be based on a working relationship among members in a group, in which the leader acquires special status through active participation and demonstration of his or her ability to carry tasks through to completion.

Yet the search goes on, to somehow find the magical attributes that will transform us to esteemed positions as leaders (Hall and Williams, 1971). There are a variety of theories as to why the romantic conception of the leader with magic attributes persists. Some relate it to the common childhood experience of having a father, a prestige figure, magically endowed (Knickerbocker, 1961). Many find security in that figure. Since each of us continues to need security, perhaps we carry with us from childhood that symbol, the Leader. If such an assumption is accepted, it would be more readily understandable that the leader, or the person we conceptualize as the leader, should be larger, more intelligent, more cultured, more impressive than we. The leader represents the symbol, the ink blots onto which people project their desires for security, dependence, glamour, and power. It perhaps makes more understandable the persistent search for leaders who will arrive full blown, without an abracadabra or seven-league boots, but who by their presence can remove difficulties, overcome obstacles, and attain the goal. And yet most of us realize that this is fantasy.

How familiar is the disappointment we fleetingly feel when we discover that the bigger-than-life hero is really very small, almost insignificant physically, but we in our own minds and with the aid of the media have blown him or her out of realistic proportion. We know from the avidly read "inside stories" by leaders' friends, assistants, even valets, that they are very human. To those who know them well, they can be infuriating or charming, opinionated or receptive, verbose or taciturn—in short, human beings working at their jobs. We know the leader is not the person for all seasons; great leaders have been repudiated at the polls, viewed as anachronisms—throwbacks to earlier eras who have lost symbolic favor with a fickle public. So far, research findings seem to indicate that leaders do not possess qualities that demarcate them from nonleaders; that who will be the leaders depends upon a complex of forces that do not permit prediction; and finally, that leadership potential is widely diffused.

POTENTIAL LEADERSHIP

Another type of response might be evoked by the question, "Who is a leader?" A leader is a pope, the president of the United States, the chairperson of a corporation, a superintendent of schools, or a general in the army. At another level, he or she is the mayor, the district manager, the school principal, the lieutenant. Yet at another level, he or she may be the foreman, the head of the English Department, the shop steward, the corporal. By now it is evident that in answering the question, we have responded with a series of titles representing positions. We call the person a leader if the person holds a particular position in an organization (Jenkins, 1956). It's that circle (or dot) at the top of the organizational chart, or farther down the chart if it represents a lesser leader.

Leaders could have attained their positions in a variety of ways: by official choice of the organization, by appointment or selection by some higher authority, by succession, or by seizing control. Regardless of how they got there, the position grants them authority or influence over people. Often, when we talk about leaders, we are talking about certain people who hold positions that carry authority. We say they can influence the group because of their position, or because they have control over the group. Think of the power the legendary Vince Lombardi had over his players as he prescribed what they would and would not do, not only how they would play on the field, but what time they would eat, what time they would go to bed, and even how long their hair could be. Or consider the influence an ordinary coach has over his or her team and the compliance he or she demands. Consider how important is the president's feeling on war, or the role of the press, or advocacy of human rights. Or, think how we feel when the boss does not listen to our suggestions, or hear our side of an issue.

A position gives the occupant power to influence (Abrahamson and Smith, 1970). One theory of leadership states that leadership lies with the position rather than the person. Leadership is conceived as what the person in a position does; by definition of that position the person is a leader. Thus the announcement of a new superintendent of schools (which is a rather usual occurrence) produces apprehension among the staff. They wonder what the new superintendent will be like in a working relationship, what he or she will favor, which programs will be cut; the apprehension is directly related to the degree of power and control he or she has over them. The staff does not know the new superintendent as a person, nor do they know his or her style. They are, however, consciously aware of his or her position and its implications for them. This anxiety recurs whenever a new supervisor or principal or "position above" is filled, in terms of what it will mean to those in "positions below." One source of tension relates to how the new appointee will use his or her authority, and for those below him or her this is an area of central concern.

Moreover, the position may be a source of influence even for those not on the staff. We have expectations for how an occupant of a position should behave, and these expectations influence both our perception and our behavior to the position occupant. Imagine yourself with a bank president: envision the office, appearance, manner of dress, education, race, sex, religion, style of speech, manner in greeting a business executive It seems quite simple to conjure up an image. The banker you imagine will probably be male, white, Anglo-Saxon, middle-aged, conservatively dressed, well-spoken, and lodged in a well-appointed office.

Now imagine yourself with a Marine officer. What is his appearance as compared with the bank president's? What is his manner of dress, his education, his style of speech? How would he speak to a recruit? Imagine a party committeewoman in an urban district. Consider her office as compared with the others, her physical appearance, her style of dress, her education, her manner of speech. How would she greet a constituent? Finally, consider a nun-principal. Imagine her office, dress, style. In each case, the naming of the position projects onto our personal screens a series of specific images of specific positions, and we respond to these positions in terms of our expectations. We speak formally and intellectually to the bank president, and our manner is subdued; his position seems to require it. We subconsciously straighten up our posture, tuck in our stomachs, and speak tersely with the marine officer. With the committeewoman, we are preoccupied, busy; she will be informal, ingratiating, and laugh at our jokes. With the nun-principal, we become conscious of our language and censor any possible vulgarities, act very respectful, and listen more than we talk.

Each position produces for us expectations of how the occupant of that position should act, and frequently we respond to our expectations for the role occupant rather than to the behavior of the individual. In fact, an area of considerable debate centers on whether position occupants are selected because they meet role expectations, that is, because they have the attributes of the "part"; or whether they develop the behaviors expected in a certain role. For example, when we say someone is "school-teacherish," are we saying that this was her or his behavioral style before she or he became a teacher, or did she or he develop it from teaching over a period of time? The question has no clear answers. Although the sequential pattern is not clear, it is abundantly evident that an individual's behavior is influenced by his or her position, and also by how others define the position, as illustrated below.

There is a children's story of a good, kind king who lived modestly, was concerned for the welfare of his simple farmer subjects, and found time to hunt, fish, and enjoy his family. One day an ambassador from a distant kingdom came. He was shocked at what he saw—no palaces with turrets, no counting houses for storage of money and treasure, no army or soldiers—and worst of all the king in simple raiment. The ambassador clearly doubted that this man was really a king. Even fairy tale kings have identity crises. The king had never before wondered about whether he was real, or whether he performed his kingly duties properly. Obviously, he was not playing the king role as it should be played.

Therefore, he set forth to levy taxes, raise armies, build counting houses and win wars. With each success, he added a new turret and a new flag pole to his palace, a new set of kingly raiment, and a new counting house from which to count and store his new wealth. He conquered and added, with no time left for his subjects, his family, or his former interests.

One day, while counting his riches, he slipped from the topmost turret and fell, hitting each flag pole until he reached the ground—broken, tattered, unrecognizable. He pleaded for help, but no one (except finally, his daughter) even knew him or would bother with him. He did not look like the king; he was not acting like a king; and so, of course, he was not treated like a king.

Our expectations of how the occupant of a position should behave frequently influences our interaction with him or her as well as influencing his or her behavior in the position.

Let us return to the prime question. Is the person in the position of leadership automatically the leader? If any behavior in a position is leadership behavior, we are in a clearly untenable position. History is replete with actions of kings of very limited intelligence who had tremendous influence and whose whims were law. Are we placed in a position of saying such behaviors are appropriate? We all know from our experience that there have been position leaders whom we saw as excellent or outstanding and others who were failures and about whom we raised questions concerning their connections. In our own experience, we have worked under the same position (boss, supervisor) in which there have been a number of occupants. Each occupant was different, despite the same title and job description; each emphasized different aspects of the job, each had different relationships with the staff, and each was often viewed as "good" or "bad". We get into evaluations of "good leadership" when the leader is doing things that help the group, or when he or she does things we like, or "bad leadership" when the leader does things that hinder the group, or things that irritate us.

It is necessary to draw a distinction between leader and leadership (Holloman, 1968). A leader may be a person in a position of authority; he or she is given the right to make decisions for others—as a teacher is given the right to teach the class, or the foreman the authority to assign work for his or her unit. From that position, the leader may influence others who look to him or her for clues or seek to emulate him or her.

From another perspective, it might be said that whoever influences the group is the leader; that is, any person who influences the group (whether in the position of leadership or not) exhibits leadership behavior. Leadership behavior is distinguished from the leader position; leadership behavior has to do with influence on the group regardless of the position. An illustration may be helpful as a means of clarification.

After-school meetings are a chore; they are usually a tiresome extension of an already long day. Representatives of the faculty and an equal number of representatives from the student council were to meet to plan a workshop to explore possibilities for improving faculty-student relations and to clarify the duties of the student council. These were the explicit agenda items; many others were obviously present.

The room had the institutional pallor of a chamber impervious to change in time, generation, or subject matter. The drab walls, nondescript flooring, and windows with shades precisely drawn on an invisible line were all reminiscent of school rooms built for the grandparents of the present school population. The participants sauntered in: the students were casually dressed in bright shirts or velours, jeans, and slacks; the teachers were conservatively dressed, except for a few who regarded themselves as young, but none wore jeans. Some of the women wore brighter colors and shorter skirts, and some of the men had more informal attire and long sideburns. Student teachers had not yet declared their visual allegiance; some dressed as students, and others identified with teachers.

The principal called the meeting to order in her most informal style. The small talk stopped; she was given attention. A teacher was asked to state the previous duties of the student council. Those sitting close to him listened; the others renewed their subconversations.

One red-haired student talked about how insincere everyone was toward her since her election. Students now saw her as one of the most popular girls in the school, someone who had influence, someone they should get to know. She felt as if she was some kind of superstar; as if she now had obligations to be a certain kind of person in school—the perfect person. Other students responded with laughter; it was ridiculous. But they then got into a discussion of what is expected by students, by the school, by teachers or by leaders in the student council. Are these expectations mythical? Do they influence their behavior? Is it helpful or harmful?

A black faculty member, the only one, arrived and attempted to be inconspicuous by sitting in a chair closest to the door. Two black students quickly stood up, greeted him, and suggested that he sit in a chair by them. He did. More talk. One of the black students indicated that he could not come to the workshop. People looked at one another with silent messages. A tall student, very casually, asked why. The black student replied that he had a job on Saturday mornings, a new job, and he could not jeopardize it by not showing up the second week. There was an obvious sigh of relief among participants. One of the teachers suggested that she would be glad to call his employer and explain the situation. Someone suggested that the teacher do it immediately so that they could be assured of everyone being able to attend the workshop.

The above illustration is a brief sample of a meeting, one not very different from many meetings. But, who was the leader? Was it the principal who opened the meeting? What were the issues the group was working on? Who influenced the discussion; how?

An effort to understand the meeting in terms of the leader or his or her position, in this illustration, would yield little in recognizing the issues or those who influenced the group. A study of leadership in terms of position is limiting, then, on several counts. It accepts any behavior by a person in a position as leadership behavior, which can result in terming contradictory behaviors in the same position (i.e., as principals) leadership behavior, thus escalating the confusion. Second, it loses the dynamics of what is happening in a group—who is influencing the members, how, and on what basis. If the positional leader says little, does that mean that by definition nothing is happening?

Finally, there is the problem of the "leader behind the throne." We are aware of persons who occupy the position. Well-documented accounts of boss rule in politics are generous in their details of mayors or governors who were handsome, mellifluous-voiced errand boys for the "boss," who himself held no official position. Each of us knows of occupants of positions who are given the name-on-the-door trappings of office; but in reality they must check almost everything with someone who may be in a higher position, or someone who has retired from office but who still must be consulted prior to any move. Study limited to occupants of positions obscures who influences the decision making, the processes in that group, how they develop, and with what consequences.

SITUATIONAL LEADERSHIP

In addition to traits theory and position theory there are other theories of leadership (Barnard, 1938; Cattell, 1951). A situational theory states that leadership is a function of the situation rather than the person or what he or she does. Here the type of leader needed depends primarily on the job. For example, Lincoln was an outstanding president, because he was the right person for the job to get us through the Civil War. Had it been another time, Lincoln might not have been elected, or at least would not have become a person of such profound influence. In other words, the situation creates an environment that produces leadership.

The difficulty with this theory is that the emphasis is weighted primarily on the environment and inadequately on the individual and what he or she does. Although the captain of a football team may not have the vocabulary skills or coordinating skills needed to be a discussion group leader, are those the only skills needed? Isn't there more involved than finding out who has the most experience as a discussion group leader? Or who speaks best? Aren't there other factors such as whether one wants to be a discussion leader and to what extent? What competition is there? How familiar is he or she with the subject of discussion? What year in school is he or she; what year are the others in the group? What is the sexual composition of the group? What is expected of the leader? What status does the group have in its context? What would enhance that status? While the situation is a factor in who becomes the leader, the nature of leadership involves considerably more than knowledge of the situation.

FUNCTIONAL LEADERSHIP

Our understanding of leadership has changed dramatically over the last half-century (Golembiewski, 1962; Vacc, 1975; Stogdill, 1974). It is changing and will continue to change as research and practical experience modify our present understanding of leadership.

However, as we have come to understand leadership as behavior in a situation, and as a dynamic relationship, we have viewed the subject differently. Leadership can be viewed from a group perspective. What actions are required under varying conditions to achieve the group's goals, and how do members take part in these various group actions? From this perspective, we define leadership as those acts that help the group achieve its goals.

When the question is recast, leadership can, in principle at least, be performed by any member. Leadership acts are those that help the group set goals, or that aid in movement toward achievement of goals, or that improve the quality of interpersonal relationships among members, or that make needed resources available to the group. When the leadership question is reframed in terms of behavior that furthers the goals of the group, it is difficult to distinguish leadership from fellowship.

There are some who feel that the word *leadership* impedes understandings of group processes and that the term would be better eliminated. Regardless of the term, many agree that the functional approach to both leadership and membership permits increased understanding of the processes and dynamics of groups. This approach recognizes the uniqueness of each group. Actions required by one group may be quite different from those of another, and subsequently the nature and traits of the persons having influence will be different in each group. In addition, a variety of factors affect leadership: the goals of the group, its structure, the attitudes or needs of the members, as well as expectations placed upon the group by an external environment (Bass, 1960; O'Brien, 1969).

For example, a pilot may be the acknowledged leader of his or her group while in the air. He or she may have the skills of coordinating the crew, knowledge of the mission, and ability to cope with the difficulties with which the crew is faced. What if the plane crashes and the crew is faced with the goal of surviving in the wilderness and finding its way back to safety? The skills needed are now very different. Who knows the terrain on the ground? Who knows about "survival training"? What plants are edible? Who can calm the members and reduce squabbling? Who can organize the group for utilizing their resources? The person who will be the crew leader on the ground may be very different from the leader in the air, and the functional approach recognizes that a variety of behaviors by any member may be helpful in the achievement of the group's purposes. The functional approach stresses behaviors that influence the group. If a group is threatened by conflicting subgroups, members who engage heavily in mediating functions may be expected to be influential. If, however, a group is faced with low prestige in the community and members are leaving, quite different behaviors will be required to be influential.

Types of Functional Roles

Leadership is basically the execution of a particular kind of role within an organized group, and this role is defined essentially in terms of power or the ability to influence others. Conceived as a role, leadership may be more or less specific to the particular structure of a particular group (McDavid and Harari, 1968). A leader in one group does not automatically emerge as the leader in another group. As membership changes, the leader may change, or if the purpose or activities change, the leader may change then too. Leadership implies followership. One person exerts influence or social power and others are influenced. Leadership is defined as the frequency with which an individual in a group may be identified as one who influences or directs the behaviors of others within the group.

Studies by Bales (Bales, 1950; Bales, 1970) and others (Benne and Sheats, 1948; Rieken and Homans, 1954) have identified three types of leader-member behaviors (hereafter referred to as member roles, since they can be performed by any member). They are:

Group Task Roles Roles of members here are to help the group select and define the common goals and work toward solution of those goals.

Suppose representatives of school and community groups decided to work together to "do something about the drug problem."

Members whose actions would be categorized in the task realm would *initiate* discussion of what could be done, or how the problem might be approached, or they may suggest different methods for getting teenagers involved. Someone may offer *information* on what other groups in the city are doing and what official agencies are available for further help.

Another may offer *opinions* on the subject. Others may *elaborate* from their experience or reading. With this variety of opinions and suggestions, some can *coordinate* or clarify the various suggestions in terms of which are appropriate for this group to work on and which are more appropriate within the province of other groups. One person may summarize what has happened, perhaps point out departures from the original goals, and raise questions as to whether the group can proceed as suggested or whether it lacks the resources needed. (That person would be orienting the group.) There may be *critics* or *evaluators* who question the facts as presented, or the effectiveness of such a volunteer group. An *energizer* may prod the group to reconsider its potential and stimulate members to greater activity. There may be a *procedural technician*, who knows where materials on drugs can be obtained inexpensively; that person may have access to means of distribution of leaflets or a speaker who could help clarify some of the technical questions being raised. A *recorder* may be writing down the suggestions, or keeping a record of group decisions on what has been assigned for the next meeting.

These are functions that members undertake to help the group accomplish its task.

Group Maintenance Roles Although task roles focus on the intelligent problem-solving aspects of achieving movement toward a goal, equally important but at a different level are the roles that focus on the personal relations among members in a group. These are known as group maintenance roles.

At a meeting, members may sound as if they are giving information or opinions, or evaluating ideas; yet a newcomer may sense that an opinion is attacked before it is fully expressed. The group keeps getting bogged down with inconsequential points, members subtly attack one another on a personal level. Newcomers may feel that to present an idea means being open to attack, and they will therefore remain silent; there is such bickering that more must be going on than can be understood. Perhaps there are interpersonal rivalries or alignments being settled, and they should stay out of the crossfire; they may sense that those who get listened to are determined by their outside status or position or education rather than their ideas, and the newcomers reevaluate their own status in terms of whether it is worthwhile to remain in the group. Although it looks as if a group is working in the task area, since these are seen as legitimate roles, members are concurrently accumulating data on, "What kind of a group is it, how will I fit in, will I be accepted or rejected?" Members acquire personal data and formulate feelings that significantly influence their behavior.

Just as task roles are helpful in aiding a group to achieve its explicit goals, maintenance roles are helpful in aiding a group to work together. These behaviors help the group maintain itself so that members will contribute ideas and be willing to continue toward progress on the group task. Both kinds of roles are needed, and each complements the other. The chapter on problem solving (Chapter Six) examines these relationships in greater detail, but it can be briefly noted here that the relationship is not only complementary but spiraling—successful work on the goal increases members' feelings for the attractiveness of the group and for liking each other. Liking each other and having worked out interpersonal relationships frees energy to put into the task and its accomplishment.

To return to the illustration of the group studying the drug problem, as opinions are given, the *encourager* may ask for additional examples or ask if others have similar opinions. The *supporter* may agree with suggestions of others and offer commendations. The *harmonizer* may attempt to mediate differences between members or points of view, or relieve tension with a joke. Someone who previously felt that public health agencies and not citizens' groups should work on the drug problem may, after hearing the discussion, come around and agree to a compromise whereby the coordinating group sponsors a series of meetings in which public health officials describe their efforts in drug statement. Someone, a *gatekeeper*, may notice that the representatives from one community group have not spoken, and may ask if they have any ideas on the subject. These roles help a group maintain itself in order that work on its task can proceed without becoming immobilized by inappropriate social behaviors and so that individuals are brought effectively into the emotional sphere of the group's life.

Individual Roles Another set of behaviors has been identified in which members act to meet individual needs that are irrelevant to the group task and are not conducive to helping the group work as a unit. In fact, this individual-centered behavior frequently induces like responses from others. An attack on one person leads to a response of personal defense; a joke may escalate to an hour of "I-can-top-that" jokes; and blocking by one member may lead to retaliatory blocking. The goal and the group are forgotten; the individual acts primarily to satisfy his or her personal needs.

To return to the representative citizens' group, the *aggressor* may question, with thinly veiled sarcasm, the competence and veracity of the person giving his or her opinion. The aggressor may imply that the speaker does not have the foggiest notion of what he or she is talking about. Or the aggressor may sneer that this "half-baked" committee hasn't the competence to solve a serious problem; why not sponsor a day for the kids at the ball park and accomplish something? It might have been agreed that drug addicts would be referred to the local community mental health center; however, the *blocker* persists in stating that unless a center is opened especially for addicts, the committee is useless. The blocker may turn every request for suggestions into a renewed attack on present plans or a renewal of advocacy for a treatment center. The *self-confessor* may use the audience to express personal problems and gain sympathy through catharsis. The self-confessor may reveal the problems he or she is having with a son who is disrespectful, shocks the neighbors with his late hours, and in general is a disgrace—to a parent who has worked hard all his or her life, in order that the children may lead better lives than the parents did. In an emotional voice, the self-confessor despairs that family relations and respect are not what they once were. The *recognition-seeker* may respond with his or her own personal advice and describe in glowing detail how it was successful in numerous other instances; or he or she may remind the group of the paper he or she has just delivered at the convention, or other important committees on which he or she has served. The *dominator* attempts to take over with an assortment of strategies by interrupting others, flattery, or asserting superior status. Consider the following example.

A rather prestigious community committee had been appointed. The members were prominent business people, well-known lawyers, a sprinkling of university faculty, several executive directors of health and welfare agencies, and a few upper-middle-class community representatives. One woman, descended from a prominent family and with a college education, was the wife of a member of the state Supreme Court. Despite her decades of experience serving on community committees, she felt inadequate and lacking in status in this committee. However, she did have strong feelings about what positions the committee should take and the proper direction of its work. It was important to her to gain the attention for her proposals she felt they deserved. Whether consciously or not, she prefixed each statement or opinion with, "Just the other day I was discussing this with my husband, the Justice, and he said . . ." or, "My husband, the Justice, feels. . . ." On the one hand, the members of the committee inwardly smiled, but, on the other hand, they were not quite sure of how influential she really was—so they listened.

There is the *playboy* (or girl) who jokes, keeps bringing up unrelated subjects, and is cynical about what meetings like this accomplish; he or she may read a newspaper for part of the meeting in a conspicuous effort to indicate his or her lack of involvement in the problem. There is the young person who states he or she "speaks for youth, and young people want . . ." or the parent who stands up "representing the concerned parents of the community, and we want. . . ." In both cases the *special-interest pleader* cloaks his or her wishes in "representative" terms. (Note that this is different from the delegate who represents a neighborhood youth group, or a specific businessperson's association.)

There are others. A high incidence of individual-centered as opposed to group-centered participation is an indication that the group will probably have a difficult time reaching its goal.

Both task and maintenance rules are needed by the group, but even though role functions may be identified, it is not always easy to separate them. Any member behavior may have significance for both goal achievement and maintenance. When a member offers information that the group sees as just what was needed (the missing piece in the puzzle) and now at last they can proceed, the information may not only have served a task function but also a maintenance one in that members feel encouraged and anticipate movement. In a similar way, the group working on a difficult problem may increase solidarity, participant interest, and efficiency simultaneously.

In the functional approach to leadership, who is the leader? Is it the person who has the greatest variety of behaviors, the person who excels in the task areas, or the person who is the social-emotional, maintenance leader?

It seems that much like in the traditional family in which the father is the task specialist and the mother the social-emotional specialist, differences appear in groups between the individuals who press for task accomplishment and those who satisfy the social and emotional needs of members. Over time, groups develop one or more leaders in each category, and when faced with a choice, most group leaders tend to give up the instrumental role in favor of popularity. Some searchers define the "hero" (Borgatta et al., 1955) as the leader who pushes the group to higher achievements, or tries to, at the cost of his or her own immediate popularity. One way to look at great individuals (Borgatta et al., 1955) is to consider them both simultaneously and over some period of time as those satisfying the best-liked and the best-ideas requirements.

The distinction between leadership as a position and leadership as a functional role is an important one. The first session of a T-group (training group) illustrates this distinction.

The session is held in a typical classroom. Participants enter as if arriving at any other class. They arrive with notebooks, pens, and perhaps some reading material on sensitivity training that they hope will be properly impressive. They arrange themselves in the circle of institutional chairs, and wait. Some stare blankly; some appraise or stereotype the others by sex, age, or clothing clues. Some avoid any contact. Some make a fleeting contact by discussing the weather or offering a cigarette; they wait with growing impatience and anxiety. Where is the "teacher"; did she forget, is she there? Who is the teacher, which one? In due time, the facilitator introduces herself, probably to the extent of her first name.

In casual tones, she indicates that the participants will create their own learnings, that feelings are important, and that participants can share their "here-and-now" experiences. This is followed by silence, a long silence. Very anxious members are thinking, what kind of a leader is this; what happens now; isn't she going to do something? Others decide that she has probably never run a group before, and resent getting stuck with her. Still others may be thinking that this is some kind of game, that they are being tested.

One rather confident male member begins cautiously by saying, "We're going to be spending the next couple of months together; we might as well begin by getting to know each other."

"How shall we do that?"

"Maybe we could go around the room and have each person introduce himself or herself."

"I have trouble remembering names, and besides that doesn't tell me anything about someone—who they are."

More silence. Glances toward the facilitator. More silence.

"How shall we proceed?"

"Does anyone have any ideas about how we could get started?"

And so it goes, with the facilitator taking on a behavioral role. She is not acting in the directive, lecturing position students expect of a teacher. The distinction is especially clear as to the difference between position and process. The facilitator's behavior, in this situation, indicates that an invitation to lead is open to all—to examine expectations, and to develop goals and means for implementing them. Acceptance by one member of a leadership function does not preclude others from participating, and it may even encourage others to begin to assume some responsibilities for the direction of the group. By her behavior, the facilitator indicates that leadership in the group is to be shared; she does not see herself as responsible for all the functions of the group. In this group, any member, regardless of position, can perform leadership functions.

WHO ATTEMPTS TO LEAD?

If any member can perform leadership functions, who attempts to lead? Who attempts to influence the group? In the illustration of the T-group, who is the person who first suggests an action; how is it that he or she ventures forth? Who are the others who offer suggestions and try to move the group in their direction?

There are two basic conditions for a member to take initiative: first, a member must be aware of what function (for example, information, opinion, encouragement) is needed; secondly the member must feel able to perform that act and that it is safe to do (Hemphill, 1961; Pepinsky et al., 1968). Therefore, prior to action, a member needs to have diagnosed the group situation and determined what behavior is needed to help the group move ahead; then he or she must make the difficult appraisal of whether he or she has the skills, the influence, or resources to act; and finally he or she weighs the consequences or repercussions. Some members of the T-group might have felt that the facilitator was sitting it out, but they had no idea of what could be done. For them, it was the facilitator's responsibility to lead the way; if she did not, there was very little they felt they could do or knew how to do. There might have been others who knew it was up to the participants to start, but they were not quite clear about how to go about it, or on what topic. Others might have known what they would like to do, or how it might have started, but they would not be first. To be first means acting without clues, without knowing the "lay of the land." What if this attempted initial behavior revealed some inadequacy that would be better unrevealed? It is better to play it safe and keep quiet. These are the basics. A person must know what functions are needed, and have the skills to perform them.

Our illustration has been a T-group with the characteristic ambivalence about the role of the facilitator and the role of members. However, to some degree, a predictable ambivalence about leadership is found in many non-T-groups. There is the conflict between a desire to be dependent and a need to be independent. On the one hand, we admire strong leaders who chart a definite course of action and give orders as to what needs to be done next; on the other hand, we resent taking orders; we value using our own initiative, and developing our own plans. A nondirective leader may be viewed as weak and lacking in leadership abilities, and we may wish he or she were more directive; a directive leader may be viewed as authoritarian, and we wish he or she were more open to suggestions from others. From childhood, we are taught to be independent, to develop our own skills and resources; and yet as human beings we are frequently nurtured and dependent well beyond the time of physical maturity, frequently well into the third decade of life. Is it any wonder that our ambivalence about leaders, our expectations of them, our conflicts between being dependent and being independent are a perennial source of stress in many groups?

But it is important to return to the prime question of who attempts to lead. One series of experiments attempted to develop some tentative answers. The researchers found that although some people enjoy exerting influence at any and all occasions, this was not typical. Rather, they found that the most potent source for attempting an act of leadership was generated out of the situation (Gershenfeld, 1967; Hemphill, 1961). When someone found himself or herself in a situation in which he or she had expert information required for the solution to the group's problem, he or she was strongly motivated to lead and influence the group in how to solve the problem. For example, a statistician could readily determine the procedures needed for a survey problem and would try to influence the group on how to approach the problem.

Then, too, tasks have different requirements. In a discussion task, many people may give opinions, yet there may be few efforts to change the opinions of others or suggest a given course of action. In a discussion, it may not be seen as appropriate to attempt to have others change their opinions. However, in a situation where a complicated product must be assembled, more leadership efforts will be made because suggestions for coordination of efforts are necessary for development of a constructive plan. In other words, the number of leadership attempts is influenced by the degree of coordinated effort required by the task. A finding of particular significance, and wide applicability, is one that we frequently feel but infrequently admit. Whether or not a person attempts to lead depends upon how his or her previous efforts were received (Gray et al., 1968). If a person offers a suggestion, and finds others nodding in agreement, he or she is likely to offer more suggestions. If, however, he or she offers a suggestion and gets disapproving glances or negative comments, he or she picks up the clues and desists from further leadership attempts. In one research study (Ring and Kelley, 1963) sets of four-member teams were assigned a task, and two members, by design, were cast in rejecting roles. Every time one of the members attempted to lead, a rejecting member expressed disapproval. There was a marked reduction in attempts to lead. However, when two confederates were assigned accepting roles, there were many more attempts to lead; participants who had rarely led now did so without hesitation.

Another factor that influences motivation to lead is how important the task is to the person, and whether he or she is in a position to help the group solve the problem. Thus, many of us may feel that certain goals are important—for example, greater flexibility of curriculum in the high school. But as parents or outsiders, we may feel our position to influence is extremely limited if not hopeless, and we will therefore not exert energy toward influencing the curriculum committee. However, if we are appointed to a Parents' Council committee to examine school curricula, we would feel now that the task is important *and* that we are in a position of influence, and exert more attempts to lead. Frequently, those who perceive themselves in higher or more important positions are more likely to attempt to influence the group.

In summary, when leadership is conceived as a function that can be performed at least theoretically by any member, the question arises as to who attempts to lead. It would seem that this is dependent upon many factors; the degree of coordination required by the task, the ability to diagnose, the ability to perform the needed behavior, the importance of the task, the possibilities of action leading to success and the feelings of personal acceptance or rejection.

WHOSE ATTEMPTS TO LEAD ARE REWARDED?

Leadership is defined as the frequency with which an individual may be identified as one who influences or directs the behavior of others within a group. A first step in this process involves an attempt to lead, but the crucial step is whether the group accepts the influence attempt. Why does a group do what one person suggests over what another suggests? Why are some suggestions listened to closely and considered, while others are dismissed even before the speaker finishes? How does the person whose suggestions are frequently accepted attain this influence? Why do members do what he or she advocates?

Power

Simply put, when one person does what another wants him or her to do, we would say that the influencer has power over the other. Leadership clearly involves power, that is, the ability to influence other people by whatever means necessary (McDavid and Harari, 1968). A person may be very influential and have a great deal of power in one group, and he or she is considered the leader because the group frequently accepts his or her direction. In another group, he or she may have little power; his or her suggestions are infrequently accepted by the group, and he or she would not be identified as one of the leaders. It is not unusual in community groups that a person who is a clerk in a business may be a powerful board member in a Boy Scout Council. The reverse, although less usual, also occurs. The chairperson of a university department—high power in the department—may only be window-dressing (low power) in a community association. Power is not a universal; it is limited by the person being influenced. A powerful person only has power over those whom he or she can influence in the areas and within the limits defined by the person being influenced. In other words, you only have the power those being influenced let you have.

Discussions of leadership sooner or later evolve into a discussion of power. The word itself evokes visions of manipulation, the omnipotent "big brother," and personal feelings of powerlessness. We think of Machiavelli's *The Prince* and his strategies of power; the dictum of Lord Acton, "Power corrupts, and absolute power corrupts absolutely" in relation to the centralization of power; and politics is defined as the ultimate power game. Perhaps we fantasize and begin to think of flower power, black power, and power to the people as also being power concepts. What do we personally think of power? Would we rather be powerful or powerless?

READER ACTIVITY

How would you rate yourself on a power dimension at present?

|__|__|__|__|__|__|

Not at all powerful	*Moderately powerful*	*Very powerful*

Where would you like to be on a power dimension in ten years?

|__|__|__|__|__|__|

Not at all powerful	*Moderately powerful*	*Very powerful*

On what basis do you see yourself attaining more power?

Perhaps it is the influence of American history, with its nation of egalitarianism in the ideals of "life, liberty, and the pursuit of happiness" for each person that is the basis for our ambivalence about power. Perhaps it is a distortion of middle-class values, which consider each of us as equal—none higher or lower. Perhaps our unease is related to the social connotation that being assertive or pushy are other words for being obnoxious. Perhaps we associate power with illegality or with conniving, shrewd maneuvers, which if not illegal are shady and despicable at best. Perhaps we are fearful of being thought pathologic—to need power is definitely viewed as "sick." We want power—to be the decision maker, to be the controller, to have things go our way. It is what each of us wants, yet we are ashamed to admit, even to ourselves, that we desire power. To be "power hungry" is to be despicable; yet to have power is valued.

And why wouldn't we want to be powerful (except for our confusions on the subject)? The more powerful members of a group tend to be better liked than low-powered members and are imitated more (Polansky, Lippitt, and Redl, 1950; Lippit, Polansky, and Rosen, 1952). They speak and are spoken to by the other higher powered members more than are lower powered members (Stogdill, 1974). They participate more, exert more influence attempts, and their influence is more accepted (Mulder, 1971; Rubin and Lewicki, 1973; Rubin et al., 1971; Hoffman, Burke, and Maier, 1965; Gray, Richardson, and Mayhew, 1968). Groups tend to be better satisfied when more powerful members occupy leadership positions (Stogdill, 1974) and those in positions of power enjoy being in the group more (Kipnis, 1972).

And further, being in a position of power is predictably an antecedent of higher self-concept. In an experiment with groups of Harvard undergraduates, Archer (1974) found that those with high power over the experimental period changed in the direction of more positive self-concepts and those with low power changed in the direction of more negative self-concepts.

If there is any area in which we are ambivalent and confused, it is in thinking and behaving in relation to power. We think of it as conniving and "dirty," and at the same time secretly strive for it, knowing the acceptance and satisfaction associated with it.

Power, then, is a multifaceted concept. Social power is defined as an influence relationship. Studies of power are concerned with the means, and the extent to which one individual can influence others. Power represents one aspect of role differentiation (but it is not synonymous with leadership).

What determines who has power? One conceptual scheme (French and Raven, 1960) distinguishes five different kinds of power.

1. *Referent Power* First, there is the kind of influence we do not even think of as power. We may emulate the clothes of someone we consider fashionable, we may espouse an argument we first heard from a brilliant intellectual with whom we identify, we may buy a book because someone whose opinion we value commented favorably on it. These people have referent power over us; we identify with them in certain areas, and they influence us without our feeling manipulated. In smaller groups, we hear the suggestions of those whom we perceive as having good ideas, or as being "with it" quite differently from those whom we categorize as pedestrian and "out of it." We hear the member who speaks for us, who represents our point of view, who sounds as if he or she understands our position, and we are much more influenced to act in accord with his or her suggestions. We may be influenced by those of higher status, a position we regard as important, a personal style, or a charisma. In each situation, the powerful person has power because we accept his or her influence and do it voluntarily. Obviously, this power exists only as long as that person is a referent for us. Fathers are powerful referents for children until they are teenagers, and perhaps for a period become powerful negative referents; later their referent power usually diminishes.

2. *Legitimate Power* A second kind of power is legitimate power. This is an authority relationship in which one person through his or her position is given the right to make certain decisions for others. It is the congressperson who represents our voting preferences, the department chairperson who represents us at the faculty executive committee, the foreman who supervises our work. It may also be the person we elect as president of an organization, the members of the committee we agree will make arrangements for a banquet, the observer we ask to process our behavior at today's session. The legitimacy may be derived from a number of sources. It may be from a higher level of the organization, it may be by law, it may emerge from the group. However, the recipients of influence see it as legitimate that the powerful person has a right to make decisions for them (Goldman and Fraas, 1965; Julian, et al., 1969).

3. *Expert Power* Frequently allied with legitimate power is expert power. Over time a person may become expert in an area, for example, a congressperson may point out to a citizens' group a strategy that will be carefully considered, because he or she will be seen as both an ally through his or her legitimate power and also an expert in terms of being knowledgeable on the machinations of Congress. Expert power may also exist independent of position. It is based on the person's specialized knowledge, information, or skills. When preparing a brochure, we call in a printer to give us an estimate, since from experience we know that our guessing will not be helpful in planning a budget. In the same way we seek the services of a plumber, a TV repairperson, a counselor, a psychiatrist; they have power over us because we see them as experts. In a group, those with expert power are asked more questions, they participate more, they are less likely to be women, and they are more likely to attain power in ongoing groups than in spontaneous groups (Richardson et al., 1973). As with all classes of power, we determine who the experts are, for how long we will be influenced by them, and the area of their expertise. In each of the first three mentioned types of power, we are voluntarily influenced. Somehow, because they are voluntary actions, we hesitate to think of these influence situations as power. We are more likely to conceive of power in the emotional "kid glove" or "iron fist" dimension; we normally think of power as reward or coercion.

4. *Reward Power* In the reward situation, the powerful person gives carrots, promotions, gold stars, or *A*'s to the recipients for complying. It may be the "bribery" of a parent as he or she entreats a child to finish dinner with a reminder of ice cream for dessert, or the parent who tells his or her teenager that if he or she makes the honor roll, he or she can have the car on weekends. Usually, reward power is situational, that is, determined by position. The parent rewards obedience in children, the boss gives rewards to workers. Often, the recipients of the rewards feel controlled. It means compliance, running the rat race, playing the "company game," or carefully following the rules. We do, however, tend to accept reward power, because the reward is more pleasant than the unfavorableness of the task or the other alternatives; we go through the task in anticipation of the reward. We enjoy the nod of approval, the smile, the support for our suggestion, the increased responsibility, the fatter paycheck. However, when the reward is not one that we perceive as more favorable than the other alternatives, the person who can administer the rewards has no power over us. If the child does not like ice cream, the inducement to finish dinner is clearly lacking. The student who does not care about grades considers the rewards of more

time with his or her friends and other alternatives more attractive. Reward power can only be exerted if the recipient values the offered rewards.

5. *Coercive Power* If the reward does not bring compliance, those in authority frequently resort to coercion. The child who does not finish dinner will be told he or she cannot leave the table until he or she does, or is sent to his or her room. The teenager who does not bring up his or her grades is threatened with having his or her allowance cut off or revocation of any rights to the family car (Kahn and Katz, 1966). The student who continues to create a disturbance is coerced with threats of detention; the employee with dismissal; the committee member who does not perform on a high-status committee with threat of not being reappointed. Coercive power, however, is not just the opposite of reward power. Whereas with reward the individual does what the powerful person desires in hope of attaining the reward, in a coercive situation the individual usually will first attempt to escape the punishment. Coercive power invokes not only coercion but no way to escape what the powerful person wants.

These are the types of power. Acts of leadership, if they are to be effective, must rely on some basis of power. Referent power, where the person identifies with the other and respects him or her, often will have the broadest range of influence. With reward power, there is increased attraction because of the promise of reward and low resistance. Coercive power is likely to produce increased resistance, although the more legitimate the power, the less the resistance. At the right time, when it is functional, each type may be very powerful. In many situations the leader's influence is based on a combination of sources of power, as we see in the example below.

Students mechanically attend the first session of class in a required course. They carefully scrutinize the teacher for clues as to how much reading will be required, how much work, how often examinations will be given, and how interesting the lecturer sounds. They also look for clues on attendance requirements and the possibilities for getting a good grade. Simultaneously, they acquire data on how expert the instructor seems to be, and over time determine how they feel about the instructor as a teacher, as a scholar, as a human being.

The university gives the instructor legitimate power to teach the course and administer rewards or coercions in the form of grades. The students also perceive the legitimacy of this power. The students, however, determine the degree of expertness they attribute to the instructor; the students determine the extent of their being influenced by the instructor as someone to emulate or relate to. How much influence the course will have on the students will to a large degree depend upon how much power the students attribute to the instructor. It may be only coercive power, and in a hostile environment the students "get through." Or the course may have a profound influence; the students may relate to the instructor as a personal model at least in some aspects, consider him or her genuinely knowledgeable, and find the course personally rewarding in adding to insights or skills. Although legitimate power is the basis for influence of an instructor, and while some students even question this and drop the course, other bases for power develop and determine the extent of influence.

For some, legitimate power is enough. The right to be in a position to make decisions over others is everything. Wielding power gives them enormous satisfaction and a sense of prominence. There is something about being able to control others which is overwhelming. Consider an example.

The lobby corridor was wall-to-wall with about twenty elevators. As the woman came around the corner, she noted one with the door open and the "up" light on; she immediately darted in, pleased that she didn't have to wait. Then, she "caught it." From someplace down the hall she heard, "Hey, where do you think you're going? I'll tell you which elevator is going up; I'll tell you where you stand to get the elevator. You want to stand there? Fine. But it isn't going up—not 'til I say so." It was incredible; in just seconds the woman felt like a bad child. She meekly got out, waited until he told her which one was going up, and got on feeling intimidated (he might not have liked her and then she would never get to the 32nd floor). All the while the woman was thinking, "Can you believe it? I got into a power hassle over getting into an elevator? His bit of legitimate power is determining who gets into which elevator, and he's power mad."

For some, wielding power in the form of rewards or sanctions is everything. Being liked is unimportant or, at best, secondary to having power. Being in a position of legitimate power, of being able to influence decisions and the "lives of others," is the insatiable quest.

READER ACTIVITY

Earlier in the chapter, we cited some people with power. Consider:
Pope John Paul II. What is his basis of power for you?

Al Capone. What was his basis of power?

John Kennedy. What was his basis of power in the early sixties?

Name a couple of others who are people with power to you. What is their basis of power?

STYLES OF LEADERSHIP

Most of us concerned with leadership, however, face the dilemma of wanting to be liked but also of wanting to get the job done. We may think: I know it will make the staff feel good if they think they are making the decisions, but what if they make the wrong decisions? I should want their help, but I have been here longer and know the realities; they may go off on some idealistic notion that would not be practical. Sure, I want to be popular, but I also want my job, and ultimately I am responsible for the decision made. Leaders at every level—teachers in a classroom, presidents of organizations, supervisors in business, parents raising children—face these dilemmas and ask these basic questions phrased appropriately for their own situation.

Conceptions of Other People Influence Leadership Style

How a leader answers these questions and resolves the dilemmas rests on his or her assumptions about other people. The assumptions usually are not conscious or formally defined, but a leader's conception of human beings has implications for his or her leadership style (McGregor, 1960; Schein, 1969; Maslow, 1954). There are two images or theories of how to lead. In the first (Theory X according to McGregor; rational-economic man to Schein), people are seen as having little ambition, a reluctance to work, and a desire to avoid responsibility. People are motivated by economic competition, and conflict is inevitable. Without managerial effort, basically men and women do nothing. The leader operating under these assumptions must motivate, organize, control, and coerce. He or she directs, people under him or her accept and even prefer it, because they have little ambition or desire for responsibility. The leader bears the responsibility and burden of his or her subordinates or followers' performance. This represents the traditional theory of management, especially business management. Another theory (Theory Y for McGregor, self-actualizing man for Schein) holds that people are motivated by a hierarchy of needs. The assumptions in this theory are that as basic needs are met, new emergent needs become motivating forces. Each of us has a desire to use our potential, to have responsibility, to *actualize* ourselves. The theory assumes that men and women enjoy work as well as play or rest. This conception suggests that individuals will exercise self-direction and self-control toward the accomplishment of objectives they value. Furthermore, they can be creative and innovative. They will not only accept responsibility but also seek it. That potential for imagination, ingenuity, resourcefulness is widely distributed within the population but poorly utilized in modern society. In this theory, the leader creates challenge and an opportunity for subordinates to use their abilities to a greater extent. There is no need to control or motivate; the motivation is waiting to be unleashed. A leader's conception of men and women will greatly affect his or her style of supervision and the bases of power he or she will implement. The first theory is more likely to use money as a motivating reward and coercion to compel compliance. The second theory is likely to induce intrinsic rewards of self-satisfaction and pride in achievement; coercion is used infrequently.

Does Leadership Style Make a Difference?

A classic research, the Lewin, Lippitt, and White studies (Lewin et al., 1939; White and Lippitt, 1968), investigated the following questions: What difference does style of leadership have on the group? Is the group more productive if the leader is autocratic, democratic, or laissez-faire? Does it make a difference in how members relate to one another? Is there a difference in the social climate?

In each experimental situation three leadership types were established: the autocratic leader, the democratic leader, and the laissez-faire leader. Each leader had legitimate power as he worked with ten-year-old boys on basically similar craft projects. The findings were dramatic. The results indicated that demonstrably different group atmospheres developed. And further, in each experimental group, there were readily perceived differences in relations among members and their ability to handle stress, as well as in their relations with the leader. The findings were widely disseminated and convincingly demonstrated in that particular situation, that the best leader was the democratic leader. Yet, the implications for leadership are not that clear.

Choosing a Leadership Style

We label leaders, and the problem is: Which is the "best"? It all depends on how we perceive the label. If a person is described as an autocratic leader, an image occurs in which he or she is allied with demagogues, dictators, and coercive administrative processes. Yet autocratic can also describe a person who is directive, who stands firm in his or her convictions, who accepts the responsibilities of supervision and ultimate responsibility for his or her decisions—in short, one who has the necessary attributes of leadership.

To be labeled a laissez-faire leader is to be viewed as in a fog, incompetent, fearful of making a decision, and shirking responsibilities. This is clearly an offensive label. Yet, on the other hand, "Creativity must be given free rein"; "he who rules least, rules best." Shall the leader supervise closely, or "trust his or her people"?

To be labeled a democratic leader usually suggests that the person is well-liked. As for his or her behavior, does it mean that he or she shares all decisions with others regardless of the consequences? Does it mean that the staff members are one big happy family, that they talk in terms of *we* rather than *I*, and that all relationships are collaborative rather than competitive? Do all decisions have to be group decisions; is giving up power the price of popularity? Is the "big happy family" the goal to strive for no matter what; is any aspect of competition to be avoided at all costs?

At one time this labeling was important as we sought to understand the continuum that went from the laissez-faire leader who was minimally involved to the autocratic leader who arbitrarily made decisions based solely on his or her own style. At that time, the democratic leader was in the middle, neither the "abdicrat" nor the autocrat, and there was a desire (especially in an era of World War II dictators and totalitarian governments) to reinforce our conviction that democracy is "best." The entire concept of experimentally inducing three different leadership styles, analogous perhaps to governments, was powerful. The results generated increased understanding of the problem and limitations of each style.

The studies have been so effective that they continue to be replicated (Koch, 1978; Bernstein, 1971; Sargent and Miller, 1971; Scontrino, 1972; Sudolsky and Nathan, 1971); researchers find the early hypotheses continue to be valid. Yet, today, the question of whether one is an autocratic, democratic, or laissez-faire leader is no longer meaningful; we have come to look at leadership styles in a different way (Hersey and Blanchard, 1969). We no longer view leaders as being in a box that can be labeled, whether by their detractors, their friends, or even their own dilemmas.

Effective Leadership

The question to ask is: Which style or combination of styles will help a group arrive at its goals in a manner consistent with the values of that organization? What leadership style is effective?

Researchers have emphasized leadership traits, the effects of the situation, the position, and functional roles on leadership, and leadership styles. Certainly with all of this information and research, we should be able to accurately determine what is needed in a given organization, and train a person to be an effective leader. Fiedler (1973) makes a strong point when he says we know little. "Research has failed to show that leadership training makes organizations more effective. No one has established a consistent, direct correlation between the amount or type of a leader's training and the performance of the group he leads" (p. 24).

In a series of training programs, Fiedler and his associates took populations who had no training. Each population was divided into small teams and given a series of tasks to perform. Fiedler found that those trained performed no better than those who had no training. He found that those who had been in T-group or sensitivity training performed no more effectively on their jobs than those who had not. In another series of studies, he found that experienced leaders (most training is on-the-job training) tended to be no more effective than managers with little experience.

Although Fiedler is commenting on the ineffectiveness of leadership training programs, others are commenting on the mixed findings related to goal productivity. Some studies indicate that quality and productivity are higher with a high-task leader, while others indicate that productivity is greater with a leader who has a high social-emotional or interpersonal relationship orientation. Some report greater member satisfaction with a high-task leader; others with a high social-emotional leader.

The mixed findings on productivity, the studies that show leadership training is not significantly effective, and the desire to produce more effective managers have all led to a reexamination of leadership theory. The result has been a series of leadership models, typologies, and theories (Fiedler, 1967; Yetton and Vroom, 1978; Blake and Mouton, 1976; Hersey and Blanchard, 1969), all of which are integrative or contingency models—in other words, attempts to avoid the either/or dilemma.

LIFE-CYCLE THEORY OF LEADERSHIP

All of the theories, concepts, and empirical research presented earlier in the chapter have made a contribution to the field of leadership and management. These viewpoints have often appeared like threads, each thread unique unto itself. Hersey and Blanchard, in their *life-cycle leadership theory*[2] (1969, 1975, 1977) have created a synthesizing framework that allows these varying viewpoints to be woven into a fabric that attempts to explain why results of leadership training have been mixed and why efforts to produce effective leaders have been so limited.

Task and Relationship Dimensions

To begin, there has been general agreement for some time that there are two central dimensions of any leadership situation, that is, a task (goal, production, product, or initiating structure) dimension, and a relationship (social-emotional, consideration for others, interpersonal relations) dimension. Most typically, these are depicted on a managerial grid (Blake and Mouton, 1969). The figure at the top of page 268 illustrates the central dimensions of a leadership situation.[3]

In this formulation, task behavior is illustrated on the horizontal axis. Task (production) becomes more important to the leader as his or her rating advances on the horizontal scale. A leader with a rating of 9 has a maximum concern for production.

Concern for people is illustrated on the vertical axis. People become more important to the leader as his or her rating progresses up the vertical axis. A leader with a rating of 9 on the vertical axis has a maximum concern for people.

[2]Originally, when the theory was first reported, it was called *life-cycle theory of leadership;* more recently it has been called *situational leadership theory.*

[3]From Hersey, P. and Blanchard, K. H. "Life-cycle theory of leadership." Reproduced by special permission from the May 1969 TRAINING AND DEVELOPMENT JOURNAL. Copyright 1969 by the American Society for Training and Development, Inc.

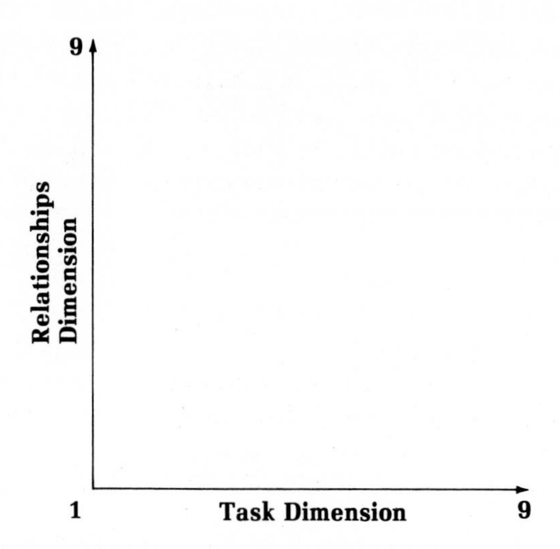

The four quadrants produced by the grid describe leader behavior (five in the Blake-Mouton formulation, the fifth being at the cross of all four quadrants). Quadrant I represents a leadership style that is high on task and low on people; Quadrant II represents a style high on both task and people; Quadrant III is high on people with little concern for the task; Quadrant IV is a style low on both task and people. See the figure below.

Relationships		
	High people *Low task* **III**	*High task* *High people* **II**
	Low people *Low task* **IV**	*High task* *Low people* **I**

Task

After identifying task and relationships as the two central dimensions, some management writers have suggested a "best" style. Most of these writers have supported an integrated leader behavior style (high task and high relationships) or a people-centered, human relations approach (high relationships). However, some of the most convincing evidence that dispels the idea of a single "best" style was gathered and published by Korman (1966). Korman reviewed over twenty-five studies that examined the relationship between dimensions of "task" and "consideration" (people) and various measures of effectiveness, including group productivity, salary, performance under stress, administrative reputation, work group grievances, absenteeism, and turnover. He came to the conclusion that ". . . very little is now known as to how these variables may predict work group performance and the conditions which affect such predictions. At the current time, we cannot even say whether they have any predictive significance at all" (p. 360).

Korman saw no predictive value in terms of effectiveness, as situations changed. This would suggest that since situations differ, so must leader style. Fiedler (1967) in his studies of leadership over fifteen years came to similar conclusions. He found that both directive, task-oriented leaders and nondirective, human relations-oriented leaders are successful under some conditions. Other investigators have also found that different leadership situations require different leader styles.

Yet, empirical studies tend to show that there is a "best" style of leadership; that successful leaders are those who can *adapt* their leader behavior to meet the needs of their followers and the particular situation. Effectiveness is dependent upon the leader, the followers, and other situational determinants. Effective leadership requires the leader to diagnose his or her own situation in the light of his or her environment.

The Effectiveness Dimension

To measure more accurately how well a leader operates within a given situation, Hersey and Blanchard added a third dimension—effectiveness—to the two-dimensional model. The effectiveness dimension cuts across the two-dimensional task/relationship factors and builds in the concept of a leader's style, integrated with the demands of a specific environment. When the leader's style is appropriate to a given environment measured by results, it is termed effective; when his or her style is inappropriate to a given environment, it is termed ineffective.

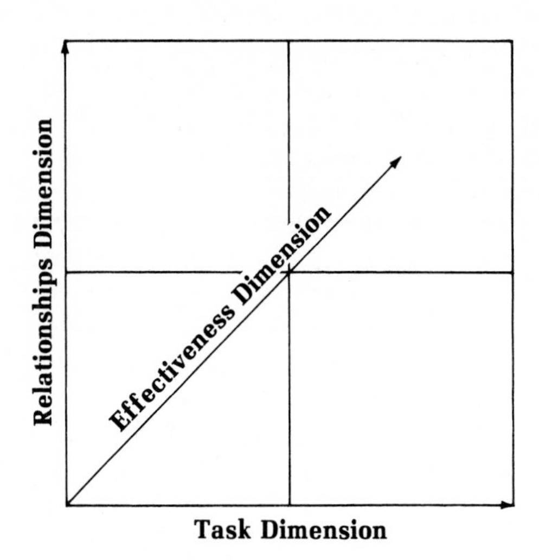

If a leader's effectiveness is determined by the interaction of his or her style and environment (followers and other situational variables), it follows that any of the four styles (defined by the grid quadrants) depicted in the model may be effective or ineffective depending on the environment. Therefore, there is *no* single ideal leader behavior style that is appropriate in all situations. In an organization that is essentially crisis-oriented, like the police or the military, there is evidence that the most appropriate style would be high-task style, since under riot or combat conditions success may depend on immediate response to orders. Studies of scientific and research-oriented personnel show that they desire or need only a limited amount of social-emotional support. They know what they are doing and "want to get on with it." They view meetings as "wasting time." Under these conditions, a low-task and low-relationship style (leave them alone for the most part) may be the most appropriate one.

In summary, an effective leader must be able to *diagnose* the demands of the environment and then either adapt his or her leadership style to fit these demands, or develop the means to change some or all of the variables.

The Maturity Factor

Hersey and Blanchard, in their life-cycle theory of leadership (situational leadership theory) have, in response to the vague concept of *situational determinants,* sought to create a theory with a systematic conceptualization of what situational determinants might be and how they are related to leadership behavior (task and relationships).

The primary element in the theory is that an effective leadership style is related to the level of maturity of followers. The emphasis is on followers. Followers are vital not only individually as they accept or reject leaders, but also as a group, because they determine what power the leader may have.

According to life-cycle theory, as the level of maturity of one's followers increases, appropriate leader behavior not only requires less structure (task), but also less social-emotional support (relationships). This cycle can be illustrated in the four quadrants of the leadership effectiveness model.

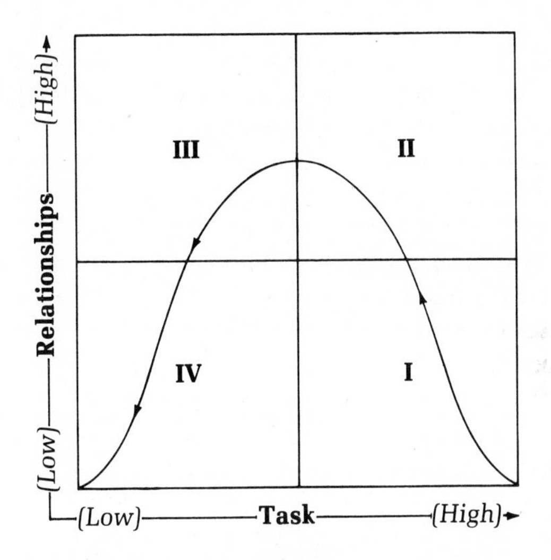

Maturity, in life-cycle theory, consists of several components. First, people have the capacity to set high but obtainable goals and a desire for task-relevant feedback (how well am I doing?) rather than task-irrelevant feedback (how well do you like me?). Second, they are willing to take responsibility, which involves willingness (motivation) and ability (competence); the highest level of responsibility would be taken by those high on ability and competence. Third, maturity can be thought of as involving two factors: job maturity—the ability and technical knowledge to do the task; and psychological maturity—a feeling of self-confidence and self-respect as an individual. These two factors seem to be related; high task competence leads to feelings of self-respect; the converse also seems to be true.

Finally, although the maturity concept is a useful one, it should be remembered that diagnostic judgments may be influenced by other situational variables like a crisis, a time bind, or your superior's style.

According to the theory, as the level of maturity of followers continues to increase in terms of accomplishing a specific task, leaders should begin to reduce their task behavior and increase their relationship behavior (beginning at Quadrant I with immature, inexpereinced followers, moving to Quadrant II and then III as the individual or group moves to an average level of maturity). As the individual or group reaches an above-average level of maturity, it becomes appropriate for leaders to decrease not only task behavior, but also relationship behavior (moving now to Quadrant IV). Now, members are not only task- but also psychologically mature, and so can provide their own strokes or psychological reinforcement, needing much less supervision. An individual at this level of maturity sees a reduction of close supervision and a delegation of responsibility by the leader as an indication of positive trust and confidence. This theory, then, focuses on the appropriateness or effectiveness of leadership styles according to the task-relevant maturity of the followers.

This cycle can be illustrated by a bell-shaped curve going through the four leadership quadrants.

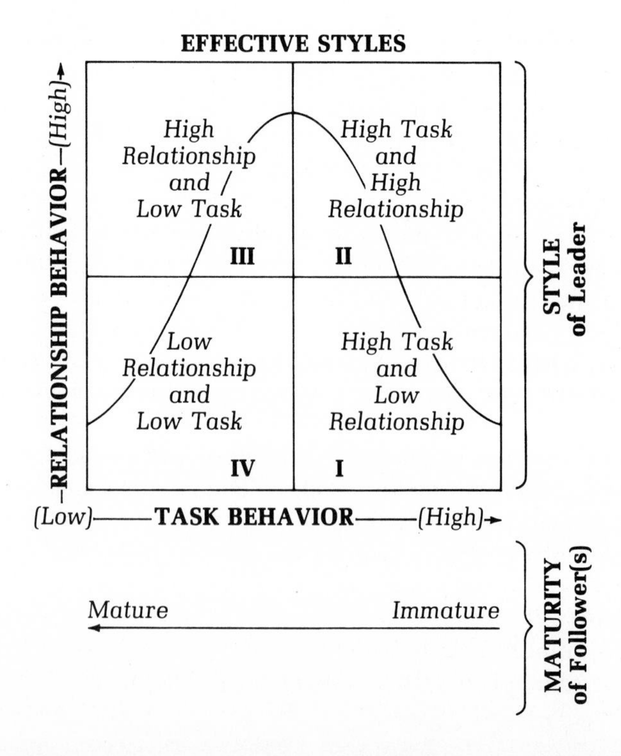

Determining Appropriate Leadership Style

What does the bell-shaped curve mean to a leader with a specific task to accomplish with a given organization or group? It means that the maturity level of one's followers develops along the continuum from immature to mature, and the appropriate style of leadership moves accordingly along the curvilinear function.

To determine what leadership style is appropriate in a given situation, a leader must first determine the maturity level of the individual or group in relation to a specific task that the leader is attempting to accomplish through their efforts. Once a leader identifies this maturity level, he or she can determine the appropriate leadership style. If the followers are of low maturity, the leadership style of Quadrant I would be most effective; the style of Quadrant II second most effective; Quadrant III next, and Quadrant IV least effective. Using the four maturity levels as a basis for diagnosis, a leader can determine which style is most appropriate.

The theory has broad applicability, not only to leadership situations in organizations, but even in the family.

A Parent-Child Example

Young children, up until three years of age, clearly need structure, support, affection, and a high degree of control as they wander through an environment fraught with dangers. There is little question that a little Quadrant I behavior of the benevolent type would be useful. Control and structure, wrapped in understanding and care, are just what the doctor ordered. The problem is that it is all too easy for parents to become addicted—habituated to a particular parental behavior. If they do, they will surely pay a very heavy price at some time in the future. For example, as children become more mobile and independent, which is one of our goals for them as parents, they also push against parental authority, and become less easily controlled. At the very point when some structures should be taken away, some parents put more on because they are personally uncomfortable with the child pushing out against the world, often against their authority. The easiest behavior for parents is to push back hard—increase control—use more of the same behavior they have been using except without so much loving and understanding as before. Since it is rather easy to intimidate a three- to four-year-old, the parent sees that their high control behavior, their intimidation, works. It takes little thought and is immediate. What the parent puts off in the present in relation to patience, understanding, and the time needed to explain, or the effort needed for greater collaboration is paid for in spades about ten years later. Children faced with unidimensional leadership on the part of the parent leader respond by being passive or by continually pushing against the authority (which forces the parent into more reactive, unidimensional, authoritarian behavior). Similarly, the overly permissive parent, who lets children set limits of when to go to bed and what to eat sets in motion a pattern of expectation that will haunt them as the children (subordinate) grow. Like any leader, the parent needs to have an array of leadership behaviors that are appropriate according to the situation.

It may be that parents, like most leaders, have more power to require others to conform than they need. Often there are many more imposed rules than make sense, and they act more as a means of creating barriers (to be torn down or in other ways resisted later) than anything else. Hersey and Blanchard believe that the basis for making decisions concerning one's leadership is dependent, to a large degree, on the maturity of the individual or group that is being led. Thus, the sooner children are brought into household decision making and problem solving, the sooner they sense their personal influence in the home and the sooner they will respond in a mature manner. Again, the problem with parents is that when they feel the growing sense of independence being generated by their maturing children, they often revert to more controlling behaviors, imposing more structure and becoming even more arbitrary. This reversion to Quadrant I behaviors will, of course, be resented and resisted by the children. The cycle continues unabated as the parent, resenting the children's resistance, then comes down even harder.

By the time a young adult is fifteen or sixteen, leadership should, as much as possible, be shared or delegated. The Quadrant III, or consulting mode, is probably most acceptable with people maturing into adulthood. In fact, if parents are wise they will begin to move completely out of their involvement in many areas and give total responsibility to the child. Sad as it is to say, many parents never can let go of Quadrant I and II behaviors, even when their children have become mature adults. Whether in a business or a family, there is a constructive and useful place for Quadrant IV behaviors. Perhaps the toughest role for a leader is letting go of old responsibilities that others can take on. Unless this letting go occurs in most systems, subordinates will continue dependency-building behaviors and often develop passivity and resentment.

QUESTIONS FREQUENTLY ASKED ABOUT LEADERSHIP

What Are Some Leadership Findings That Can Be Generalized? As must be abundantly evident by this point, there is no single right style of leadership. But there are findings which can be generalized in our efforts to understand the complexities of leadership. The following seem to hold true:

1. What at one time was the crux of a major theory—that leadership results from some personality trait, or was determined situationally, or was dependent upon culture, or arose from interaction of particular forces—today is seen as but one of many items considered in the complicated leadership equation.
2. The kinds of assumptions a person has about people will determine how he or she will lead or supervise them. Becoming aware of one's own assumptions is therefore an important factor in a personal leadership style. (There may be certain behaviors that are natural, and others which, because they are appropriate, need to be learned—perhaps with great difficulty.)
3. Leadership is closely related to followership. An act of leadership involves acceptance of influence. Leadership can be understood in terms of the influence attempts that succeed in a group, and which theoretically can be carried out by any member.
4. When a person is in a position of leadership and needs to decide how to supervise those under him or her to accomplish the task, the maturity of his or her subordinates is a significant factor in diagnosis prior to action.
5. Effective leaders are those who can behave comfortably in each quadrant, depending on the stage of maturity of their subordinates. They can adapt their behavior to the requirements of their members, the problem at hand, and the factors in the situation. They are flexible enough so that their behavior helps the group (organization) move toward its goal.
6. Frustration and confusion generally ensue when a leader verbally states one position (quadrant) and actually behaves from another position.
7. Leaders have a number of choices about how tightly to centralize control and about how flexible they can be from one situation to the next. By diagnosing forces in themselves, in the members, and the situation, they can increase their flexibility and their effectiveness.

How Is Leadership to Be Judged? Frequently a leader asks: How did others act? How do we generally solve this problem? What are the rules? However, to be effective, leadership is to be judged in terms of the goals and purposes of the group and the maturity and experience of the members (Jenkins, 1961). The leader, or others in the group who would like to be influential, ask themselves the following questions: What is needed by this group at this time to move toward the goal? How can I help movement toward it? This is a difficult task because sometimes the goals are not clear; sometimes we are not sure of a diagnosis; frequently we question our skills and resources, or fear the consequences of attempting behavior that may have negative implications.

What Kind of Leadership Is Best? There is no best kind. Groups working on different tasks need different kinds of leadership. Sometimes groups want a close personal relationship between members and the leader; at other times they prefer a more formal relationship. Some groups need to consider the alternatives and come to a decision themselves; others may need firm direction from the person in charge.

Groups need help in progressing on their tasks as well as remaining in healthy working condition. A leader must be sensitive to both aspects. Typically a leader is only concerned with the task role, but he or she also has to pay attention to increasing ways for members to communicate and work together. High task behavior is appropriate with immature (new, inexperienced) members; as the members mature, there are other behaviors needed for effectiveness.

How Is Success or Failure of the Group to Be Judged? Success or failure, in most groups, cannot be attributed to the leadership. Usually the entire group is responsible for the success or failure of the group. Rarely can it all be blamed on the leader; nor can he or she be given all the credit. Leadership functions can be shared, and sometimes members abdicate their responsibilities.

However, leaders can impede efforts of members to be helpful to the group. Leaders may not permit members to voice their suggestions and thus limit resources and information available to the group. They may limit too narrowly what they will permit members to do and thereby limit their potential contributions.

Groups can impede the effectiveness of their leadership. They may have antiquated norms that hamstring leadership. They may repeatedly reject leadership attempts and cause leaders to stop trying. On the other hand, encouragement and support are likely to produce increased leadership attempts.

EXERCISE 1 A Series of Nonverbal Experiences in Leading and Being Led

The following exercises focus on our feelings of leading and being led, and provide data to examine both our behavior and attitudes in being a follower and being a leader. The exercises are appropriate as a beginning micro-lab in a longer workshop. They are also useful to members as a means of examining their typical roles and understanding areas in which modification or reevaluation may be needed.

The exercises lead to nonintellectualized discussions of leadership in terms of personal satisfactions or conflicts, areas of skill and ineptness, and enjoyment of authority or fear of responsibility. They also allow participants to see the complementary nature of leader-member relations in trust, openness, spontaneity, communication, and dependence-independence.

The exercises are of varying lengths and, depending on the purposes, can be expanded or reduced. They involve varying degrees of risk and should be used with an understanding of both the purpose of the exercise and the type of group involved. For some groups, any touching is beyond the bounds of propriety, and these exercises should be eliminated. Because touching is counter to the norms of the group, much tension is generated and effectiveness in examining leader-follower relationships is severely curtailed. Some groups who consider themselves serious and work-oriented may frown upon games that call for a childlike spontaneity; they may consider games as contrived and not a legitimate basis for learning. The values and resistances of the group should be seriously considered so that the exercises used will help members achieve their purposes.

The procedure is as follows: The exercise would be named, as illustrated below, and the facilitator would give instructions so that the reader has both a feel for the exercise and how it is to be carried out.

1. Connectedness by Rubber Bands

The facilitator begins by saying: "Pick someone. Put your hands out in front of you. Almost touch the hands of your partner. Pretend now that your hands and his or hers are connected by rubber bands. Move—feel what happens." (5 min.) "Pretend your feet are also connected by rubber bands." (5 min.) "Talk about what happened." (10 min.)

The objectives of the exercise are to increase awareness among pairs of participants as to their own boundaries and those of others; a body awareness; control of self, control of another person. It also presents a situation of leadership-followership. Did one person lead all the time? Did they reverse roles? When? At what signals? Which role was more comfortable: leading or following? Did one partner see the other as being more comfortable leading or following? Did it make a difference whether it was the hands or feet that were connected?

Not all of these questions are asked. A few may be suggested, depending on what clues the façilitator observes. He may mention a few as basis for discussion, but it is important not to spend too much time in discussion; rather, participants should express their response to this experience and let their impressions build with subsequent exercises.

2. Communication by Clapping

With this approach, the facilitator gives the following instructions: "Pick someone who has not been your partner previously. Clap a message. Then let the other person respond." (The facilitator should select a member of the audience and demonstrate. The two stand facing each other, and the facilitator claps the phrase "How are you?" The other person claps back, "fine," or "angry," or whatever.) "You see how it is done. Remember that it involves restructuring meaning from familiar sounds. Now try having a conversation through clapping." (10 min.) "Between you, discuss what happened. What was expressed?" (5 min.) Then the entire group discusses what they experienced. (5 min.)

At one level, this exercise allows participants to experience expression through tactile and auditory senses, to create and hear rhythm and sound as expressive of emotion and sequence. At another level, communication is seen as a process requiring a sender and a receiver; one side is insufficient for communication. At yet another level, it is essential to recognize who initiated the "conversation," who "talked more," who led, who followed, who was frustrated and withdrew. The exercise can also be used to examine functional roles of members.

3. Leader-Follower Trust Walk

The facilitator begins: "This exercise focuses on being a leader or a follower. Half of you will be blindfolded. Those who are not blindfolded will select a partner. This is a nonverbal exercise, so you may not speak to your partner to tell him or her who you are. Let's begin by counting off to two." (Participants count off "1, 2, 1, 2," etc.) "Will all of the ones come to this side of the room? Here are handkerchiefs to use as blindfolds. Put them on and adjust them so that you cannot see." (The ones arrange their blindfolds and wait.)

Now, the facilitator talks to the twos quietly so that the others cannot hear. "Each of you will select a partner from the group of ones. Stand beside the partner you choose so that we can determine who still needs a partner. Remember, you and your partner may not speak, but by all means try to develop a nonverbal language between you. You will be the leader. How can you help your partner experience his or her world? Can you enlarge his or her world? Be aware of how you see your role; is it to protect him or her, to get him or her through safely? Is it to be with a minimum of effort on your part? Is it to be serious; is it to be fun? (Pause) Now select your partner."

"Explore your world, nonverbally, of course. I will see you back here in 15 minutes." (If this occurs in a building, 15 minutes is adequate time. If it is outdoors and time permits, allow up to an hour. It is frequently a moving experience to see the partners develop their own signals, an increased sensitivity to each other, a trusting relationship.)

The facilitator alerts the group two minutes before time is up. If the setting is outdoors, he or she simply hopes that they will straggle back reasonably on time.

When they return, he or she has the "blind" remove their eye covers to see who their partner is. (This produces tension, anxiety, even fumbling, as the "leader" wonders if his or her follower will be disappointed when identities are revealed. There is also the anxiety of returning to the "real world," which does not encourage the closeness and trust some of the partners felt. Now the mood changes; the "uncovering" produces laughter, squeals of recognition or surprise.)

The facilitator proceeds: "Would you share your feelings in the experience? What did you find out about yourself that was new? What did you find typical of yourself? How did you feel about your role—as a leader, or a follower?" (15 min.)

"I am sure each of you wants to experience the other role. Will all of the twos come to this side of the room? Now, it is your turn for the blindfolds. You know what to do."

He or she then talks to the ones as previously to the other group. Although they have been through the experience and know what they want their partner to experience, nevertheless it seems helpful to remind them, through the questions, of a variety of possible relations they may have with their partner in the leader role. Once more, those not blindfolded select a partner and stand beside him or her. Frequently, the choosing partner will select his or her former partner so that he may "repay" him or her for his or her felt interest. For a variety of reasons, a person may prefer to select a new partner. After selections are made, the exercise continues as in the first pairing. The groups return, see each other, share their feelings.

Sometimes the participants feel their reactions are significant and relatively private; they may only want to share with their partners. Sometimes participants are quite eager to share their new understanding of themselves with the whole group. If there is a group sharing, one of the questions the facilitator should ask is how it felt to be a leader and how it felt to be a follower. What was learned?

This exercise has usually been considered primarily an experience of trust; however, it is striking how often members report on their relations with authority. Students frequently note that it helps them understand their parents or gives them understanding on what kind of parent they would like to be. Men and women discuss their societal sex roles in which a man is expected to be the leader, and their feelings when leadership roles are reversed. Some who usually see themselves as leaders are surprised at their reactions to being followers and gain a different perspective of the relationship; those who are usually followers give similar reports. Participants will also talk about clues they pick up from each other—being tired, bored, excited—which greatly influence the other person and the relationship; new insights are reported on complementary relationships

As an aside, if a supply of handkerchiefs is difficult to obtain, paper towels as they come from a dispenser and masking tape are equally effective as blindfolds.

4. Follow the Leader—A Musical Variation on the Children's Game

Select an instrumental record with a diversity of moods, tempos, and sounds (Vivaldi's *Four Seasons*, for example). Participants are divided into groups of eight to ten and stand in a circle facing each other; in some groups participants take off their shoes to enhance a feeling of movement.

Then the facilitator says, "I'm going to play a record. Listen to it, get a feel for the mood. If it reminds you of something, or if you want to express what you feel, come into the center of the circle and do it. Those of us who can feel it with you will follow you in our places. When the mood of the music changes, return to your place. Someone else, who is feeling something he or she would like to share, will go into the center. O.K.?" (10–15 min.)

"How was it? How did you feel about what happened?" (10 min.)

One of the questions to be raised is, Who initiates leadership? It is someone who knows what is needed, feels he or she has the skills or resources to do it, and feels it is safe to try. Discussion will relate to all these issues. Some will say they cannot even think in terms of what that music could mean in a public place; their minds went blank immediately. Others had a mental image of what the music evoked but did not seem to have the skill or resources to transfer that image into body movement. Still others hesitated to come into the center for fear of appearing clumsy, or childish, or not very original. Another area for brief discussion is: How did the situation change? When did you start to feel comfortable, or enjoy it? What happened?

5. Building a Group by Music

This exercise is done with an instrumental record that has an easily discerned tempo, rhythm, and mood; ethnic folk dances are especially suitable (African dances, or Irish, or Israeli folk dance records work well).

1. *Pair* The facilitator gives the following instructions: "Select a partner. When the music starts, one of you move to the music, as you feel it. The other person will be your mirror image; he or she will do what you do. If it helps to be more realistic, pretend each of you is touching the mirror with the palms of your hands. The person who is the mirror will try to follow facial expressions as well as body movements. Change who is mirroring when you want to." The record begins. (About 3 min.)
2. *Quartet* "Add a pair to your group. Continue to move to the music but do it as a group." (About 3 min.)
3. *Octet* "Add a quartet to your group. Continue to move to the music as a group of eight."(About 3 min.)
4. *One More Time (A Group of Sixteen)* "You're right. Add an octet to your group. Stay with the music. Move to the music as a group."

(About 3 min., preferably to the end of the record so that there is a natural feeling of closure.) This produces exhilaration but also a good supply of creaking joints, and a surge of business for the water fountain. There should be no discussion for at least 15 minutes.

The objectives of this exercise are to examine leadership-membership relations in varying group sizes. Who follows, who leads? Is it easier to lead in a small group than a large one? Do some leadership patterns remain? Why? How is leadership determined? What role do members have? Do they have inputs that are listened to? Who was the leader (or leaders) at the end? How did he or she get influence? On what was it based? The facilitator asks how some of the members felt in this experience. He or she expands on some of the answers given to the above questions or raises a few new ones. Once more, the discussion should be brief and informal.

Conclusion

The facilitator in closing might summarize by reminding the group that they have been through a battery of nonverbal experiences that explore leadership-membership relations. They may consider what new understanding they have about themselves, for example, how they acted when they were asked to pick a partner. Did they choose, or more typically wait to be chosen? Why? What new insight do they have on the subject of leaders, members, followers?

Participants may be asked to write a log, a self-report, verbalizing what they have learned. They can be divided into small groups to discuss some of their experiences. A large group discussion leaves too many as listeners without an opportunity to participate; at best it might be used briefly to begin the discussion.

The facilitator should be particularly sensitive to norms of the group with whom he or she is working. These exercises should only be used where appropriate. As noted earlier, they should not be used with groups who consider touching inappropriate, nor should they be used unless the goals of the training will be enhanced by these exercises.

EXERCISE 2 What Is My Role in a Group?

Objectives

To learn who is perceived as leader of the group

To understand the many roles in a group

To develop data on what contributes to being the leader in a particular group

To increase data to participants on their perceived roles

To increase data to participants as to relationships between their perceptions of their roles and others' perceptions of their roles

Materials

Copies of the Behavioral Description Questionnaire (included with this exercise) are needed; also newsprint to record the data. If it is a large group that has been working in smaller units, data can be recorded within each unit. If there are ten members, each person requires ten questionnaires.

Rationale

This exercise is appropriate after members of a group have been working together and know one another. It provides information to members on their roles as others perceive them, and as they themselves perceive their roles. This information may be congruent, or at great variance. It is a valuable feedback to the member on his or her influence in the group. It also permits data on who is seen as the leader and with what dimension leadership in this group is allied. Data are also available on the degree of commonly shared perceptions. Do many see the same person as the leader, or are several people seen as leaders?

Action

A questionnaire is distributed listing fifteen statements. Under each is a 5-point scale. Each member is given enough questionnaires to fill one in for every member of the group. The facilitator may begin by briefly stating that each group develops in a unique manner depending upon the composition of the group, its task, its situation. He or she continues, "Each of us takes on a unique role within a group. Sometimes it is our typical role, sometimes a blend of roles from many experiences. Sometimes we see ourselves taking one role, and others see us quite differently. Perhaps by developing and sharing this data, we can learn more about ourselves and our group." It is important that the facilitator explain this rather seriously. Sometimes groups are apprehensive about getting or giving feedback and need support from the leader that such data are both legitimate and helpful.

Each person lists the name of the person he or she is describing at the top and then proceeds to indicate how typical that behavior is of that person. (30 min.)

An easy way to collect data is to give one person all the sheets for another person (not his or her own). He or she then charts the data on a blank questionnaire. Finally he totals points on each question (a rating of 3 is 3 points). (15 min.)

The facilitator asks for a report from each person on the behaviors seen as most typical for the person he tallied; or he asks for the highest points on each question. The person who scores the highest points on question 9—who is the real leader of the group—is the leader. In what other areas did the people who score highest on that question also score high? What are the behaviors required for leadership in this group?

Following discussion, all the papers are given to the person whose name is on top. The person then has all the reported data on himself. He can examine its congruence or variance with his own perceptions. (15 min.)

BEHAVIORAL DESCRIPTION QUESTIONNAIRE[4]

Indicate how well each statement describes each group member (including yourself) on a 5-point scale running from "very true of him or her" (5) to "not true of him or her" (1).

1. He prodded the group to complete the task.

---

1 2 3 4 5
Not true of him *Very true of him*

2. He was the real "idea man" in the group, suggesting new ways of handling the group's problems.

---

1 2 3 4 5
Not true of him *Very true of him*

3. He is a creative person.

---

1 2 3 4 5
Not true of him *Very true of him*

4. He was concerned only with his own ideas and viewpoint.

---

1 2 3 4 5
Not true of him *Very true of him*

5. He influenced the opinions of others.

---

1 2 3 4 5
Not true of him *Very true of him*

6. He interrupted others when they were speaking.

---

1 2 3 4 5
Not true of him *Very true of him*

[4]Morris, C. G., and Hackman, J. R. "Behavioral correlates of perceived leadership." *Journal of Personality and Social Psychology,* 1969, 13, 350–361. Copyright 1969 by the American Psychological Association. Reprinted by permission.

7. He criticized those with whom he disagreed.

1 2 3 4 5
Not true of him *Very true of him*

8. He was an aloof sort of person.

1 2 3 4 5
Not true of him *Very true of him*

9. He was the real leader of the group.

1 2 3 4 5
Not true of him *Very true of him*

10. He worked well with others in the group.

1 2 3 4 5
Not true of him *Very true of him*

11. He was disruptive to the group.

1 2 3 4 5
Not true of him *Very true of him*

12. He was in the forefront of the group's discussion.

1 2 3 4 5
Not true of him *Very true of him*

13. He kept the group from straying too far from the topic.

1 2 3 4 5
Not true of him *Very true of him*

14. His attitudes hurt the group's chances of success.

1 2 3 4 5
Not true of him *Very true of him*

15. He seemed to be a tense, nervous person.

1 2 3 4 5
Not true of him *Very true of him*

Scoring. For each person, the scores which he/she gave himself/herself and those he/she received from other members are summed to provide an overall indication of how characteristic each of the fifteen statements was of his/her behavior.

Discussion

The data gathered and presented are the basis for discussion. Any issues considered appropriate should be developed.

Who are the most influential members of this group? Why? What behaviors are valued?

What behavioral roles seem to be missing? How does this affect your group?

What difference was there between your perceptions of your role and how others see you? How do you feel about that?

EXERCISE 3 A Task-Maintenance Exercise

Objectives

To increase understanding of task-maintenance roles

To increase skills in observing a group; to learn to categorize by functional roles of members

To increase learnings about the difference between intended behavior and perceived behavior

Rationale

This exercise permits participants and observers to understand that intent is not enough. Attempting certain behaviors does not mean that a person will be perceived in the role he or she intends. In addition, intending to behave in a certain way does not indicate a person's skill in this role, factors that may interfere with his or her intentions, or the fact that more conscious behaviors are called into play. Participants come to understand some of the complexities in the interpersonal and group processes that go on in meetings.

This exercise is not appropriate unless the members have become familiar with the task-maintenance concept and functional roles and are willing to consider the implications of the learnings. It can be followed by skill sessions in practicing each of the task and maintenance roles. The role play may be repeated after discussion, or skill session, with members attempting to be more congruent or skillful in their roles, and observers attempting to focus more clearly on behaviors.

Materials

The following materials are needed: two copies of the role-play situation described below and six role-play names; six copies of the Explanation Sheet of Task and Maintenance Roles which are to be given to role players only; newsprint, magic markers, and tape; and a table and six chairs for the role play.

Design

The facilitator introduces the design as a role play. He or she explains that in a role play a situation is concocted that is not real but could be. Participants do not enact their usual roles but rather the roles that they are instructed to play. The role play permits an examination of behavior without embarrassing anyone and helps develop increased skills in the situation.

The facilitator asks for six volunteers to participate in the role play. He or she selects six people and briefs them in private. He or she distributes six pieces of paper (usually face down, and the role players select one of the papers). The papers should read as follows:

Association President—assume number 1 task role and number 1 maintenance role.

Teacher Representative—assume number 2 task role and number 2 maintenance behavior.

Association Negotiator—assume number 3 task role and number 3 maintenance role.

Superintendent of Schools—assume number 4 task role and number 4 maintenance role.

School Board Representative—assume number 5 task role and number 5 maintenance role.

School Board President—assume number 6 task role and number 6 maintenance role. You are somewhat hostile to the teachers and their position on the salary increase.

The association negotiators, the board, and the superintendent are given a few minutes to discuss as individual groups what they plan to do. While they are doing this, observer sheets are distributed to others in the group.

The facilitator instructs observers to tally every time any of the role players behave either under Task or Maintenance, as described on the Explanation Sheet. He or she asks them to record words or acts that will help recall and support their observations and also to write the names of the role players at the top of the sheets. The facilitator answers any questions observers may have. Then he or she calls the role players back into the room, seats them at the conference table, and reads the role play to the entire group. He or she gives one copy to the role players to keep before them.

The Role-Play Situation

A negotiations impasse over salaries between the Teachers Association and the school board is imminent. The association's original proposal called for a beginning salary of $12,600 with a $200-a-year increment for 12 years. This would have given the association the highest beginning and ending salary in the country.

The first counterproposal by the board maintains the existing salary of $12,600, but with $100 per year increments for 12 years. This was refused by the association. A second counterproposal by the board maintains the existing salary increment of $200 per year for 20 years, but begins at $6,400. This would give the association the lowest starting salary in the country but the highest maximum salary. A special levy has already been called for, and the amount, which includes no salary adjustment, has been earmarked. As the scene opens, the board members say that they cannot change the earmarking of the special levy funds and that no other funds are available. The teachers say that they cannot accept the last proposal, maintaining that it will decimate the professional staff of the district. The association negotiators have agreed that if they cannot get the board to utilize the reserve funds (7½ percent of the total budget), they will declare an impasse at the present meeting.

The facilitator continues the role-play situation for 10–15 minutes. He or she cuts it when it appears that there will be no further movement, before interest lags. Then he or she asks observers to meet in trios and compare their perceptions and tallying. The trios do this and arrive at a joint report indicating what role each player was primarily playing (about 10 minutes). The facilitator returns to the general session and places the names of role players on newsprint. He or she asks for reports from trios, and next to each name he or she writes the role the trio thought each person was taking. Then he or she asks each role player to reveal the role assigned to him or her and what he or she was attempting. That information is written next to his or her name on the newsprint. The prediction is that there will be a discrepancy between the way the role player understood his or her role and the way he or she actually acted it out. Also, it is predictable that the observers will vary in their reports of how the role players were acting.

EXPLANATION SHEET OF TASK AND MAINTENANCE ROLES[5]
(To be given to role players only)

Task Roles

1. *Initiating:* Proposing tasks or goals: defining a group problem; suggesting a procedure or ideas for solving a problem.
2. *Information or opinion seeking:* Requesting facts; seeking relevant information about a group concern; asking for suggestions and ideas.
3. *Information or opinion giving:* Offering facts; providing relevant information about group concern; stating a belief; giving suggestions or ideas.
4. *Clarifying or elaborating:* Interpreting or reflecting ideas and suggestions; clearing up confusion; indicating alternatives and issues before the group; giving examples.
5. *Summarizing:* Pulling together related ideas; restating suggestions after the group has discussed them; offering a decision or conclusion for the group to accept or reject.
6. *Consensus testing:* Sending up trial balloons to see if group is nearing a conclusion; checking with group to see how much agreement has been reached.

Maintenance Roles

1. *Encouraging:* Being friendly, warm, and responsive to others; accepting others and their contributions; regarding others by giving them an opportunity or recognition.
2. *Expressing group feelings:* Sensing feelings, mood, relationships within the group; sharing own feelings or affect with other members.
3. *Harmonizing:* Attempting to reconcile disagreements; reducing tension through "pouring oil on troubled waters"; getting people to explore their differences.
4. *Compromising:* When own idea or status is involved in a conflict, offering to compromise own position; admitting error, disciplining self to maintain group cohesion.
5. *Gate-keeping:* Attempting to keep communication channels open; facilitating the participation of others; suggesting procedures of others; suggesting procedures for sharing opportunity to discuss group problems.
6. *Setting standards:* Expressing standards for the group to achieve; applying standards in evaluating group functioning and production.

[5](Based on Benne and Sheats, 1948)

Discussion

Role players are asked how comfortable they were in their roles. Was it like their usual roles?
Which players enacted their roles most faithfully—were they perceived by others as being in their assigned roles? Which players had the greatest discrepancies? Why? Which behaviors caused difficulties in attempting to attain observer agreement? Are there generalizations that can be made about our observations?

The facilitator might briefly dwell on the several factors that influence the interpersonal and group processes going on at meetings. At least two factors appear prominent as a result of this role play:

1. *A perceptual factor.* Not all of us perceive the same thing when we are watching the behavior of another person. Some perceive a person giving feelings, while others hear him or her giving information or expressing opinions.
2. *Our intentions are different from our behavior.* Participants might have intended to play a given role but might have given out mixed messages. Their words said one thing, but their nonverbal behavior in body or tone emitted another message.

Sometimes we intend to play roles at which we are not skilled; we mean to behave in a certain way but get "off the track." Sometimes cues from another person evoke responses we had not intended. We sense that he or she does not like us, and we become defensive or see him or her as an opponent when we previously assumed he or she was a friend. How we feel about others personally affects our behavior.

EXERCISE 4 Functional Roles of Membership

Objectives

To increase the understanding of functional roles of membership

To see leadership emerge

To practice observing types of behavior

To develop an increased understanding of leadership requiring followership

Materials

The materials needed are the same as for the role play or tower exercise, enough for all participants. Also needed are copies of

Group Building and Maintenance Roles and Group Task Roles included with this exercise.

Rationale

Any of the role-play exercises in the book or problem-solving exercises in this chapter are appropriate as a situation to examine functional roles of members. The goal is to develop skills in observing types of behavior and increasing awareness of the functional roles.

Design

The facilitator establishes the role-play included with this exercise or uses tinker toys and asks each group to create a symbol of its group. Or he or she can use newsprint, crayons, and other materials and ask each group to create a collage for their group. Two observers are assigned to tally for functional roles. Each time someone behaves in a role a tally is made. One person can tally for task roles, the other for maintenance roles. A better procedure is to have two observers for task roles and two for maintenance roles so as to check reliability. Following the activity, the observers feed back their findings in the emerging roles.

Sometimes a questionnaire can be distributed to participants immediately following the situation, in which each person is asked to state who he or she felt was the leader of the group, who was most influential, and with whom he or she would like to work again. These results are determined. There is discussion about the amount of agreement or variety of responses and why. This discussion gets at the various bases of power. Following the questionnaires the observers feed back their findings. The group examines how these results support or are different from their ratings on the questionnaires.

GROUP BUILDING AND MAINTENANCE ROLES

Categories describing the types of member behavior required for building and maintaining the group as a working unit.

Usually helpful	*Usually destructive*
1. *Encouraging:* Being friendly, warm and responsive to others; accepting others and their contributions; giving to others.	Being cold, unresponsive, unfriendly; rejecting others' contributions; ignoring them.

GROUP BUILDING AND MAINTENANCE ROLES

2. *Expressing feelings:* Expressing feelings present in the group; calling attention of the group to its reactions to ideas and suggestions; expressing own feelings or reactions in the group.	Ignores reactions of the group as a whole, refuses to express own feelings when needed.
3. *Harmonizing:* Attempts to reconcile disagreements; reduces tension through joking, relaxing comments; gets people to explore their differences.	Irritates or "needles" others; encourages disagreement for its own sake; uses emotion-laden words.
4. *Compromising:* When own idea or status is involved in a conflict, offers compromise, yields status, admits error; disciplines self to maintain group cohesion.	Becomes defensive, haughty; withdraws or walks out; demands subservience or submission from others.
5. *Facilitating communication:* Attempts to keep communication channels open; facilitates participation of others; suggests procedures for discussing group problems.	Ignores miscommunications; fails to listen to others; ignores the group needs that are expressed.
6. *Setting standards or goals:* Expresses standards or goals for group to achieve; helps the group become aware of direction and progress.	Goes own way; irrelevant; ignores group standards or goals and direction.
7. *Testing agreement:* Asks for opinions to find out if the group is nearing a decision; sends up a trial balloon to see how near agreement the group is; rewards progress.	Attends to own needs; does not note group condition or direction; complains about slow progress.
8. *Following:* Goes along with movement of the group; accepts ideas of others; listens to and serves as an interested audience for others in the group.	Participates on own ideas but does not actively listen to others; looks for loopholes in ideas; carping.

GROUP TASK ROLES

Categories describing the types of member behavior required for accomplishing the task or work of the group.

Usually helpful	*Usually destructive*
1. *Initiating:* Proposing tasks or goals; defining a group problem; suggesting a procedure or ideas for solving a problem.	Waits for others to initiate; withholds ideas or suggestions.
2. *Seeking information:* Requesting facts; seeking relevant information about a group problem or concern; aware of need for information.	Unaware of need for facts, or of what is relevant to the problem or task at hand.
3. *Giving information:* Offers facts; provides relevant information about a group concern.	Avoids facts; prefers to state personal opinions or prejudices.
4. *Seeking opinions:* Asks for expression of feeling; requests statements of estimate, expressions of value; seeks suggestions and ideas.	Does not ask what others wish or think; considers other opinions irrelevant.
5. *Giving opinion:* States belief about a matter before the group gives suggestions and ideas.	States own opinion whether relevant or not; withholds opinions or ideas when needed by the group.
6. *Clarifying:* Interprets ideas or suggestions; clears up confusion; defines needed terms; indicates alternatives and issues confronting the group.	Unaware of or irritated by confusion or ambiguities; ignores confusion of others.
7. *Elaborating:* Gives examples, develops meanings; makes generalizations; indicates how a proposal might work out, if adopted.	Inconsiderate of those who do not understand; refuses to explain, show new meaning.
8. *Summarizing:* Pulls together related ideas; restates suggestions after the group has discussed them; offers decision or conclusion for the group to accept or reject.	Moves ahead without checking for relationship or integration of ideas; lets people make their own integrations or relationships.

ROLE PLAY TO PRACTICE OBSERVING FUNCTIONAL ROLES

Second Observation—Role Play—Preliminary Negotiations between Union and Management

1. *Mediator from Chamber of Commerce.* You want both sides to be happy. Your goal is to arrive at a solution that satisfies both sides. It would look bad for business in town if there were a strike; on the other hand, if the word goes around that labor gets everything it wants, business will not be attracted to this city.
2. *Union Representative.* The workers must have a raise: too many are beginning to question why they pay dues. If you can't get a raise, get an equivalent. The president of the company is a nice guy if you treat him with respect (he goes for that), but you don't like to kowtow. You will if things get bad; the most important part is to come out with something.
3. *Shop Steward.* You want a raise for the workers. It makes you feel important to sit down with the president of the company and feel that you are his equal. You want to be sure he treats you as his equal. You are hurt easily by any slight; your voice must be heard. Remind them you can cause a strike, make trouble for them, and so forth. You want to show the workers in your plant how important you are and to get them a big "package."
4. *Company President.* You are head of a large company. You are in the midst of modernizing your equipment, which will mean more automation and laying off workers. Possibly workers can be shifted to other places in the plant but not at the same level. A small raise wouldn't be bad (profits have been substantial), but how about the layoffs next year and the possibility of a strike then? Perhaps something can be worked out, but you want them to understand that you are considering such alternatives. You are a leader in the community; you are not the equal of the working person in intelligence, education, or standard of living; you are superior.
5. *Independent Businessperson.* You are a small businessperson. An increase in this industry means an increase for every worker. You are just barely making ends meet; you can't afford a raise; you would be faced with going out of business. You like being on a par with the president of the large company, a prominent person in town. You certainly see his point of view better than the workers, who only care about their earnings; they are unconcerned with yours.

Discussion

What did you learn about your behavior in a group? Which roles do you usually take? Which do you wish you could take on more frequently? These questions can be listed in a short questionnaire and fed back as data to participants. There could then be sessions allowing participants to build skills in the roles with which they have difficulty.

What kind of leadership is helpful to the group? What blocks movement? These questions usually lead into a discussion of clarity—or lack of clarity—of goals. It becomes clear that we are open to the influence of only some members. Frequently, we have difficulty in putting out needed behavior for fear of its not being accepted.

EXERCISE 5 Increasing Understanding of Leadership and Power

The following are a series of exercises that focus on various aspects of leadership and power. They are quite simple and appropriate when a group is working on some aspect of leadership and when acquiring data on their group would be useful.

They may be used in a variety of contexts: as a warm-up for further discussion or a theory session, as a "quickie" to help the group understand its own processes at a given moment, or as a beginning for thinking about aspects of leadership.

How Many Leaders?

Generally there are several leaders who emerge in a group: the task leader concerned with goal accomplishment, the social-emotional leader whose prime behavior is in maintaining working conditions within the group. Either of these or another may have the most influence in the group. A simple sociogram will provide data for determining who the leaders are. Each person is asked to respond to the following questions: Who has the best ideas in the group? Who is best liked? Who is most influential?

The data is gathered, tallied visibly before the group, and results determined. The group discusses findings. Is the same person the leader in all three areas or are there three different people with highest scores? What are the implications for the group? What generalizations can be derived regarding leadership?

Power—In a Line

When people seem concerned about their relative influence in the group, or who is listened to and who is not listened to in terms of influencing action, this is an appropriate exercise. The facilitator simply asks the group if it would be willing to try something. If members reply, "Yes," he or she proceeds. He or she asks that all members arrange themselves in a line, with the person who sees himself or herself as the most influential at the head of the line, and the person who sees himself or herself of lesser influence further down the line, according to the degree of influence he or she feels he or she has in the group. Then the facilitator asks that members who believe someone is misplaced move that person to where they think he or she should be. Then the members look at the ranking and discuss how they feel about it. If they discuss how they feel about their influence, it elicits data on how they feel about the group, its task, and their relationships with others. It can also lead to a discussion of how influence by some members could be increased, as, for example, the silent member who ranks at the tail because he or she rarely speaks or attempts to influence.

Higher and Lower Status

This exercise is a variation on the one preceding. In everyday conversation we may refer to someone as "the big cheese" or "a prestige member." We say of others, "Oh her," or "It doesn't matter that Bill isn't here; we can start without him." We often imply in these expressions, as well as by our nonverbal cues, that some have higher status than others (even in a so-called nondifferentiated group). High status means greater ability to influence; low status, limited ability to influence, although low status may produce negative influence.

The questions then asked are: Who has the highest status in the group? Who is next? Who has least? Members are now instructed to arrange themselves so that the person of highest status stands on a table, those of lower status stand on chairs farther out from the center. Those of yet lower status stand on the floor but also farther from the center. Those lower yet crouch or lie down, still farther from the center. Once more members may rearrange one another if they feel someone is misplaced. Members complete their hierarchy, and then discuss it. How many are in the center, what are the problems? How "far out" and how "small" do low status persons feel? What problems in the group could be reduced if this were changed? How could it be done?

A variation here is not only the height and distance relations but also a pairing and cluster relation. The exercise is the same as above, but people now also group themselves as to whose influence they are most open to and cluster around their status people.

Who Are You in a Group?

Frequently we categorize ourselves and others in dichotomies: good-bad, leader-follower, talkative-quiet. We rarely seem to see the variety of behaviors required in a group, or our own roles in them. This exercise begins to help participants see their own roles, and the roles of others. The facilitator asks members to write down their typical behavior in a group on a piece of paper. He or she asks that they fold the sheets and lay them aside. He or she then divides the group in half, naming them group 1 and group 2. He or she asks each member from group 2 to select a partner from the other group. The first group is assigned a task, for example, building a creation from tinker toys, or reaching a decision on the Johnny Rocco case (see Chapter Three), or making a collage of the world as it may be in 1990. The members of the second group observe their partners' behavior specifically. They can be given the functional roles listing as a basis for observing. At the end of the task the observer feeds back all the various behaviors of his or her partner. Then members look at the earlier statement on their typical behavior and discuss it among them. Then the procedure is reversed. The second group performs a task; each member is observed by his or her partner. Descriptions of behavior are fed back again; they examine earlier sheets and discuss them. Finally a group discussion is held on the variety of group behaviors observed and the range of behaviors exhibited in a limited experience.

Who Would I Like to Be?

Most of us wish we could be more effective in a group. Some roles come easily, and we are skillful; others we rarely use. Each member is given a copy of the functional roles and is asked to list the roles he or she uses most frequently plus those he or she uses infrequently or rarely. The facilitator divides the group into trios. One person interviews another on which roles he or she rarely uses and asks why. The third person takes notes. The interviewer asks which he or she would like to be able to use more and what stops him or her, and what he or she might do to change. The interviewer only asks clarifying questions; responses come from the person being interviewed. The trios reverse and continue until each person has discussed the roles he or she takes, those he or she rarely takes, and those he or she wishes he or she could become more skillful in. Each person is given the notes on his or her interview as a basis for future planning or skill sessions.

EXERCISE 6 Decision Making Along a Continuum —Problems

Objectives

To experience some of the roles people play in decision making

To experience ambiguity with a leader and note how it affects both the process and the product

To understand the choice-of-leadership continuum, and its implications for members

To increase understanding that a leader-member relationship is a dynamic one, each previous relationship affecting the present one

To increase observational skills of a dyad in change

Rationale

This exercise has very dramatic learnings. It becomes obvious that the leader-member relationship is a fundamental one in understanding group process and organizational functioning. The leader and member roles are defined, much as they frequently are, and the leaders have more direct access to instructions from the top than the members. The information members receive is based on how the leader decides to transmit it—whether limited, ambiguous, or honest. Although the leader becomes more member-oriented in his or her decisions, the member may not perceive himself or herself as having a more significant role. He or she may be reacting to previous behaviors of the leader, and he or she might have developed a lack of trust and an apathetic response. Participants become aware of the dynamic relationships in a group, that is, that members' trust of the leader influences the degree of work in decision making, and that clarity in decision making is an important factor in leader-member relations. It also permits members to understand the effectiveness of multilateral versus unilateral decision making.

Materials

There should be three lists of innocuous items to be rank-ordered. (The lists are not of importance in themselves; they are simply a basis for discussion and decision making.) Some examples are: What are the most important qualities of a good teacher? Ten qualities are listed, with a dash as a place for ranking before each item. Qualities might be: education, initiative, creativeness, persistence, fairness, a sense of humor, diligence, love of teaching, interest in travel, experience. Other similar questions are: Rank these ten presidents as to their importance in history. Then the names of ten presidents are listed. Another example: Rank these people in order of the importance of their contribution to mankind: Charlemagne, Julius Caesar, Socrates, Martin Luther, Galileo, Darwin, Shakespeare, Queen Victoria, Karl Marx, Adam Smith.

Three different lists are required, and there should be two copies of each list, preferably on differently colored paper. One color (let us say green) should always be given to the leader; the other color (white) should always be given to the member. There should be enough lists for two-thirds of the participants. There should also be sheets printed for observers. The sheet is headed Observer's Sheet and should list questions to direct the observer's attention to the data to be gathered (see sample Observer's Sheet included with this exercise). These sheets are needed for one-third of the group.

Action

The group is divided into trios. One person is the leader, one the member, and one the observer. Each trio will be asked to perform three ranking tasks and to turn in their decision on each, according to specific instructions. Following the third exercise, observers report back their findings, and leaders and members respond with their feelings on the situations. This is followed by a general discussion.

The group is again divided into trios and told that they will be asked to perform three tasks, each of which involves reaching a decision. In the tasks, the participants are to maintain the *same roles* throughout, that is, the person who is the leader will continue in his or her role as leader in all three tasks; the person who is the member will occupy that role for all three tasks; the observer will maintain that role in all the tasks.

Instructions are given at the beginning of each task. However, prior to giving instructions, the members are asked to leave the room. Only the leaders receive instructions, as frequently happens in many groups and organizations. Instructions are then transmitted from the leader to his or her member or subordinate. It is important that the members leave prior to instructions so as to avoid the psychological effect of feeling left out, and because they should not know the changed rules for each task.

TASK I

Instructions are given to leaders; members are not in the room. The leaders are instructed that they and their paired member will each be given a list of items to rank-order. (The leader is given the green sheet, the member the white sheet.) Each person will rank the items individually. Then the leader and member will discuss the lists between them. No attention is to be given to the member's list. At the completion of this task, the leader will only turn in the list he or she made. No mention is given to the member's list. Members are not told about the special instructions. The leader may discuss his or her list with the member in any way he or she likes, but only his or her list represents the final selection. Observers will not participate. They will be concerned with gathering observational data. Members are called back in. The leaders and members take their sheets and rank the items individually. The facilitator announces that the group has 20 minutes to work before each group is asked to submit a list. The groups work on the task; at the end of 20 minutes the leaders turn in their sheets.

OBSERVER'S SHEET

Pay particular attention to the following:

How does the leader act toward the member (friendly, cordial, patronizing, and so forth)? Attempt to note specific behaviors—verbal or nonverbal.
How does the member act toward the leader?
How open are they to each other's influence in this situation?
How is this situation different from the previous one for the leader? For the member?

Task I:

Task II:

Task III:

Discussion

The facilitator should be aware of how members feel when they are asked to leave the room and how they feel about only the leaders receiving instructions. They feel left out, apprehensive. They feel that "this is how it always is"; the members are given the information secondhand and to the extent the leader wants to relay information to them. They also feel "used" since the leader paid no attention to their suggestions and submitted only his or her own ranking. Leaders often feel that they should discuss issues with members, frequently leaving members with expectations of influence. It is a disillusioning experience to find suggestions were not considered. This discussion is not brought before the group. It is developed here so that the facilitator may understand some of the dynamics developing in the relationship.

TASK II

Members are asked to leave the room; instructions are given to the leaders. The leaders are instructed that they and their paired member will each be given another list of items to rank-order. However, this time they are to use a fair and equitable decision-making process. They are told to adequately consider the member's list so that a collaborative ranking is developed. They are told not to mention their special instructions to the members. The members are called back in. The leaders and members take their sheets (a new listing but maintaining the same colors for leaders and members) and again rank the items individually. The facilitator announces that the groups have 20 minutes to work before each group is asked to submit a list. The groups work on the task; at the end of 20 minutes joint sheets are submitted.

Discussion

The member once more feels left out, but he or she feels it more acutely following the unilateral decision making just experienced. His or her feelings about the leader may range from strong anger at being manipulated to "you can fool some of the people some of the time, but you won't fool me again." He or she will be more guarded and less willing to be involved in the task. He or she may feel the leader is pressuring him or her; he or she may be verbally antagonistic to the leader. He or she may see the leader turn in a joint list, or he or she may still feel the leader is turning in his or her list. It is important to note that the member's subjective reality greatly influences the climate of the group as well as his or her relationship to the leader.

TASK III

The members are again asked to leave the room; as previously, instructions are given only to the leaders. The leaders are again instructed that they will be given a list of items to rank, as will the members. Once more the new lists are distributed on the same colored and white paper. The instructions to the leaders now are, "The list you turn in really does not matter, because we will throw it away. Let the members turn in whatever they want; their list will be the one representing the group. Once more do not tell the members of these instructions."

Members are asked to return. The members and leaders take their lists and rank them. The facilitator announces that each group will work on its task and will be asked to submit a list for the group. Usually this takes about 10 minutes. The members submit their lists.

Discussion

The members become even more annoyed at being left out of the instructions. Attitudes toward the leader are mixed. In some groups they are beginning to reconsider their relationship, partially and somewhat grudgingly. In other groups, the relationships are strained and viewed as a "personality clash." However, now the leader acts either apathetically or seems to be pressuring the member in order to influence his or her list. The relationship is ambiguous. The level of involvement is low, effort on the task is greatly reduced, and the climate is characterized by minimal communication and maximal suspicion. Although the member has more influence in making the Task III decision, he or she neither knows nor is involved at this point. The actions of the leader leave him or her confused; he or she is increasingly less sure of his or her own role.

General Discussion

Following the three tasks, the group observers report back on the behaviors they have seen. Leaders and members respond with their feelings on each of the situations. There can be a discussion with the entire group related to their reactions in each of the situations, the level of involvement, and their feelings about the group as a whole. Discussion can be in terms of how participants felt after each task, and perhaps they can generalize some of the learnings.

The following questions might be the basis for general discussion: How do people feel about multilateral versus unilateral decision making? Both the leader and the member experienced a task in which each made decisions irrespective of the other; they also experienced a time when they shared the decision making. How does it affect both the process (the degree of involvement) and the product (how good do the participants think their list was)?

How did the members feel when they were asked to leave? Why? (Allow time here since feelings are generally quite strong.) How did members feel about the leader? Why? What are the implications for working with ongoing groups? What are the implications of this exercise for decision making in a group? What behaviors are helpful? Which hinder?

BIBLIOGRAPHY

Abrahamson, M. and J. Smith. "Norms, deviance and spatial location." *Journal of Social Psychology,* 80, No. 1 (1970), 95–101.

Archer, D. "Power in groups: Self-concept changes of powerful and powerless group members." *Journal of Applied Behavioral Science,* 10 (1974), 208–220.

Bales, R. *Personality and Interpersonal Behavior.* New York: Holt, Rinehart, and Winston, 1970.

Bales, R. *Interaction Process Analysis,* Reading, Mass.: Addison-Wesley, 1950.

Barnard, C. I. *Functions of the Executive.* Cambridge, Mass.: Harvard University Press, 1938.

Bass, B. *Leadership, Psychology, and Organizational Behavior.* New York: Harper and Row, 1960.

Bass, B. M. "Some observations about a general theory of leadership and interpersonal behavior." In *Leadership and Interpersonal Behavior.* Ed. L. Petrullo and B. Bass. New York: Holt, Rinehart and Winston, 1961, pp. 3–9.

Benne, K. D. and P. Sheats. "Functional roles of group members." *Journal of Social Issues,* 2 (1948), 42–47.

Berelson, B. and G. A. Steiner. *Human Behavior.* New York: Harcourt, Brace and World, 1964.

Bernstein, M. D. "Autocratic and democratic leadership in an experimental group setting: A modified replication of the experiments of Lewin, Lippitt, and White, with systematic observer variation." *Dissertation Abstracts International,* 31, No. 12A (1971), 6712.

Bird, C. *Social Psychology.* New York: Appleton-Century, 1940.

Black, T. W. and K. L. Higbee. "Effects of power, threat, and sex on exploitation." *Journal of Personality and Social Psychology,* 27 (1973), 382–388.

Blake, R. and J. S. Mouton. *Consultation.* Reading, Mass.: Addison-Wesley, 1976.

Blake, R. and J. S. Mouton. *Building a Dynamic Corporation Through Grid Organization Development.* Reading, Mass.: Addison-Wesley, 1969.

Bohleber, M. E. "Conditions influencing the relationships between leadership style and group structural and population characteristics." Dissertation Abstracts, 28, No. 2A, 766–777. Unpublished dissertation, University of Wisconsin, 1967.

Borgatta, E. F., A. S. Couch, and R. F. Bales. "Some findings relevant to the great man theory of leadership." In *Small Groups: Studies in Social Interaction.* Ed. A. Paul Hare et al. New York: Alfred Knopf, 1955, pp. 568–574.

Butler, R. P. and C. L. Jaffee. "Effects of incentive, feedback, and manner of presenting the feedback on leader behavior." *Journal of Applied Psychology,* 59 (1974), 332–336.

Cattell, R. "New concepts for measuring leadership, in terms of group syntality." *Human Relations,* 4 (1951), 161–184.

Chow, E. and C. Billings. "An experimental study of the effects of style of supervision and group size on productivity." *Pacific Sociological Review,* 15 (1972), 61–82.

Davie, J. S. and A. P. Hare. "Button-down collar culture: a study of undergraduate life at a men's college." *Human Organization,* 14, No. 4 (1956), 13–20.

Dunno, P. "Group congruency patterns and leadership characteristics." *Personnel Psychology,* 21, No. 3 (1968), 335–344.

Elder, G. H. "Group congruency patterns and leadership character. Parental power legitimation and its effect on the adolescent." *Sociometry,* 26 (1963), 50–65.

Fiedler, F. E. *A Theory of Leadership Effectiveness.* New York: McGraw-Hill, 1967.

Fiedler, F. E. "The trouble with leadership training is that it doesn't train leaders." *Psychology Today,* (February 1973), 23–26, 29–30, 92–93.

Finch, F. E. "Collaborative leadership in work settings." *Journal of Applied Behavioral Science,* 13 (1977), 292–302.

French, J. R. P., Jr. and B. Raven. "The bases of social power." In *Group Dynamics.* 2nd ed. Ed. D. Cartwright and A. Zander. Evanston, Ill.: Row, Peterson, 1960, pp. 607–623.

Geier, J. G. "A trait approach to the study of leadership in small groups." *Journal of Communication,* 17, No. 4 (1967), 316–323.

Gershenfeld, M. K. "Responsible behavior under conditions of threat and non-threat." Unpublished doctoral dissertation, Temple University, 1967.

Gibb, C. A. "Leadership." In *Handbook of social psychology,* Vol. II. Ed. G. Lindzey. Reading, Mass.: Addison-Wesley, 1954, pp. 877–920.

Goldman, M. and L. A. Fraas. "The effects of leader selection on group performance." *Sociometry,* 28, No. 1 (1965), 82–88.

Golembiewski, R. T. *The Small Group.* Chicago, Ill.: University of Chicago Press, 1962.

Gray, L. N., J. T. Richardson, and R. H. Mayhew Jr. "Influence attempts and effective power: a reexamination of an unsubstantiated hypothesis." *Sociometry,* 31, No. 1 (1968), 245–258.

Guyer, B. P. "The relationship among selected variables and the effectiveness of discussion leaders." *Dissertation Abstracts International,* 39, No. 2A (1978), 697–698.

Hall, J. and M. S. Williams. "Personality and group encounter style: A multivariate analysis of traits and preference." *Journal of Personality and Social Psychology,* 18 (1971), 163–172.

Hemphill, J. K. "Why people attempt to lead." In *Leadership and Interpersonal Behavior.* Ed. L. Petrullo and B. Bass. New York: Holt, Rinehart and Winston, 1961, pp. 201–215.

Hersey, P. and K. H. Blanchard. "Life-cycle theory of leadership." *Training and Development Journal,* 23, No. 5 (1969), 26–34.

Hersey, P. and K. H. Blanchard. "A situational framework for determining appropriate leader behavior." In *Leadership Development: Theory and Practice.* Ed. R. N. Cassel and R. L. Heichberger, North Quincy, Mass.: The Christopher Publishing House, 1975, pp. 126–155.

Hersey, P. and K. H. Blanchard. "*Management of Organizational Behavior: Utilizing Human Resources.* Englewood Cliffs, N.J.: Prentice-Hall, 1977, pp. 106–107; 162–183; 307–324.

Hoffman, L. R., R. J. Burke, and N. R. F. Maier. "Participation, influence, and satisfaction among members of problem-solving groups." *Psychological Reports,* 16 (1965), 661–667.

Holloman, C. R. "Leadership and headship: there is a difference." *Personnel Administration,* 31, No. 4 (1968), 38–44.

Jenkins, D. H. "New light on leadership." *Adult Leadership,* June 1956.

Jenkins, D. H. "New questions for old." In *Leadership in action.*

Washington, D.C.: National Training Laboratories, 1961, pp. 23–25.

Jones, R. A., C. Hendrick, and Y. Epstein. *Introduction to Social Psychology.* Instructor's Manual. Sunderland, Mass.: Sinauer Association, 1979, pp. 121–126.

Julian, J. W., E. P. Hollander, and C. R. Regula. "Endorsement of the group spokesman as a function of his source of authority, competence, and success." *Journal of Personality and Social Psychology,* 11, No. 1 (1969), 42–49.

Kahn, R. and D. Katz, *The Social Psychology of Organizations.* New York: John Wiley and Sons, 1966.

Kipnis, I. "Does power corrupt?" *Journal of Pesonality and Social Psychology,* 24 (1972), 33–41.

Knickerbocker, I. "Leadership: a conception and some implications." In *Leadership in Action,* Washington, D.C., National Training Laboratories, 1961, p. 69.

Koch, J. L. "Managerial succession in a factory and changes in supervisory leadership patterns: A field study." *Human Relations,* 31 (1978), 49–58.

Korman, A. K., "Consideration," "Initiating Structure," and "Organizational Criteria—A Review," *Personnel Psychology: A Journal of Applied Research,* No. 4 (Winter 1966), 349–361.

Lewin, K., R. Lippitt, and R. K. White. "Patterns of aggressive behavior in experimentally created social climates." *Journal of Social Psychology,* 10 (1939), 271–299.

Lippitt, R., N. A. Polansky, and S. Rosen. "The dynamics of power: a field study of social influence in groups of children." *Human Relations,* 5 (1952), 37–64.

Mann, R. D. "A review of the relationships between personality and performance in small groups." *Psychological Bulletin,* 56 (1959), 241–270.

Margulies, N. "Organizational culture and psychological growth." *Journal of Applied Behavioral Science,* 5, No. 4 (1969), 491–508.

Maslow, A. H. *Motivation and Personality.* New York: Harper & Row, 1954.

McDavid, J. W. and H. Harari. *Social Psychology.* New York: Harper and Row, 1968.

McGregor, D. *The Human Side of Enterprise.* New York: McGraw-Hill, 1960.

Mulder, M. "Power equalization through participation." *Administrative Science Quarterly,* 16 (1971), 31–38.

O'Brien, G. "Effects of organizational structure, leadership style, and members' compatibility upon small group creativity." *Proceedings of 76th Annual Convention.* Washington, D.C.: American Psychological Association, 1969.

Pepinsky, P., J. H. Hemphill, and R. Shevitz. "Attempts to lead, group productivity and morale under conditions of acceptance and rejection." *Journal of Abnormal and Social Psychology,* 57 (1968), 47–54.

Polansky, N. A., R. Lippitt, and F. Redl. "The use of near-sociometric data in research on group treatment processes." *Sociometry,* 13 (1950), 39–62.

Reddin, W. J. "An integration of leader-behavior typologies." *Groups and Organization Studies,* 2, No. 3 (1977), 282–295.

Richardson, J. T., J. R. Dugan, L. N. Gray, and B. H. Mayher, Jr. "Expert power: A behavioral interpretation." *Sociometry,* 36 (1973), 302–324.

Rieken, H. W. and G. C. Homans. "Psychological aspects of social structure." In *Handbook of social psychology,* Vol. II. Ed. G. Lindzey, Reading, Mass.: Addison-Wesley, 1954, pp. 786–832.

Ring, K. and H. H. Kelley. "A comparison of augmentation and reduction in modes of influence." *Journal of Abnormal and Social Psychology,* 66 (1963), 95–102.

Rubin, J. Z., and R. J. Lewicki. "A three-factor experimental analysis of promises and threats." *Journal of Applied Social Psychology,* 3 (1973), 240–257.

Rubin, J. Z., C. T. Mowbray, L. Collette, and R. J. Lewicki. "Perception of attempts at interpersonal influence." *Proceedings of the 79th Annual Convention of the AOA, 6 (1971),* 391–392.

Sadler, P. J. "Leadership style, confidence in management, and job satisfaction." *Journal of Applied Behavioral Science,* 6, No. 1 (1970), 3–20.

Sargent, J. F., and G. R. Miller. "Some differences in certain communication behaviors of autocratic and democratic group leaders." *Journal of Communication,* 21 (1971), 233–252.

Schein, E. H. *Process Consultation.* Reading, Mass.: Addison-Wesley, 1969.

Scontrino, M. P. "The effects of fulfilling and violating group members' expectations about leadership style." *Organizational Behavior and Human Performance, 8* (1972), 118–138.

Smith, R. J., and P. E. Cook. "Leadership dyadic groups as a function of dominance and incentives." *Sociometry,* 36, No. 4 (1973).

Sorrentino, Richard M. "An extension of theory achievement motivation to the study of emergent leadership." *Journal of Personality and Social Psychology,* 26 (June 1973), 356–368.

Stang, D. J. "The effect of interaction rate on ratings of leadership and liking. *Journal of Personality and Social Psychology,* 27 (1973), 405–408.

Stech, E., and S. S. Ratliffe. *Working in Groups.* Skokie, Ill.: National Textbook Company, 1976, pp. 214–225.

Stogdill, R. M. "Personal factors associated with leadership: a survey of the literature." *Journal of Psychology,* 25 (1948), 35–71.

Stogdill, R. M. *Handbook of Leadership: A Survey of Theory and Research.* New York: The Free Press, 1974.

Sudolsky, M., and R. Nathan. "A replication in questionnaire form of an experiment by Lippitt, Lewin, and White concerning conditions of leadership and social climates in groups." *Cornell Journal of Social Relations,* 6 (1971), 188–196.

Tindall, J. H., L. Boyler, P. Cline, P. Emberger, S. Powell, and J. Wions. "Perceived leadership rankings of males and females in small task groups." *Journal of Psychology,* 100 (1978), 13–20.

Vacc, N. A. "Cognitive complexity: A dimension of leadership behavior." In *Leadership development,* Ed. R. N. Cassel and R.L. Heichberger, North Quincy, Mass.: The Christopher Publishing House, 1975, pp. 278–288.

White, R. and R. Lippitt. "Leader behavior and member reaction in three social climates." In *Group dynamics.* 3rd ed. Ed. D. Cartwright and A. Zander. New York: Harper and Row, 1968.

Yetton, P. W., and V. H. Vroom. "The Vroom-Yetton model of leadership: An Overview." In *Managerial Control and Organizational Democracy.* Ed. B. King, S. Streufert, and F. E. Fiedler, New York: John Wiley, 1978, pp. 133–149.

Zigon, F. J., and J. R. Cannon. "Process and outcomes of group discussions as related to leader behaviors." *Journal of Educational Research,* 67 (1974), 199–201.

Six

The Incredible Meeting Trap

At any given hour during virtually any working day there are a million meetings going on throughout the United States. We have become a society of meetings. There are meetings to plan, to problem solve, to dream, to organize, to resolve crises and to create them, to explain things, to make us feel better, to punish, to reward, to build and to desolve, and to give hope. There are meetings to plan meetings to plan meetings. The cost is staggering. Perhaps a billion dollars worth of people time each day. A day-long meeting of eight to ten executives can easily cost a business $1600 to $2000 in salaries alone. Often the value is hardly worth the cost or effort.

Meetings are meant to be sources of stimulation, support, and solutions, and should fulfill any number of personal and organizational needs. Some of them are and do. But if one stands by the door of most meetings, one will inevitably hear such comments as "What a waste of time," "I could have just as easily stayed home," "What a meaningless bore," or "Why did they have us there anyway? They already made the decision before we came . . . such a fuss." These comments are not the exception; they tend to be the rule. For most of us, meetings are an annoyance and a waste of time. Many times, people feel they are walking away with more problems than they had when they started. Just why meetings fail, why they become traps, and what we can do to improve them is the thrust of this chapter (Antony, 1976; Bradford, 1976; Miller, 1972; Schindler-Rainman and Lippitt, 1977).

Meetings, of course, can be incredibly valuable. They provide the possibility for resolving an enormous number of both organizational and individual needs. As a people bound in the principles of the democratic process, our meetings are in many ways an obligation. With all good intentions, they are meant to draw people together, to help create teams, to involve people in decisions that affect their lives, to share ideas, to express feelings and support for each other, to communicate new ideas and review old ones, and to establish a commitment among those that have to live with various decisions. There are few experiences that can equal the feelings of goodwill, satisfaction, and camaraderie at the end of a meeting when participants have felt heard, appreciated, and have accomplished a mutually important goal.

READER ACTIVITY

The first step in becoming either an effective leader or participant in a meeting is to become keenly aware of what is occurring that either facilitates or hinders progress. Consider for a moment the last meeting you attended. Now, utilizing a number of the concepts discussed in the text to this point, diagnose the effectiveness of the particular meeting. For example:

1. What individual and group goals influenced the group?
2. What type of communication patterns occurred?
3. Did any norms act to block the group?
4. Were roles of individual members clear?
5. Was real membership accessible to those present?
6. How was leadership gained? Was it effective?

7. To what degree did members of the meeting seem interested and involved and does it appear that they would be motivated to attend another (an informal measure of group cohesion)?

As you know by this time, much of what occurs in a meeting that affects the life of the group passes unnoticed by the participants. Often they are too busy reacting to what is happening or simply do not know the questions that must be asked in order for a comprehensive picture of the group process to be understood.

If there is so much potential value for meetings and they can be so beneficial, then why are they disastrous so often? On the one side of the ledger we can place the blame on the fact that many leaders simply are not trained in the art of conducting stimulating and productive meetings. On the other side is the less optimistic view that many leaders are quite satisfied with the meetings as they are, whether those attending like it or not. Many meetings are held for the wrong reasons, and it is not uncommon that the reasons stated have nothing to do with the real purposes.

In many organizations people are "meetinged" to death. Individuals literally become burned out from meetings. Thus, when groups are regularly called together out of courtesy, habit, or formality and little of constructive value occurs, individuals tend to become passive and a sort of group lethargy takes over. Passive resistance can become more overt when participants are feeling overburdened with other work and have no recourse but to attend. Unresolved authority issues suddenly have a common point of focus and the informal system buzzes with rumors, complaints, scapegoating, and other less than constructive behaviors. Other sources of frustration result when a meeting is billed as participative but those present sense quite early that the meeting is really designed to elicit support for a decision previously made and the leader's goal is to appear democratic without having to relinquish any real control.

For many individuals, meetings become predictable. Predictably boring, predictably nonproductive, predictably noninvolving. Because it is not within the goals of many leaders to make meetings creative, stimulating, and participative, their behavior as well as those of other participants become repetitive. This is true even though there are a wide variety of meetings, each of which ideally would demand a different type of structure and format—a different design. The failure to utilize different designs for meetings with different purposes results in a high degree of failure and, ultimately, frustration on the part of many of the participants.

TYPES OF MEETINGS

The great majority of organizational meetings probably fall under one of the following categories:

Information sharing—communication meetings—such meetings are usually directed at providing common information across diverse sections of an organization. They are motivated by a desire to keep people abreast of what's happening, because one of the major sources of alienation within organizations is a sense of isolation or impotence many people feel because they are outside of the formal communication channels.

Diagnostic or factfinding meetings—effective organizations spend considerable amounts of time identifying problems, establishing priorities, and generally attempting to take the pulse of the organization. Such meetings can be focused around a particular topic or have a more general thrust, but the primary purpose is to generate information that can provide a better understanding of existing problems or conditions.

Brainstorming meetings—increasingly, problem solvers agree that idea building and actual decision making should be separated in time and place. Having people consider a wide range of alternatives can best be done when decision making can be delayed while those ultimately accountable consider the cost and benefits of the various ideas developed.

Decision-making meetings—in these meetings, participants consider which of the variety of alternatives they should actually take. In each individual case, the consequences of such action deserve special attention.

Planning meetings—in this instance, we identify planning as a particular function whose primary purpose is to establish the means of implementation for decisions made.

Coordination and monitoring meetings—because delegation and accountability are two of the critical foundation stones upon which effective management is built, managers need to conduct well-organized and executed periodic meetings of those involved in the implementation phase to insure that all aspects of the program are being covered accurately.

Ongoing business meetings—many organizations rely on committees that meet regularly to work with issues of what could be called system maintenance, including selection of new members, discipline problems, promotions, employee grievances, marketing, and many others (Antony, 1976; Bradford, 1976; Maier, 1963; Strauss and Strauss, 1951).

It is obvious that almost all of the types of meetings identified here are interdependent rather than independent of each other. The point is that each type of meeting warrants careful planning and design and has its own peculiar problems that deserve special attention. Thus, if an organization is considering new policies and if the process is to be one that is thoughtful and intellectually honest, then a number of meetings should probably occur, each building from the goals and progress of the last. Meetings can provide the foundation for rationality, as long as each has a purpose or theme that is clearly understood and a framework that will facilitate its execution. Even meetings with more general purposes that are less structured (for example, an operations meeting that occurs every day in a plant during which participants review problems and procedures) can be looked at in light of the various categories established previously. Later, we will talk more specifically about the concept of design, which is the keystone to the success of any meeting for which specific goals or purposes have been defined.

THE INTERPERSONAL-PERSONAL DIMENSION

Over the years, professionals have become increasingly aware that clear organization and group goals directed at the achievement of particular products are not enough to sustain a healthy climate of work. Meetings are simply group vehicles for accomplishing various tasks. As groups, they represent a developing social environment that can provide a constructive and positive experience for the individuals involved or, as often happens, can result in frustrating and dissatisfying interpersonal relationships. The point here is that the group—the committee—the team—the staff—can also be a vehicle for improving the overall life of the organization and can help meet the kinds of individual needs that people have whenever they get together in a group context. The failure of most meetings does not lie in the task-product domain, but rather in the interpersonal relationships and the personal needs that provide sources of energy, commitment, and dedication that eventually spell success or failure for the group itself.

Potential Benefits of Meetings

Following are a number of personal and interpersonal needs that can be satisfied in a well-functioning meeting and that justify the occasional use of a group even where the delegation of a task to individuals might be more time efficient and cost effective:

1. Meetings give individuals the opportunity to belong or have membership in a number of groups that support the organization. Belonging gives individuals further identity beyond a single job role. This is true at any organizational level but of particular importance when job roles tend to be monotonous or lack variety.
2. Meetings make individuals feel that their ideas are being sought and there is a mutual building from these ideas to new procedures, policies, or programs.
3. Beyond the creation of the ideas, being part of a meeting gives the psychological satisfaction of being able to identify oneself with visible outcomes that have value for the organization. This can be like the pride construction workers have when passing a house or building that they have had a role in constructing.
4. People are psychologically committed to ideas that they help generate in meetings, which goes beyond the product itself. Thus, to have a personal point of view forged into a final declaration of a group can be incredibly rewarding even though the idea is insignificant when compared to the impact of the entire statement.
5. Those involved in meetings have a chance to experience opposition and conflict and then to resolve these in a positive manner. The level of trust within the group may increase and be transferred to the organization because mutually agreed-upon solutions require the sacrificing of personal self-interest or gain. It is the giving up of one's personal vested interest that often increases one's stake in the organization itself.
6. Accountability within a group is often clearly defined in meetings. A commitment of the group or other individuals to action and the monitoring of areas of responsibility within the group can provide important sources of success, reinforcement, and support necessary in any healthy organization.
7. The meeting allows the differentiation of roles according to skill and provides a healthy source of status depending on the nature of the role that is taken.
8. People learn from each other by working together in meetings and enhance their social or technical skills through the observation of other people's performance.
9. Meetings help build an overall sense of mutual accomplishment in having succeeded in a mutual goal, having risked certain new

ideas or innovations, and having experienced success and overcome failure.

10. Meetings build personal relationships through the legitimate work process where little time is provided for social niceties. Individuals begin to know each other within the organization on a different and more meaningful level because of shaved goals, shaved points of view and the satisfactory resolution of conflict. Put simply, people "get to know each other" by acting together.

11. Meetings provide the opportunity for fun and social rewards that can only occur spontaneously when people work together with common purpose and under some pressure of obligation to each other (Bradford, 1976; Burke and Beckard, 1976; Marrow, Bowers and Seashore, 1967).

Obviously, the eleven characteristics suggested here could be expanded or reduced. The list is imperfect and is meant only to be a guide to some of the personal and interpersonal needs that can be met by effective work in organizational meetings. The problem is, of course, that in most meetings few of these potential benefits are actualized. More often than not, the experience leaves people with bad feelings even if the stated goals are accomplished. Worse than this are the feelings of impotence or uselessness resulting from a lack of success that occurs in both the "product" and "personal" domains.

READER ACTIVITY

Use the eleven criteria to measure the value of the last important meeting you attended. Simply imagine a one-to-ten scale for each item, with *one* representing that the particular criterion was not reached at all and *ten* indicating total accomplishment. If you haven't been to a meeting of consequence lately, use the criteria to measure several classes you attend or your department at work. Clearly, we are talking about some rather universal qualities of effective organizations that are directed at meeting participant needs as well as organizational goals.

BLOCKS TO MEETING EFFECTIVENESS

The primary reasons that committees fail is that leaders and members alike refuse to see the meeting as a social system and often as a microcosm of the larger organization itself. Thus leader and participant style, member interest, the degree of interaction, the amount of interdependence or dependence that exists, the norms, membership criteria, and communication patterns all will tend to reflect the kinds of messages being given off by the parent organization. If those in the meeting are not asking serious questions relating to both task and process, much of what is occurring that blocks the group from being productive and enjoyable will simply not come to light. A group not aware of its own functioning will inevitably run into troubled waters, lose its sense of direction, and tend to flounder. This may be revealed by fights with each other, avoidance of the task at hand, or in a variety of ways that undermine the purpose of the group without the participants even knowing it. Following are five examples of ways in which success is blocked in meetings. In every case, had the group stopped to take stock, to look at its own process, the course of the meeting could have been changed and the outcome might have been more productive.

The Domination of Single Members

One committee of ten was representative of high school students and teachers who had the task of establishing a new discipline policy for their school. There were six faculty and four students on the committee. Asked to observe the group for a period of an hour and then comment in a way that might be helpful to their development, an outsider simply observed the distribution of "airtime" that was occurring in the group. At the end of one hour, four of the members had talked 80 percent of the time about a subject that everyone in the school lived and breathed. Not only this, but three of the four people were faculty. By the end of its first hour as a group, the committee was beginning to show telltale signs of participant withdrawal, passivity, and disinterest in a topic that was vital to them all. The four who were involved were highly involved and simply assumed that anyone who wanted to get into the conversation would do so. There didn't appear to be any overt power struggle, simply insensitivity to the degree to which all members felt able or welcome to participate. Not only were good ideas being lost to the group, but energy was being drained away and passive hostility and disillusionment were slowly getting a foothold. A two-minute commentary on the observed data along with five minutes talking about the implications reestablished the committee on the right track. The members immediately developed certain procedures for involving everyone's ideas and by the end of two hours, a lively debate was ensuing in which people were listening and participating with vigor.

Critical Norms

A meeting had gathered to generate new ideas in marketing for a rapidly expanding market. The session was designed to develop new ideas in a positive manner and not to come to decisions. During the first fifteen minutes, discussion was lively, ideas were batted about, there was considerable laughter and nearly everybody was participating. By the end of forty-five minutes, the meeting had become obviously hard work for everybody. Instead of all eight members participating in a rather restrained manner, the flow of ideas came almost to a halt. On this particular occasion, not only did the observer watch which individuals talked, but he also noted the number of positive, neutral, and negative statements made by people as they were throwing out ideas. During the initial fifteen minutes, positive ideas outnumbered the negative ones by 4 to 1, with neutral responses making up 15 percent of the conversation. By the end of thirty minutes, the negative comments, or "yes, but" statements, and divisive critical remarks had increased to almost an equal share. Finally, by the end of forty-five minutes, the group found itself in a myriad of negative statements. Virtually every idea that was raised was met by two or three statements that negated or questioned its validity. The better the idea the more resistance that was generated. Much of the tension was generated by a group of young, aggressive, and competitive participants among whom good ideas were seen as trophies, thus, making others appear like losers. The group was not even aware of its own pattern and if asked would have found it difficult to explain how almost every meeting it had would tend to degenerate into debates and haggling. As strange as it may seem, the group had created a norm that, although destructive to the group itself, was perceived as protective of individuals by deterring the success of others.

Vested Interest

A group was composed of seven executives of a representative organization that bought and sold fuel. During the period of initial fuel shortage in the United States, the company found itself in a position of having less fuel to sell than it had commitments to buyers. For the first time in its history, their sales force was told actually to reduce the volume it sold. This meeting was established on an ad hoc basis to explore how this was to be done. Appropriately, the first hour and a half of the three-hour meeting was a review of the present situation from both the group and individual perspective. The leader then directed the group to discuss overall corporate goals in relation to fuel allocations. Again, there were no problems and many creative ideas. After about eighty minutes, the topic changed to focus on short-term realities and the sacrifice individual sales personnel would have to make if the company was not going to be embarrassed by not being able to meet its fuel commitments. Almost instantly, the tone of the meeting changed. From an atmosphere of listening, support, and goodwill, the air became charged, defensive, and resistant. People stopped listening and began to defend their own postures. One hour later when the meeting ended, the morale of the group was gone and any optimism with which the meeting began had disappeared. The reality was that the decisions had to be made and the group appeared ready to make them up until the time that each individual's vested interests were attacked. Salespeople who had always been regarded for selling suddenly found themselves in a position of letting down clients and appearing ineffective. Because the decision-making process had not been established prior to this point of the meeting, it suddenly became impossible to make any decisions to which the group would commit itself. Any attempt to establish a fair decision-making process would simply be perceived as an attempt of one individual to maintain control over others.

The Dilemma of New Members

An eight-person committee was one of the three ongoing committees of the hospital. Its purpose was to establish and review policies that cover all hospital employees. For the sake of continuity, two new members were rotated into the committee every year and two others left. The committee had the reputation of being amicable, businesslike, yet friendly, and was somehow run without the petty problems of many committees that can bog down and are unable to work effectively because of hidden agendas, authority problems, or some of the other "sticky wickets" of group life. This year, however, shortly after two new members were elected and rotated into the committee, the committee began to experience a wide variety of problems and tensions that had not existed previously. It wasn't even possible to link the problems to the behavior of the two new members, who were generally well liked and respected. Nevertheless, things went from bad to worse as individuals began to listen less and talk more, to resist each other's ideas, and in many ways to act like a committee of adversaries rather than colleagues. The problem, of course, was that the loss and addition of members simply created a new group. Suddenly, who had membership and on what basis went up for grabs, norms that somehow had been a stable part of the committee were opened to question and to the pressure of new behavior from the entering members. In addition, roles and goals and expectations of people's behavior in the group suddenly changed and individuals began to vie for both authority and attention in a manner that would not have been acceptable several weeks before. Only by the group taking the time to assess the changes in roles and expectations, goals and norms, and membership criteria would the group begin to deal with the blockages that were being created. The interesting part is that in this particular group the resolution of these new tensions would probably not be very difficult because most of the participants enjoyed their place on the committee, worked in the spirit of goodwill with each other, and saw their participation as a real opportunity. It was very likely that just the raising of the issues in such a supportive group would be enough to reestablish some of the previous behavior patterns as well as some of the goodwill that still remained in the group.

Problems of Physical Structure and Space

Virtually everything one does as leader of a group, in this case a committee, can influence the life of the group. Most of us have been at committee meetings as the members of the group sit scattered around an auditorium in rows, looking at the necks of their colleagues with little opportunity to share ideas or collaborate within the meeting itself. Similarly, most of us have been part of committees where we have sat around a long wood table with some fifteen or twenty people having all eyes directed at the primary source of influence at the table or those few individuals who share the wealth of influence through either structural proximity around the table or status gained through other sources of power and control. One committee observed recently was an almost "textbook" version of a group governed by how they sat around the table. The two individuals who were most antagonistic toward one another sat on opposite sides of the table. Individuals closest to the leader in authority and trust tended to be on the same side of the table, and those with less influence and involvement tended to fade off toward the ends and corners and were physically far from the appointed leader. The committee, which met once a month, virtually always moved into these positions and thus immediately reinforced each other as to interest, power, influence, and general involvement. The patterned routine of the group made certain antagonisms almost inevitable and provided members with easy "turn offs and turn ons" in terms of the expected behavior that was so often fulfilled.

These five brief examples are by no means the exception. In every instance the meetings were meant to be purposeful and effective.

None were a case of people being forced together to do things they didn't want to do. Even though the apparent causes may have differed drastically in each situation, results were similar, with morale, productivity, and interest reduced in each situation. Each represented a misuse of people' time and energy, in which participants inevitably left the committee meeting less satisfied than when they entered. The troubling part is that each of the committees would predictably become worse or less effective as time went on simply because the members and leaders were not aware of the causes undermining their effectiveness. Nor are the solutions necessary to change the situations terribly demanding or complex. In most instances, a simple awareness of the problem would make the solution rather self-evident. The problem, of course, is that without an awareness that the problem exists there can be no solution. It's because the leaders and members fail to ask the right questions or are unwilling to stop and look that the problems in groups tend to remain over time, causing increasing frustration and tension among those involved (Auger, 1972; Maier, 1963; Prince, 1972; Schindler-Rainman and Lippitt, 1977).

READER ACTIVITY

Most of us are involved in organizations (school, work, social groups) that conduct meetings, and often more meetings. Pick a meeting of several hours duration that you will be in the position to observe. Select two of the five brief topical vignettes you have just read and observe the meeting with those in mind. For example, you might select *the domination of a single member* and *critical norms*. Then, first analyze the meeting in light of whether one or two individuals tend to dominate the meeting and the impact of their behavior on the group, the product, and overall atmosphere of the meeting. At the same time, try to determine whether certain norms exist that are helpful and whether there are others proving detrimental to the goals of the meeting. If you had the power to be constructive in altering the meeting to cope with problems you see arise, what would you do? As members become more aware of the factors that influence the outcome of a meeting, they find that there are things they can do to change the situation.

RUNNING AN EFFECTIVE MEETING

Assessing the Needs of the Group

In most cases we learn parenting from our parents and they from theirs and they from theirs. Since people seldom attend courses in parenting, the mistakes of one generation are passed generously on to the next. Certainly, there are books and new theories offered from one generation to the next, but the fact is that most of what we learn comes from what we have experienced directly.

Meetings are the same. We learn to run meetings from our previous experience in meetings. Just as few parents are trained to be parents, few leaders are trained to chair groups. For the most part we learn by osmosis, filtering into our repertoire of behavior those models or approaches that seem somehow appropriate in our society. Rarely are these approaches questioned, seldom if ever do average people have the opportunity to express their opinion about the nature of the meetings they have just attended. As a result, those running the meetings seldom receive necessary feedback and rarely are they stimulated to do anything differently. In most instances, even if the feedback were available, the norms or patterns are so habitual and ingrained that it would take a terribly dedicated leader to change the way it's always been done.

Just as it is impossible to prepare each of us to deal with every different personality that we meet, it's also impossible to teach those running meetings solutions to all the potential problems they might face. Even in the five brief examples we discussed previously, it can be seen that these are only five of thousands of situations that one might expect to occur since each new group of people coming together takes on a life of its own, which in turn creates a multitude of new and interesting problems. The best we can do is provide people with the ability to be aware of the problems that do exist and develop the skills to design what might be called "nonprescriptive" solutions for the problems that are identified. Thus, one cannot carry around a bag of solutions to be dealt out in any particular meeting; rather, solutions must be uniquely designed for the particular situation. The following are some rather simple, direct, but tough questions that should be asked prior to virtually any meeting that is ever held. Although these are not absolute or all-inclusive, they represent the essence of how we believe a leader should psychologically approach a meeting.

Knowledge of the Participants Although it seems obvious to ask the question "who's coming" to a particular meeting, we are interested in knowing everything that we possibly can about the participants because it's from the participants that trouble will occur if it's going to occur. Thus, we are interested in much more than questions of simple professional affiliation or status. We are interested in as much information as we can gain in a reasonable amount of time and in a traditional fashion. These questions include:

1. Whom does each of the participants represent within or outside of the organization?
2. Why have these individuals been chosen and not someone else?
3. What do we know about each individual's organizational interests and needs? What do they want from the meeting? What hidden agendas might be influencing the life of the group?
4. What is known about each individual's personal goals and needs? Is his or her ego on the line? Do any have a stake in certain outcomes in addition to the organizational needs that they represent?
5. Do the participants come with individual biases toward the leader, other members, the task at hand, or the general format of the meeting?
6. What are the personal skills and strengths available in the group that can be utilized, if necessary?
7. What are the personal limitations and idiosyncrasies present in the group that might block the task at hand?

History of the Group It is essential in planning any meeting that the leader be keenly aware of unfinished business left in the group. This unfinished business can relate to task issues or psychological interpersonal ones. The questions to be asked include:

1. How did the participants leave the last meeting? Were they pleased? Did they experience success? Or failure?
2. As a result, what are the expectations of this meeting?
3. What were the sources of pleasure experienced at the last meeting, of accomplishment for other rewards?
4. What were the sources of tension, frustration, or conflict?

Realistic Goals A crucial element of planning an effective meeting is goal setting. Planners must have an idea of the realistic goals that can be reached by the group, given the limitations of time and human resources. To set such goals, planners might ask:

1. What are the real task priorities of the members?
2. How is it possible to best utilize the resources available in meeting these priorities?
3. What needs to be done to ready the group to work on the task so they can hear each other and focus on the job to be done?
4. What do the individuals know when coming to the group?
5. What information do they not know but which they need to have in order to work effectively on the priorities at hand?

It is not uncommon for planners to pass informally through these kinds of questions prior to entering a meeting. The problem is that the answers to these questions are rarely reflected in the design of the meeting. Meetings tend to be very similar because even when we have diagnostic information we fail to apply it in a creative manner to meet the needs of the group. We are also quite capable of fooling ourselves by seeing what we want to see from impressionistic information, rather than seeking out more descriptive data from which to build a plan for the meeting.

In one interesting study (Amidon and Blumberg, 1967), the views that school principals had toward the meetings they ran tended to differ significantly from the impressions of teachers at those same meetings. Although teachers' views may be distorted as much as principals', the fact remains that the principal must deal with the teacher perception of reality. Thus, principals tended to see themselves as accepting and encouraging within the meetings while the teachers generally sensed, at best, mixed reactions and often a noncommital response to the ideas suggested by them. While principals viewed teachers as being encouraging of each other, the teachers felt themselves to be more critical. Furthermore, principals liked to believe that their teachers felt free to say anything they wished, but teachers saw themselves as cautious and rather careful. Similarly, although the principals said they felt open and free to say anything, teachers saw the principals as cautious in what they said and how they conducted themselves at meetings.

It becomes apparent from this information that leaders may carry distorted views of how their constituents feel about the meetings they run. Although in this particular study the differences were not dramatic, there were enough differences to assume that in the eyes of the teachers the meetings were not nearly as open and honest and effective as they might be. In one sense, the principals as a group may be deluding themselves into believing that the meetings are not in need of improvement. By not seeking more objective data than their own impressions, there is little reason that many would be motivated to change. But, by maintaining a diagnostic mentality, leaders who conduct meetings would always be aware of the changing needs of the participants as they enter a meeting, what left over feelings or business remains at the end of the meeting, and how the process of the meeting itself was helpful in achieving the goals of the particular session (Burke and Beckhard, 1976; Marrow et al., 1967; Schindler-Rainman and Lippitt, 1977; This, 1972).

The Concept of Design

Once leaders are aware of the needs of those who will be attending the meeting, they will realize that certain elements of business need to be accomplished, and that feelings and interpersonal relationships exist that one way or another will influence the outcome. With these in mind, it becomes time to design the actual meeting itself. It is not unusual for a typical leader to spend as little as ten minutes rushing together an agenda at the last minute before a particular meeting. Because of the structured and predictable pattern of most meetings and the low level of expectation on the part of the participants, this kind of preparation may serve to get the leader through another mediocre session. If, for example, each participant of an eight-member group put in three hours at a session, it is flabbergasting to realize that twenty-four person hours of time have been subjected to all of ten minutes of planning. Since very few people have ever experienced well-designed meetings, the standard of expectation is low indeed.

For our purposes, *design* means the building of a series of activities or events that move the group in an integrated manner toward the accomplishment of certain goals. Effective designing is a learned craft or skill that assumes that the leader has a rather thorough understanding of group process, at least minimal diagnostic skills, as well as a clear understanding of group problem-solving and decision-making practices. Most leaders have a very limited repertoire of activities from which to draw and tend to utilize the same design components for virtually every meeting. In fact, many leaders perceive meetings to be a necessary evil because they have little faith that the utilization of a group can be an effective means of sharing ideas or solving problems. One cannot blame them for such a jaundiced view because, again, effective meetings are by far the exception and not the rule. It is our major assumption that without setting aside the appropriate amounts of time to design meetings, all the reasons for ineffective meetings described earlier will occur. Below is a case in point.

A TYPICAL MEETING

The meeting was scheduled to last the usual 2½ hours. The group met regularly (every two weeks) to discuss its department goals and problems. The twenty members of the department usually straggled into the meeting somewhere between 4:15 and 4:30 PM, even though first arrivals began appearing at 3:55 for the 4:00 PM starting time. The same ones were always there on time and the same members always late. The sidelong glances of annoyance among the early arrivals as the stragglers arrived were all too apparent and humorous digs floated out occasionally to greet them.

Once the meeting began at 4:18, the leader handed out the agenda, prepared minutes before the meeting and looking much the same as every other agenda during the past five years. People glanced at it dutifully and put it aside. The agenda read as follows:

1. Review of the minutes
2. Old Business
   A. Report of the Committee on filling the departmental vacancy
   B. Report of the Christmas Party Committee
   C. The issue of preferential parking places for senior members of the department.
   D. Review of budgetary cuts discussed at meeting of November 1, 1980.
3. New Business
   A. Yearly review of departmental programs
   B. Organization of the department basketball team
   C. The condition of the lunchroom facility
   D. Report of the Financial Department on new guidelines for expense reimbursement
   E. Cost-of-living increases versus organizational pay schedule
   F. Other business from department members

Five minutes were spent reviewing the minutes of the previous meeting. Nitpicking comments were made by several of the early arrivals while most of those present ignored this ritual and either carried on conversations or conducted activities of their own, including reading the organizational newsletter or the help-wanted section of the daily newspaper, knitting, needlepointing, and doodling.

There was no discussion of the agenda and the leader simply began with the first item on the agenda under "Old Business." The first two reports took a total of ten minutes, with a few questions of clarification from the participants. At this point, the leader casually opened the topic of preferential parking places for senior members, which had been raised at the last meeting but not discussed. To put it mildly, "all hell broke loose." It seemed that everyone had an opinion and issues of equality, tenure, loyalty, service, and favoritism sailed around the room. Somehow what appeared to be an inconsequential issue devoured forty minutes of the meeting and the exasperated leader in frustration decided to table the problem until further study had been given the problem prior to the next meeting. A committee was appointed to review the issues that had been raised.

At this point, feeling the meeting getting away, the leader requested to move to item *D* under "New Business" because a representative of the Financial Department was present and it would be polite to utilize him at this time. The new guidelines for expense reimbursement were explained and because it was a mandated decision, the individual simply asked for questions of clarification. Warmed up by the previous discussion and frustrated by it, the members of the department attacked. How was the decision made? Why were they not represented? Did they realize the hardship the new guidelines created? Twenty minutes of harassment, defensiveness, and rationalizing left the group further antagonized with nowhere to go. The financial representative thanked the group and assured them that their ideas would be shared with the appropriate individuals. Murmurs of discontent followed him out of the room.

Sensing that the review of budgetary cuts (item *D* under "Old Business") would not be well received in the present climate, the leader moved directly into "New Business" explaining that the present issue would demand more attention than the time remaining allowed. The few weak protests from the members fell on deaf ears. The next thirty minutes were spent on the review of five departmental programs (five minutes each). The goal was to keep everyone informed of what was happening in different areas of the department. But, it was clear that the previous discussion had not resulted in a climate conducive for listening and most of the group faded into daydreaming or related activities as each presenter hurried through his or her report.

At this point in time, someone in the group requested that it might be of interest to the group to move to item *E*, dealing with pay schedules and the cost of living, rather than the organization of the basketball team. The leader, in a kindly and understanding manner, agreed except that the beginning of the interdepartmental league demanded at least some attention be given to the issue of team selection and practice so they wouldn't be embarrassed. Besides, there was less than thirty minutes left and perhaps it would be better to deal with the lunchroom issue and wait until there was more time to really get into the salary issue.

The meeting ended on time with several proposals being passed in relation to both the basketball team and the lunchroom with surprisingly little discussion or involvement on the part of the staff. Adjournment came at 6:26 PM.

For the most part, the failures of this meeting could be chalked up as unintentional and the result of poor planning or lack of know-how. The result was a staff divided, frustrated, angry, and feeling their own impotence. Worse than this was the fact that most meetings by this group resulted in similar feelings even if the circumstances were somewhat different. From the perspective of design there had been none. An agenda is not a design because in theory each item on the agenda deserves to be viewed separately and developed in a manner that reflects the realities of time, interest, need, and the probability of success. The meeting was not an exaggeration. It resembles thousands of others. It is not a fabrication, but one that occurred, attended by real people of goodwill and talent, including the leader. The problems were many and are listed below:

1. The participants were not involved in establishing the goals of the meeting.
2. The very format of dealing first with old business results in the passing over of critical issues for those of less importance.
3. People need successes to feel their ability to influence their environment. At this meeting there was almost no possibility for success, given how the issues were arranged.
4. Priorities according to the agenda seemed arbitrary or based on the hidden agenda of the leader.
5. Meetings at the end of a work day are often self-defeating. In this instance the first 40 minutes of a 150-minute meeting resulted only in frustration and hostility. The leader had lost control before the meeting even began.
6. The history of the group and personal antagonisms were working against the potential success of the meeting and there was no mechanism for altering the process to take care of such tensions.
7. The meeting was predictable, boring, and in most instances unproductive. There was little excitement (except around vested interests or scapegoating) and almost no humor.
8. Participation was almost totally reactive and not constructive.
9. The resources of the group were poorly utilized.
10. Any decision making was not defined and the total meeting was left to the whim of an arbitrary leader's arbitrary choices.
11. At least half of the agenda items could have been dealt with outside the meeting itself with greater efficiency.
12. The parking place issue would never have gotten out of hand if the leader had taken the pulse of the group, known the importance of the issue, and designed a means of problem solving that would not have opened the group to many old issues.

A MEETING REVISITED

Let's look again at the typical departmental meeting. Obviously, there is no one way in which to make the meeting successful. But, if we consider the general guidelines discussed to this point, we should gain some ground that is much more firm to walk upon than that created by the leader's agenda and overall format.

Twenty people sitting around a table are bound to be self-defeating unless a variety of activities are designed to draw them into participation. Usually in a group of this size, a majority will feel "out of it" unless a special effort is made. Ideal planning for such a meeting should include one or two members from the group plus the leader. These individuals must have the respect of the other participants. In this case, it would have been easy to have listed all of the potential agenda items and to have had the participants interviewed to determine those issues of greatest importance to them and whether there were other issues they felt deserved attention. Undoubtedly, a pattern of priorities would have evolved and issues of less significance might have been delegated to committee action prior to the meeting itself. The fact is that the agenda was so long and potentially complex that there was little hope that the leader would escape with his hide given the limited time and the intensity of feelings underlying many of the issues.

Once it is understood that only a limited amount of work can be accomplished in the time available, then specific activities must be designed to deal with each, all within a time-limited framework. It is the leader's responsibility to communicate with participants in advance to let them know the agenda, how it was decided, why the meeting is important, what they will need to prepare prior to coming, and that the meeting will begin promptly. Starting with an issue of importance to the group will inevitably encourage people to attend on time. It is up to the leader then to start on time. If meetings are important and people feel their presence is important, they will be there.

An issue of importance (emotional if not substantive) to the participants was that of parking. Clearly, it was symptomatic of other issues of favoritism, seniority, and privilege that creates adversary positions among the members. In the actual meeting, forty minutes were spent raising feelings and hostility, and then the issue was tabled unresolved. At the outset, it would be important for the leader to suggest how, after appropriate discussion and deliberation, the decision would be made. Second, it should be made clear that several viable alternatives, along with a rationale, for each would be developed. A design for this might be:

State the problem and a history of the issue to this point in time. Next, reach agreement that at the following departmental meeting a decision would be made that would be tried for six months and then reopened for discussion. Define the condition that a 60 percent vote would be required to change the present system. Randomly create some groups of three or four and give each group twenty minutes to develop two alternatives to the present situation. After twenty minutes, combine the six groups into three groups of six and ask each to negotiate a single best solution in thirty minutes. Then ask a member of each group to present the idea of his or her group to the total group. Ask the members to discuss these ideas during the period between meetings (one assumes considerable discussion would ensue) and take a vote at the beginning of the next meeting. If the group cannot reach a 60 percent majority at that time, discuss the two favorites for a limited amount of time. If a 60 percent vote cannot be attained, then use the system presently in vogue for a period of six months.

The design encourages a movement toward consensus and an exploration of a variety of issues. It insures participation and involvement in an issue that influences nearly everybody. The process tends to be one where the chances of failure are limited. One of the leader's goals is to provide the group with success experiences. Creating what appears to be a temporary solution mitigates the resistance to any change and minimizes a win-lose mentality. The initial move from groups of three to six forces a consensus process on the group with the assumption that any idea that six people could agree to would probably be acceptable to most people in the group. The opportunity for wide-open discussion enhances the exchange of information, insures interaction among people of varying persuasions, and initiates the beginning of a proactive rather than reactive approach to problem solving.

Thus, after a period of approximately the same amount of time as the original meeting, the design for the hypothetical meeting would have allowed the groups to have experienced several small successes and to have seen the light at the end of the tunnel in what appears to be a rational process.

It is almost certain that the cost-of-living versus the organizational pay schedule would be a high priority among the participants. The type of discussion that would develop and the type of design best suited to the topic would depend on whether or not there was any possibility for changing the present situation. Too often such items draw time and energy away from the group and influence morale when nothing can be done by the group. In this case, it would be inappropriate to develop alternatives until the situation had been studied carefully. This would best be done by a study group to meet with appropriate organizational personnel, to explore other realities, and to report back to the staff recommendations for next steps. Again, even this is a design since it represents a proactive approach rather than an avoidance of the issue as occurred in the actual meeting.

As a leader interested in the morale of the department, it would be wise to encourage the Christmas party committee to involve the staff in the planning process; the meeting would provide an opportunity for the polling of interest or for the delegation of certain tasks. It is important as a leader to encourage the various acting committees to utilize meetings in a positive and developmental manner so that individuals feel involved in issues influencing their lives even in small ways.

The report of the financial department on new guidelines to financial reimbursement was undoubtedly known in advance to be a volatile issue. A thirty-minute planning meeting with those visiting the department would have warned them as to the sensitive nature of the issue and helped familiarized them with potential resistances. Ideally, the financial department would not make decisions influencing those involved without their participation. For such an approach, the leader would have had to be supportive of such a process and encourage early participation.

If, as happened during the actual meeting, the department members' antagonism was directed at the visitor it could have been diffused by simply stopping the process and focusing individuals on the sources of tensions existing. It is possible that matters could have been moderated and not intensified. A design might have been for the leader to stop the attack on the visitor and ask the group to suggest two reasons for the degree of anger being generated. Had they been asked to discuss their anger for a few minutes in clusters of three, it might have allowed the group to experience some catharsis and enabled them to see that much of their anger was over the process of the mandate rather than with the substance of the mandate itself. This is an example of how stepping back for a moment during a meeting can provide perspective and new direction. Anger and frustration are realities for any group that is imperfect. It seems important, however, to deal with the real sources of anger rather than with issues raised from other unresolved situations.

To accomplish an effective design, the leader must always be conscious of both the task and the process domains of a group. If the leader is not aware of both individual and group needs, of interpersonal relationships, of the efficiency of various task activities, and of a wide range of other variables, it will be impossible to develop a coherent design. Similarly, if the individual leader is concerned with the symptoms and not the causes that are blocking group effectiveness, the design cannot reflect the necessary changes that need to occur to move the group forward. Finally, what blocks the successful implementation of a design often has nothing to do with the design itself but more to do with the "plumbing" of the meeting. So often it is the little things that need to be taken care of that can destroy the environment of the meeting or make it less effective than it might be. Information about the meeting, the language in the invitation, seating arrangements, the availability of necessary visual aids and materials, the presence of refreshments, right down to the nuts and bolts of name tags or some means of identifying the members of the group are all important considerations. The ability to anticipate obstacles that can influence the psychological atmosphere and remove them before they pose a problem is critical. Who is invited and how they are notified can affect the attitudes of participants before they even walk in the door. One need not be a nervous wreck or become overly compulsive prior to a meeting, but it seems crucial to try and step into the shoes of the participants and ask questions concerning their needs, expectations, and concerns. Following is an example of an effectively designed meeting.

AN EFFECTIVELY DESIGNED MEETING

It had been building up for weeks. Tension, hostility, and frustration from a group of thirty black parents was being directed at the faculty and principal of the small, neighborhood elementary school. Somehow, communication had broken down and the parents felt the primarily white faculty was not being sensitive to the needs of their children. Several outspoken critics of the city school system had been drawn into the battle and it appeared that a classic confrontation was about to occur.

Finally, it was agreed that a meeting would be held between the parents and the faculty. Predictably, the community group was suspicious because the school system had avoided every effort at previous meetings and the local press had labeled the parents as "radical." The consultant requested to help design the meeting knew there would be little chance that either group would hear the other initially. Thus, something had to be done to keep each group off balance and to minimize the tendency to accelerate the tension by increasing the adversarial climate. Somehow, the meeting had to reduce the fears and threatened feelings of the faculty without appearing to the black participants as an avoidance of the issues that concerned them.

White and black, parent and teacher, educated and uneducated were drawn into the gym for the three-hour meeting beginning at

6:00 P.M. It seemed that nearly everyone had their arms stiffly folded, watching, occasionally laughing among friends as if to deny the tremendous discomfort. The principal thanked everyone for attending and threw the ball immediately to the consultant. He immediately divided the group of sixty into ten groups of six. The effort had been made to invite equal numbers of faculty and community members. The six groups included:

two all-parent groups
two all-faculty groups
two groups, each composed of three faculty members

and three individuals representing the community.
Each of these groups was requested to develop a list of statements reflecting *what they as a group agreed* the school should be providing the children of the community in the way of educational services. Each of the groups fell into their tasks, which allowed concerns to be aired but minimized a polarizing, win-lose climate to develop with everyone performing for their own groups and having to hit out at the other.

At the end of forty minutes, the groups were asked to write their lists on large sheets of newsprint so that all sixty participants could review them together. Each group was requested not to identify themselves on their paper. By seven o'clock, all six sheets had been posted and faculty and parents were asked to note similarities and differences among the various groups. As individuals scrutinized the sheets there was a release of nervous energy, laughter, talking. It seemed impossible. There were almost no major differences across the different sheets. It became obvious to everyone that there was tremendous overlap, that people regardless of background seemed to want the same things for the children. Even the "radical" parent groups were not so different and certainly could not be identified from the other groups. The barriers of stereotypes, past experience, fear, suspicion, and racism had to take a back seat to the positive tone that filled the room.

Clearly, the issues were less educational than personal. The parents felt unheard, impotent, misunderstood, and that they lacked access to the principal and the faculty. The faculty felt they should be left to do their job, and were insulted at the accusations and recriminations being made. The children were a vehicle for the community to gain the access they needed.

The design allowed people to vent their concerns, helped individuals back off and gain some much needed perspective, and immediately increased levels of trust in the two groups. Each could lend some credence to the other group since both parties had some obvious good sense and showed their wisdom by agreeing with the other. Instead of a confrontation and a series of predictable justifications by the two groups, each now had the opportunity to look at the real issues. The meeting, although representing an extreme in terms of tension, required the same tough-minded and creative design process that should occur for virtually any meeting where ideas and concerns are shared and where differences should be expected to exist.

ENSURING MEETING SUCCESS

If a meeting represents a one-time gathering of individuals, it can be treated somewhat differently than a meeting that is part of an ongoing series of events among the same group of people. Nevertheless, any leader should, among other things, consider the means of incorporating the following within the context of a particular meeting. The department meeting met few of these criteria and the faculty-community meeting did. The agreement to which any one of the following points receives attention will depend on the particular goals of the meeting, but success can be enhanced if:

Participants in the meeting are, to some degree, disoriented or kept from falling into an expected routine. By continuing to be stimulated through a variety of activities and experience, their minds and spirits are kept active.

Each goal is considered separately in terms of the kinds of activities that will best insure the appropriate outcome. Creative designing by its nature does not allow for pat, routine, or stereotypic approaches to meetings, either in the areas of problem solving or discussions. Thus, meetings can be simple, straightforward, complicated, or sophisticated depending on what is demanded by a particular group and their goals.

Participants feel utilized in a meaningful manner during the process of the meeting itself. All too often, meetings represent intelligent people sitting and consuming information that can be shared in other ways, or listening to the opinions of others while their own ideas lie dormant within themselves.

The individuals in the meeting experience some feeling of success, outcomes that are visible to themselves and others and suggest that the meeting has been purposeful and worthwhile.

Individuals have the opportunity to learn something new and interesting during the meeting, either from other participants and the experience or from a structured learning activity provided by the leader.

The participants feel challenged by what they are doing so that they have to draw on their own resources and extend themselves beyond what might be called "the routine."

The members of the meeting enjoy—even have fun. The fact is that fun can be designed into a meeting just as can a serious discussion, a debate, or a problem-solving activity. The norms of seriousness and appropriateness that govern the nature of many meetings can be broken through effective designing and most assuredly will not be changed based on the simple desire of a few individuals to lighten the flavor of the meeting.

People feel they are members of the group, and that they are accepted by each other as equals, even though they may not be equal as resources to the group.

When the group is more than a one-time group, an effort is made to create a sense of a team so that members feel interdependent and supportive of each other and toward the task at hand.

If the leader is not a craftsperson of design, the chance of the individual utilizing the resources of individual members and maximizing the potential of the group as a whole is minimal. Perhaps the most important aspect of design is the diagnostic phase, by which one is able to focus clearly on where the group is and what needs to be done. From this reality, many useful and creative designs will literally fall out quite naturally. Thus, a group needing success must be provided the opportunity for success. A group needing information must be provided information. A group needing to problem solve, or requiring information, or needing to relax and have fun must have access to fulfilling these needs or the meeting itself will stand a good chance of failure. The monsters of most meetings are created from lack of an effective diagnosis and a poor design.

From this overview of meetings in general, we will turn our attention to a critical aspect of any working group, that of problem solving. Although we have explored a variety of reasons that people have meetings, it is the problem-solving and decision-making aspects of group life that tend to create the greatest sources of stress and confusion. Here, the execution of leadership is perhaps most demanding and challenging. Thus, the following chapter will relate theory concerning group problem solving and decision making to specific situations that demand the application of design skills.

EXERCISE 1 The Newspaper Interview: A Means of Discovering the "Real" Issues in a New Group

It is true that what limits many working groups is their lack of tools or methods for solving problems and making decisions. However, just as often, the problems that prove most disruptive (and cause the greatest loss of time and energy) are the result of poor interpersonal relationships and the lack of any mechanisms for improving this human relations factor. Similarly, tensions resulting from innumerable sources can immobilize a group (and yet, the destructive feelings generated find no legitimate outlet) indirectly through covert hostility and passivity. Here we are talking about the emotional or maintenance side of the ledger. Increasingly it is being recognized that if a group is not responsive to the emotional needs of its members as they work together, the task may never be completed.

Thus, it is necessary for groups that work together for any length of time to spend some of that time taking care of the problems that are bound to arise because of their own insensitivity to each other or simply because of their own lack of interpersonal skills. The problem is that people may willingly talk for hours about problems involving technical, task-related skills, but will avoid, even for a few minutes, looking at how the group operated on the human level. Were individuals in the group shut out of participation, did certain individuals dominate, was hostility suppressed, was problem solving shared, were the real issues raised? These and many other questions reveal how closely personal feelings are related to overall objectives. The aim is not to have the group solve all the personal problems of its members. Rather, it is to come to grips with procedures, fears, or behaviors that reduce the effectiveness of that group. In looking carefully at their own work process it is quite possible that individuals will grow personally and become more effective as members. Even more important, the group will develop a climate built on a base of honest communication, shared leadership, and a concern for its members. But, again, this deserves and requires time, effort, and a certain amount of risk. Following are a few exercises that can help a group begin to work with process skills at both a personal and a group level.

Objectives

To bring into the open issues that may influence how a group operates from the beginning

To establish an immediate climate of honesty and leveling among the participants

Setting

Very often when a group of people come together for the first time they bring with them an assortment of feelings, concerns, and expectations that may influence their participation for some time. It is very important to help clear the air from the beginning, to find out where the group is, and to give people an opportunity to express themselves. This particular exercise is more effective with a group of over ten (ideally twenty or more) and as many as one hundred. From the large group, the facilitator selects a random group of "reporters"—enough for one reporter in every group of seven or eight. These individuals should be brought together while the larger group is still milling around waiting for the beginning of the program (allow 10 minutes). The facilitator gives the following directions to reporters:

You each represent a different newspaper. It is your task to interview a group of about seven or eight of these people and find out as much as you can about them—as a group and as individuals. Names are not important, but you may wish to find out:

Why are they here? What do they expect to get out of the meetings?

Do they have any reservations about coming?

Outside of the stated reasons for the group being together are there any particular hidden goals different members have that they would like to share?

Do they have any doubts, suspicions, or special concerns about the meetings, and how they are going to be conducted?

If they could wish for one thing during this period of time together what would it be?

You should feel free to pursue any line of questioning you like and to use these questions as guidelines. After about twenty minutes of interviewing, you will be asked to synthesize your findings and report to the larger group. It is doubtful that there will be time to hear a full report, so please give only that information not presented by other reporters or that which is particularly important for your group.

Action

After a general introduction, the facilitator suggests that one way the members can become acquainted rapidly is by taking part in a brief exercise that will clarify the purposes and focus on the expectations of the group. He or she then quickly organizes the group into sets of about seven or eight (so as to break up any cliques). Once in groups, the participants are told that they are to be interviewed by the local press and that it will be important for them to give the press the cooperation they need. They will not be quoted by name so they should be free to express any of their concerns or feelings about this meeting. The interview will last about 20 minutes, and then each reporter will share his or her findings with the large group.

During the reporting session it is important that the facilitator does not allow one or two reporters to dominate the session. A reporter gives a piece of information on a point, then the next reporter speaks on that point. It is helpful to have the points being made recorded on newsprint so that they afford immediate visibility for the group. Most of the information will be obtained in about 15 or 20 minutes. It is at this point that a general overview of the program should be developed by those in charge. Considerable effort should be made to link this presentation to the expressed concerns of the group. It is possible that a brief discussion about an issue or two may prove necessary. Thus, the major reason for this activity is to help express a norm of openness and allow the participants to express their feelings and expectations. Once expressed, they can be related to the actual plans and may be used by the organizers to alter some of them.

EXERCISE 2 The Process Diagnosis: A Means of Altering Group Behavioral Patterns

Rationale

Most groups are suspicious about being observed or looking at their own patterns of operation. People fear "getting it in the neck" and seldom have really experienced constructive criticism that does not degenerate into personally evaluative comments. Part of the problem, of course, is that most participants simply do not know the many possible ways a group can look constructively at its own work process and not feel personally threatened. It is only after a while that most groups begin to feel better about the notion of process and begin to welcome it as an integral part of any meeting. This exercise will assist a group to diagnose the factors that are both facilitating and inhibiting members in their efforts at problem solving.

Objectives

To help establish a desire in the group to process its own behaviors and procedures, to look at what happened, how people felt, what helped or hindered movement on goals

To present explicit information about how the group is operating without becoming personal and focusing on any single individual

To help the group look at alternative operating procedures, thus introducing a degree of organizational flexibility that might not have been present

Setting

This exercise is directed primarily at relatively small (five to eighteen) working groups who need and desire to maximize the use of time and available resource personnel. This type of experience should not be imposed on a group by the well-intentioned facilitator. The diagnosis format and reviewing of data should be explained in detail before it is attempted. Usually if the group is concerned about maximizing its working efficiency and if there is the assurance that the diagnosis looks at group behavior rather than individual behavior, there will be a curiosity to proceed. Although most people fear looking at their own or group behavior, it also has a fascination for them, and they will be willing to experiment if the risks are not too high. If possible, this exercise should be conducted after a period of work. The collection and analysis of the data along with the report back to the group takes between 45 minutes and one hour. Time must also be left for discussion. In all, the facilitator should figure on 1½ hours (in this example six questions are used, but three might be enough). The key to success in this exercise appears to be:

1. Making certain that the questions are not too threatening.
2. Presenting the data in an objective fashion and letting the group sort out the implications.
3. Leaving the group with the feeling that an expert is not necessary to carry out this type of process analysis.

Action

Having agreed that they would like to participate in the diagnosis, members are (*a*) given a brief questionnaire or (*b*) asked the questions directly from newsprint charts on the board. The latter approach seems to promote a feeling of less secrecy and allows data to be transferred directly to the board. However, if trust is a real problem in the group, the questionnaire tends to give a greater feeling of security and confidentiality (the facilitator should state as casually as possible that the responses will be completely anonymous). The diagnosis is aimed at painting a more descriptive picture of the group, which is not often available in a discussion. It usually takes about 15 minutes to answer the questions. Then, while the data are being tabulated and posted for presentation and discussion, a number of options are open to the facilitator. He or she may suggest that the group break into trios and discuss a particular question on how the group operates and what implications there may be in the various responses. Or he or she could have the members (again in subgroups) discuss the kinds of behaviors in this group and others they have noticed that reduce participation and increase defensiveness on their part and in others. The posting takes about 20 minutes, and the task during that period should be process-oriented and directed in a manner that will lead to more rather than less openness in the subsequent discussion. Following are six questions that exemplify a wide range of possible questions that could develop important data. All of the data from them can be easily tabulated and visually presented to the members for their own interpretation.

Question 1

Indicate which of the diagrams best represents the relations that exist among the members of this group. Place the letter *A* beside the figure that *you* feel most represents the group. Place the letter *B* beside the figure you believe most group members will choose.

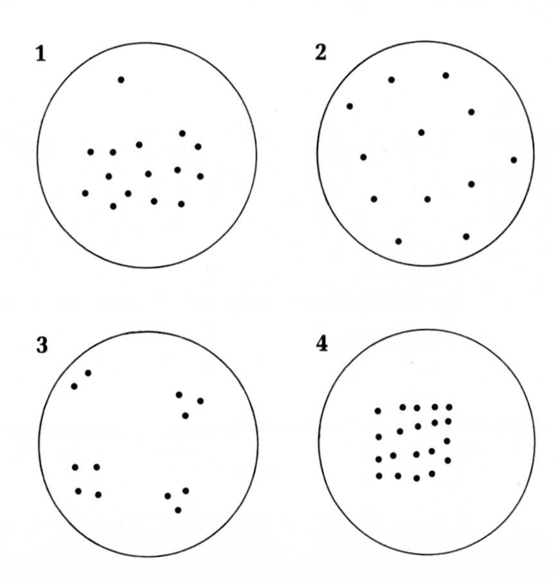

Rationale for Question 1

Most people have a definite feeling about their group and how its members relate to one another and to the group as a whole. It is also true that in many groups members feel that others see the group differently than they do. If, as it is assumed, the group wishes to maximize its working relationships and it is discovered that most of the members respond to the number that refers to subgrouping, then the group must face the implications of this in its decision-making efforts. If some members of the group see it as a group of individuals (diagram 3), it is likely that there is considerable blocking and that active or passive hostility is being generated as members attempt to win points for themselves, while failing to work for the benefit of the group. Further, a group that sees itself dominated by a single personality may well wish to explore means of reducing such dependency-building relationships (1). Finally, a discrepancy between what individual members think and what they feel others think about the existing pattern of relationships can provide an important reality test for a group.

Question 2

In this group a person who feels that he or she has the greatest possible influence and respect in the decision-making process would place himself or herself at 1 in the circle. A person who feels he or she has no influence at all in this group would place himself or

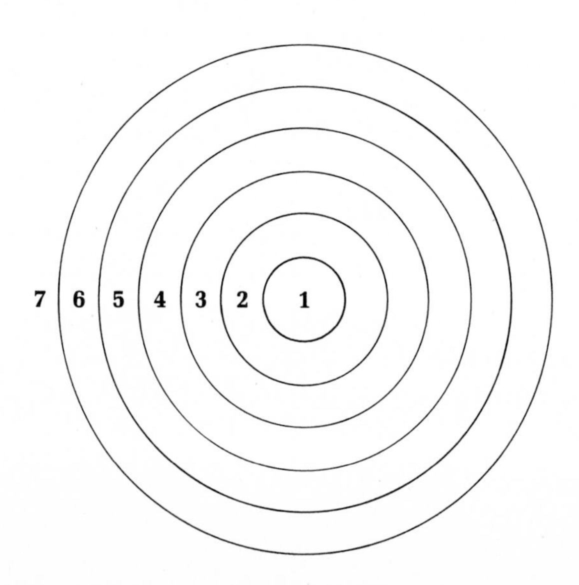

herself at 7 in the circle. Place an *A* at the point in the circle where you feel that you belong. Place a *B* where you believe the group will tend to respond to the question.

Rationale for Question 2

People need to have influence; to feel impotent and remain in the group is seldom tenable. Thus, individuals often rank themselves higher in this particular question than they see others. It is an important discrepancy, for example, if all twelve members of a group see themselves in the first four circles, while six of the members saw the group tending to have very little influence (points 5, 6, 7). Also, the second question supplements the first question. For example, in the first question, if members see the group dominated by a single person, or if they see a sharply divided series of subgroups, they would have to explain the discrepancy if most of the group felt influential. Also, if four or five members of the group feel they have considerable influence and another four or five (or even one or two) feel they have virtually none, then there is almost inevitably going to be tension and nonproductive behavior (withdrawal, defensiveness, dispute) as those being controlled try to cope with their own feelings of impotency. Again, it is for the group to interpret the data and for the facilitator to make them aware of the evidence so that they can raise the most helpful questions.

Question 3

In a perfect group a person is able to say almost anything he or she thinks or feels as long as it is not totally irrelevant or destructive. Mark an *A* on the diagram at a point that represents how free you feel in communicating with the group. Place a *B* at the point you feel most members of the group will respond to the question.

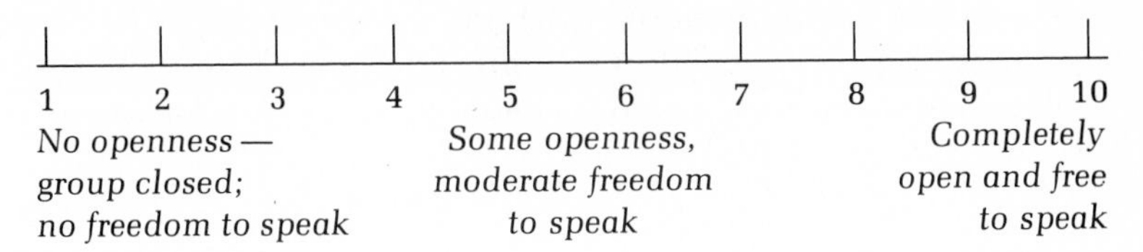

Rationale for Question 3

When the data are presented in graphic form, it is difficult to avoid certain implications. For example, a wide difference in the perception of the openness that exists immediately suggests that certain members are more open that others which, in turn, implies an issue of control and raises some of the questions suggested previously. As in the other questions, it is important to explore the difference between how individuals perceive themselves in the group and how they perceive the group responding to the question. It is possible to compare individual and group averages in this and the previous question and, if desired, deviations from the mean. However, most of the implications will be quite visible when charted.

Question 4

Suggest two reasons that keep this group from being as productive as it might be when working together.

Rationale for Question 4

All groups have problems working together, and those unable to recognize some rather specific conditions that need to be improved probably are not very involved (a problem in itself) or are not being quite honest. Human relations problems are part of the game and to be expected. The only real problem occurs when the group does not face them and try to remedy them. This question is strictly designed to get some of the issues out on the table so that the group can begin to deal openly with them. Resolution will probably not come in this session, although the mere mention of some problems will go a long way toward solving those problems. The responses to this question should be integrated to some degree so that members can see which areas seem to be of most concern. Thus, instead of recording all twenty-four responses of a group of twelve, there may be six or eight that can be identified in terms of importance (that is, the number of members who mentioned it) and noted next to the statement. Of course, a point mentioned only once may represent an important issue others were afraid to mention and therefore should be given equal time for exploration.

Note: If the participants have numbered their questionnaires or responses, there can be some interesting comparisons between responses made to the first three questions and those on this question. However, one must also be careful not to overwhelm the group with data. There is just so much that can be digested at one time.

Question 5

What are two of the best qualities of this group?

Rationale for Question 5

A person or a group grows not only by strengthening its limitations but also by maximizing its good qualities. Seldom do we take time to focus on these characteristics. The simple expression of these strengths can give a group the lift needed to deal effectively with its limitations. It may also be a first step in legitimizing praise, which is crucial in any effort to describe clearly the nature of the working group.

Question 6

What behaviors do you feel occur in groups with which you work (not necessarily this one) that you personally find annoying and disturbing to the progress of the group?

Rationale for Question 6

It is quite possible that group members are not able to face some of the issues hindering progress. By focusing upon other groups, members may indirectly raise problems that may have direct implications and that may be used as a source for diagnostic questions at a later time. Most important, the group may wish to build some preventive measures into its own operations to make certain that some of these problems do not occur.

Conclusions

The questions are diagnostic in nature, and, although some of the statements imply solutions, the data should basically be used to establish a desire to look toward solutions. If steps are not taken to resolve some of the problems raised, if the new awareness and facing up by the group is not matched by an effort to develop more effective problem-solving and communication patterns, the exercise may introduce a period of disenchantment and increased tension. Thus, again, the group and those in a position of influence must desire not only to *look* but also to *do*. As in any problem-solving situation, this will take time and energy away from other issues. However, in the long run, the time should prove beneficial to the entire work process.

EXERCISE 3 The Open Chair: A Means of Insuring Greater Group Participation

Objective

To interject new ideas and opinions into a group meeting without losing the advantage of a small number of participants

Rationale

Quite often a large group (perhaps ten to twenty-five members in this case) is faced with an issue that needs the involvement of all its members, but the facilitator is aware that an open discussion would probably only polarize positions and cloud the important aspects of the problem. Thus, while it is important for members to feel that their ideas are being represented in the discussion, they must feel some identification with and responsibility to the group as a whole. If those involved in the discussion can hold a responsibility to the group rather than the need to defend their own position, the chance for compromise and eventual consensus is greatly enhanced.

Setting

The large group selects a representative group of individuals to discuss the topic that is posing a problem. The smaller group of perhaps six or seven takes seats in a small circle and an extra chair is placed there. Those not directly involved take chairs on the outside of this inner circle.

Action

As the discussion develops, individuals on the outside of the circle may move to the empty chair. They may make an observation or express an opinion on either the issue (task level) or on how the discussion is being conducted (process level). Unless asked to remain for a few minutes by the inner group, the contributor then leaves the circle, and the chair is open to another interested person. This structure insures the outside members that their concerns are being represented and, as observers, they also have a better opportunity to note any personal factors that may be reducing the effectiveness of the group. This particular problem-solving structure assumes that the group is at a point in its development where it can delegate authority in a reasonable manner. It also assumes that participants would like those in the middle to resolve the problem being discussed. If this is not so, it is possible for disruptive persons on the outside to destroy the intent by dominating the inner group with personal biases or irrelevant points of view.

EXERCISE 4 Role Reversal: A Means of Unblocking a Polarized Group

Objectives

To help group members gain a broader problem-solving perspective

To help a polarized group with a particular issue on which certain members are especially intransigent

Setting

If, regardless of the facilitating methods used, a group is making little progress in reaching a joint decision because of strong differences of opinion, it is often helpful to pull individuals away from their defensive positions. This can be done by introducing a role-play situation where the participants are either (*a*) asked to argue from the point of view of the opposition or (*b*) design a solution that is as far as they could go in satisfying the opposing position.

Action

A group that has moved from open discussion and the presentation of ideas to one of defensive debate is doomed to failure, frustration, or both. The facilitator may wish to give the group members a short break (often merely changing the physical scene can help) so that they have a chance to move away from the positions they have taken. Sometimes members may wish to come to a compromise position, but the strength of their argument has boxed them into an all-or-nothing position. It is for the facilitator (or a skilled member of the group) to help remove this barrier. Thus, on returning from the brief break, members of the group are asked to think for a few minutes of the opposing positions being expressed. The discussion begins, and it is evident that many of the arguments on both sides have not been heard as the members argue those issues that are most difficult for them to accept.

At the end of about 10 minutes, the total group is asked to consider the role reversal and what implications it has for the group. By getting into the other person's shoes and having to argue his or her side, it often becomes easier to understand diverse points of view, and it may give those desiring to break the deadlock an opportunity to do so honorably.

People usually argue from extreme positions. There is usually (except on issues of morality) a wide range of acceptable compromise. But, again, intensive debate and argument lead to a condition in which personal image more than personal belief or idea is being defended. To not lose becomes of greatest importance. By asking members to think of a solution that they might not like, but which would incorporate as much of the opposing position as they possibly could, they are forced to do something constructive rather than just argue and defend. Just getting the group to move in the direction of being constructive may induce the compromise and resolution that are needed. Once the group is willing to make a few concessions, the process becomes easier, and it is no longer a problem of saving face.

There are many ways of going about helping groups move off what appear to be irreconcilable positions. One way is to simply get them thinking about the other person's position and not just their own, or by inducing them to build a workable solution instead of just arguing intellectually. Most action programs represent compromise merely because many people must be satisfied.

REFERENCES

Amidon, E. and A. Blumberg. *Understanding and Improving Faculty Meetings.* Minneapolis: Paul Amidon & Associates, 1967, pp. 30–35.

Antony, J. "How to run a meeting." *Harvard Business Review,* #72604 (March–April 1976), 43–58.

Auger, B. Y. *How to Run a Better Business Meeting.* St. Paul: Minnesota Mining and Manufacturing Company, 1972.

Blake, R. and J. Mouton. *Group Dynamics —Key to Decision Making.* Houston: Gulf, 1961.

Bradford, L. "Leading the large meeting." *Adult Education Bulletin,* 12, No. 3 (1948).

Bradford, L. *Making Meetings Work.* La Jolla, Calif.: University Associates, 1976.

Burke, W. W. and R. Beckard, eds. *Conference Planning.* 2nd ed. La Jolla, Calif.: University Associates, 1976.

Kaufman, R. *Identifying and Solving Problems: A System Approach.* La Jolla, Calif.: University Associates, 1976.

Maier, N. R. F. *Problem-Solving Discussions and Conferences.* New York: McGraw-Hill, 1963.

Marrow, A., D. G. Bowers, and S. E. Seashore. *Management by Participation.* New York: Harper and Row, 1967.

Miller, E. C., ed. *Conference Leadership.* New York: American Management Association, 1972.

Prince, G. M. "Creative meetings through power sharing." *Harvard Business Review,* #72410 (July–August 1972) 47–55.

Robert, H. M. *Robert's Rules of Order.* Chicago: Scott Foresman, 1943.

Schindler-Rainman, E. and R. Lippitt. *Taking Your Meetings Out of The Doldrums.* La Jolla, Calif.: University Associates, 1977.

Strauss, B. and F. Strauss. *New Ways to Better Meetings.* New York: The Viking Press, 1951.

This, L. *The Small Meeting Planner.* Houston: Gulf, 1972.

Zamke, R. "What are achieving managers really like?" *Training,* (Feb. 1979), 35–36.

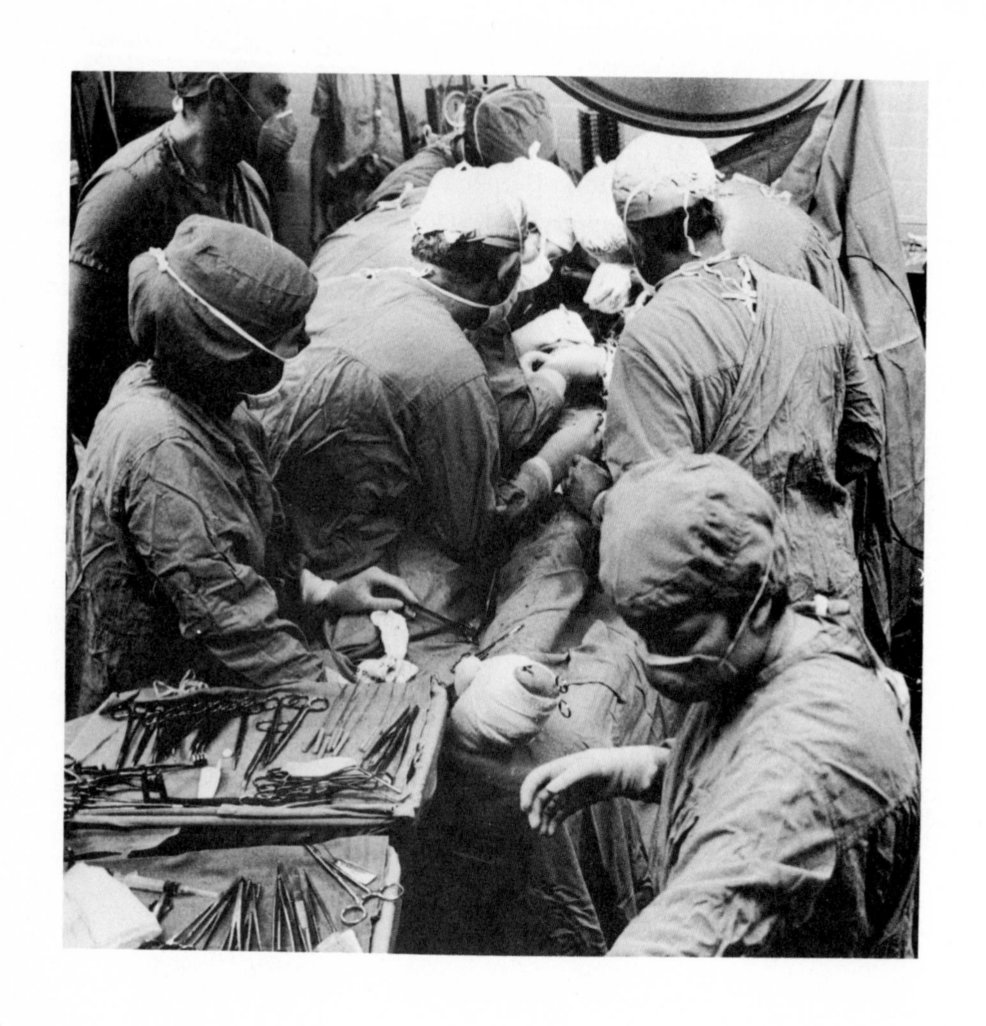

Seven

Group Problem Solving and Decision Making

Consider the poor, homely camel, the product of one of Mother Nature's efforts at group problem solving. Their goal was a horse, but the heavenly hosts involved in the project, so the story goes, were so busy compromising their personal interests that nobody really much cared for the finished product. Embarrassed by their own handiwork, the group could agree on only one thing and that was that they didn't wish anyone to see the outcome of their endeavors. Thus, for a hundred millenium, the bedraggled camel has been forced to wander the deserted backroads of the earth.

It's sad to say, but most of us have experienced all too many of this type of problem-solving group, in which the finished product is something with which we wish we did not have to live. Somehow, it is all too common that the good intentions of the group give way to vested interests, organizational politics, personality conflicts, competition, ambition, and an endless number of seemingly unpredictable variables that jam up the works, leaving us frustrated and disappointed. People want to be effective, to be proud of their efforts and to savor a job well done. Yet, rarely are we satisfied with the product, or do we believe that the group effort has spawned trust, openness, and teamwork while utilizing the resources in the group. But, don't despair. If you have never experienced a group problem-solving effort that yielded creativity, high levels of personal motivation, commitment to the outcome, and a sense of cohesion among the group's members, you are probably among a large majority. It is little wonder that many of us shudder at the thought of working with a group to solve a problem, let alone to agree on a meaningful decision. Consider an example.

One group of six executives who represent the key senior leadership in a company of ten thousand, come together once every other week for five or six hours to share information, develop new ideas, solve problems, and make decisions that affect the total organization. The salary of each individual is approximately $80,000. With benefits, the cost to the company for this group of executives is over three-quarters of a million dollars each year. The group appears professional, polite, well-organized, and serious-minded. The individuals involved are civic leaders and pillars of their churches, and not one is divorced. Good people indeed. Yet, their meetings are travesties of mismanagement, poor utilization of time, hidden agendas, bickering without owning their complaints, and competition. A careful analysis of twenty meetings held over a year revealed that 80 percent of the time of the group was not involved with meaningful problem solving and decision making, not even with discussion of critical issues. Rather, it was spent sharing information, recounting personal experiences, justifying previous decisions, and spending enormous amounts of time with inconsequential questions. Controversy, arguments, even a well-ordered debate went against the norms of the group, in which passive politeness and agreement were keyed to the carefully laid cues of the president. Everyone knew the rules, minimized conflict, and held back important or high-risk opinions until informal outside meetings. Each individual was careful not to be vulnerable, to be seen negatively, or to attack a peer directly in the group.

A change in the executive and middle-manager parking lot took four hours of the group's time, a change of the organizational logo another four hours, and more complex problems resulted in little discussion because of the implicit understanding that the group was only the figurehead for democracy in an autocratic organization where personal influence was wielded informally and on a face-to-face basis with the president. In addition, members of the group protected themselves by forming coalitions with at least one other member.

This chapter will focus briefly on the problems and stresses that undermine effective group problem solving. More importantly, we will explore how to improve the situation, and how to apply the enormous amount of good information learned in the behavioral sciences during the last twenty years to group problem solving and decision making. Twenty years of study in this area have provided great potential for improvement. But, the vast majority of players in the group problem-solving game have been raised and rewarded playing a different game. Thousands of years of dealing in a top-down, high-control, power-oriented, and nonprocess-oriented manner have resulted in enormous skepticism and mistrust for any alternative viewpoints. Nevertheless, in the face of great resistance, there is a growing body of evidence that supports involvement and collaboration in many situations and encourages the building of teamwork. This view has fostered a variety of approaches to problem solving, depending on the nature of the problem, its potential impact, and factors such as time, available resources, and the type of problem itself. The point is that there are new and better ways of problem solving being discovered nearly every day, and it will take patience and perseverance to overcome the habits, biases, and poor experiences most of us still hold dear. We appear to be at the cutting edge of new approaches that over time should help to improve the human condition and facilitate our ability to think, decide, and act.

READER ACTIVITY

Take a sheet of paper and make two columns. On the left-hand side, list the qualities and characteristics of the last group you attended in which members successfully solved a significant problem together. In the right column, list the characteristics of the last group problem-solving session you attended that ended unsuccessfully. Extend this list to include specific behaviors or events that worked against success. Thinking in this manner should bring perspective to the following pages and help you relate to the common patterns that exist in many groups.

HUMAN BEINGS AND THE DECISION-MAKING PROCESS

Decision making is at the center of our very being. A thousand times each day we make decisions, sometimes casually, almost without thought, responding to long-established routine. We are rational, irrational, spontaneous, and strategic. Depending on who we are, we can be victims of our own narrowly defined and predictable patterns or we can seem to be forever throwing off the shackles of precedent, shaking every tree we come near. We can pay a price for carefully weighing every new step, literally taking the life out of life, or we can act too quickly, being constantly reactive, without plan, and stung by consequences somehow overlooked. Who we are as decision makers is no more or less complex than who we are as people. The weave of factors influencing us can be incredibly complex: our cultural backgrounds, parents, schooling, feelings of attractiveness, social status, religion, and general level of achieved success. Add to this very special mix our willingness to risk, our shyness, our inclination toward bias and stereotypes, jealousy, fear of failure, and a hundred other variables and we begin to have some idea of how complex even our most casual or spontaneous decisions might really be.

Now place five or six, or ten or twenty such complex individuals together, attempt to develop an agreed-upon decision and the potential difficulties seem almost beyond comprehension.

Sources of Tension and Conflict

Who knows a perfectly relaxed person? Simply fulfilling our basic life needs provides us with a never-ending source of tension and the constant demand to make decisions. For every potential decision there are two potential sources of tension and conflict. First, whenever the decision involves a choice between alternatives, there is a loss-and-gain factor(s) that must be weighed. To eat roast beef means no chicken, to have another drink means feeling less alive in the morning, to buy six shares of IBM means foregoing a down payment on a piece of property, or to offer a suggestion means possible rejection. Similarly, just being in a group poses new tensions and decisions for an individual. What must I do to be accepted here? Should I relax and just be myself? They say they want my opinion, but should I risk having it rejected? I really don't like the way things are going. I wonder if things will be any better if I say something? These and hundreds of other personal decisions add an enormous reservoir of tension to any group even before its task has begun. Then when the group does begin its own process of decision making, there is further potential conflict as a result of disagreement among individual participants, as well as from the implications any decision will have upon the group as a whole (Lewin, 1948; Lewin 1951; Adams and Adams, 1967).

A second source of natural tension and conflict is created after the individual or group makes a decision. This stems from being faced with having to live with the decision that has just been made and thus having to continually justify it in the mind of the group (or individual) and in the minds of others (Festinger, 1957).

It appears natural, therefore, that tension and points of conflict should exist within decision-making groups. The question becomes one of whether or not the sources of tension are clearly recognized and dealt with in the most constructive manner possible. All too often the greatest sources of tension and conflict are completely avoided (denied or ignored). If the actual sources of tension are not uncovered and dealt with, it is highly likely that they will be diffused into other areas of the group's experience. Here is a case in point:

A number of years ago all the principals of a large urban school system were urged by the administration to leave the city and attend an intensive fourteen-day human relations workshop. The workshop was aimed at increasing leadership skills and providing an opportunity for the group to work through a number of pressing problems that they all shared. Extreme tension developed to a point that the participants were practically immobilized. On the one hand, it was clear that the opportunity to get away from the pressures on the job, reappraise their own skills, look at their own shared problems, and build a supportive network among themselves would probably not come again. On the other hand, the members felt that they were not there of their own free will and that the request to attend had been a scarcely cloaked demand. Thus, feeling manipulated, the entire group was faced with the uncomfortable question of their own potency.

Second, there was an underlying fear that it was one thing to gather answers to problems in a "mountain haven" and quite another to apply them back in the "jungle." To develop new answers and new personal skills and then not be able to implement them successfully would be worse than to continue to struggle on with no new ideas. It is one thing to blame an unworkable system, but it is quite another thing to be faced with one's own inadequacies.

Both these sources of tension had to be surfaced before the real work of the group could even begin. The issue of coercion was dealt with by raising it for the group to discuss. By allowing them to legitimately vent their anger and by helping them to see there were few feasible alternative approaches given the shortness of time, the issue died. But, the very process of becoming angry as a group helped them recapture some of their lost sense of potency. The second issue was resolved by developing a training period of one week back in the city (cutting short the time away) where the skills learned in the workshop could be tested on groups of students and community members at a time when the principal group was still together and able to share in the success or failure of the programs they designed.

In this case, the group had to decide whether to acquiesce and accept the program or to resist (in some cases by overt action and in others by withdrawing). The tension was relieved once the issues of conflict were uncovered and dealt with, but this took nearly two days of the group's time and energy.

Tensions within a problem-solving group can arise so subtly that the participants find it almost impossible to extricate the real issue. For example, in a certain factory, the division manager was constantly being pushed to give more responsibilities in decision making to subordinate members of staff. At one point, it was decided to give more of the administrative decisions over to the workers themselves. The reaction to the decision was immediately favorable, and a representative group of eight workers was elected to iron out the new administrative details to be presented later to the rest of the factory for a referendum vote. From the beginning, sources of tension were apparent among the working group, but no one could put a finger on the source of difficulty. After three weeks of virtually no progress, a pattern of behavior in the group had developed that seemed to lead to the following conclusion. Apparently, although the workers had intellectually liked the idea of more influence, a greater awareness of necessary time investment and accountability for outcomes had dampened any original enthusiasm. It became increasingly evident that the group was in conflict over the fact that they had won an argument (more influence) but were going to lose in the sense of having to expend considerably more time and energy as a result. This is the kind of issue that is extremely difficult to dredge out of a group, but, without doing so, compromises and alternative administrative patterns could not even be raised.

Also, what does a person do in a case where someone wants to spend some time looking at the problem of openness and trust within the group? The idea may be supported by other members because who can argue against such hallowed virtues as trust, openness, and authenticity? Again, individuals are more willing to talk about an ideal condition rather than work at developing it. Even those who initiate the discussion may begin to feel the growing tension between self-exposure and job evaluation, between authenticity and ascribed role, between honesty and different levels of power in the group.

Even a choice that must be made between two positive alternatives may create tensions that show themselves through increasing interpersonal conflict. Often such a choice is put off, and the anxiety of the group is taken out in the discussion of other issues.

The point to be made is that groups, like individuals, are constantly being forced into making decisions. The problem is that the process often breaks down over personal issues that are not easily raised to the surface, but which leave a residual of tension that may debilitate the entire problem-solving process. Often the issues causing conflict are personal and emotional while those being dealt with are intellectual. By asking himself or herself what it is that is causing the group to avoid the possible alternative and what are the real underlying issues of tension, a person standing apart may be able to move the group to another stage in its problem solving. Thus, if a person is really concerned with understanding why a group is functioning as it is, it may help to know the stated goals and bylaws of the group. Of even more value would be to observe the group as an objective observer, sorting out various behaviors in relation to possible causal factors. Most important would be developing an awareness of how the individual members personally feel about the group itself, their role in it, and their feeling about the issue being discussed. Too often, we remain at a technical level when the sources of conflict are occurring at an emotional one (Lewin, 1936).

Sources of Resistance

Much of the frustration that often accompanies working with a decision-making group results from an inability to understand and accept as perfectly natural many of the resistances that develop during the problem-solving process. This in itself can reduce some of the strain. Further, some of the points of resistance readily suggest procedural approaches that might result in a reduction of tensions and potential conflict.

Most people tend to organize their lives in a manner that reduces the amount of stress they must face. There are, of course, those individuals who create stress and crisis in their lives to satisfy other needs. Yet most individuals build a pattern of existence that is familiar and comfortable, a pattern in which habit, ritual, and precedent play a relatively large part. In a sense, the individual attempts to bring his or her life into a state of equilibrium where he or she is able to predict events and reduce conflict. To change this relatively stable, steady state results in a need to change accustomed patterns of behavior and creates, at least temporarily, discomfort and tension. Problem solving and eventual decision making often lead to innovation, alternative courses of action, and a disruption of a group's or individual's state of equilibrium. This is one reason we resist new ideas. It is why we sit in the same seats, tell the same tried-and-true jokes, maintain the same prejudices, and continue the same work habits. This is the safe and comfortable way; it requires little effort and helps us to feel secure and confident in what we do. A particular solution, even though it appears acceptable and useful, may, nevertheless, be met with the most subtle and resourceful resistances.

When people work with the familiar and within a framework of accustomed behavior, they build a relationship to authority and power that is based upon relatively clear expectations. When people change their relationship to power, when they take on new responsibilities, they immediately become more vulnerable and less sure of their own position. Thus, secure dependency relations are altered and personal security lessened. Again, this may occur within a particular group member or for the group as a whole if it is required to adjust to new lines of authority and new prospects of accountability. When decisions alter such relationships, one may anticipate conflicts and tension, which are likely to be expressed in overt or covert resistance to the proposal being considered.

As suggested in one of the previous examples, people fear being perceived as inadequate or impotent. It is one thing to feel inadequate and impotent; we somehow adjust ourselves to this. But to have someone else suggest we are less than competent, less than able to do what is expected, is intolerable. It is often an illusion of impotence that reduces the desire to risk, to try a new approach. This is a primary force behind the "Oh, what's the use?" or the "That's been tried before" syndrome. It is based upon experience, unfulfilled dreams, and real feelings of inability to alter one's conditions. The attitude a group has toward its role as decision maker will be greatly influenced by the combined measure of its own potency and sense of adequacy to carry out decisions. But, again, we find ourselves in the murky realm of the unmeasurable. (At a later point, the discussion will be directed at ways of reducing this and other resistances that may hinder the problem-solving process.)

Groups, as individuals, build security by establishing standards and rules of behavior. They tend to value the traditional. They withdraw from tests of their own potency and rebel against outside intrusions that may throw their own stable and tranquil world into disequilibrium. These and other resistances are generated from the emotional dimension of group life and may have little to do with the actual "goodness of fit" of the intellectual idea being considered. It is by beginning to look within the labyrinth of emotions upon which any group is built that a person will find the keys to real movement and progress. The first step is to discover what state prevails within the group at a given time and to attempt to seek validation of these perceptions. By its nature this requires looking beyond the rational scope of the problem and into the rational and irrational perspectives of both the individual participants and the group as a whole (Heider, 1958).

It is evident that unless individuals feel personally secure and relatively unthreatened within the problem-solving group, they will tend to respond with their own characteristic patterns of defense. These behaviors can themselves be important sources of diagnostic information. The obvious withdrawal of individuals from participation, signs of passive or active aggression, subgrouping, or an excessive amount of dependency or resistance to authority will often suggest emotional issues that reduce effectiveness. As mentioned previously, the conflicts and resulting tensions synonymous with these behaviors can usually be traced back to one of four areas of concern. First, there are conflicts arising from personal goals and needs that are at variance with those of the group. Second, there are problems of personal identity and acceptance (membership issues). Third, there are problems generated from the distribution of power and influence. And, finally, there is the question of intimacy that encompasses issues of trust and personal openness. Virtually all the examples explored to this point could be incorporated within these basic categories.

As long as such issues exist, much of the group's energy will be directed toward self-oriented behaviors, and the accomplishment of the task will be disrupted. Naturally, it is impossible to remove all sources of personal tension (nor would it be desirable); but the more such problems can be raised and dealt with, the more attention can be given to the substantive issues.

Cognitive Dissonance

Decisions by their very nature suggest alternatives, argument, and conflict. Basically, a decision represents the termination of a controversy with a particular course of action. However, the conflict and tension do not end with the decision. How the individual or group copes with the doubts, suspicions, and skepticism generated during the discussion has important implications (Festinger, 1957; Festinger and Aronson, 1968). It is reasoned that the longer and stronger the discussion, the more ambivalence will be created (whether or not this is overtly recognized is another question) leading to a state of *cognitive dissonance.* This dissonance is a continuous source of tension, and the individual or group attempts to reduce it in a number of ways. For example, once the decision is made there is a tendency to begin valuing it even more than before (Brehm, 1956). As with a religious convert, many of the old questions and doubts are forgotten and the decision is constantly reinforced.

The trouble is that the decision may become overvalued, and an intransigent attitude develops that closes the door on future discussion of alternatives or even a fair evaluation of the decision at a later date. A subtle process of rationalization and justification may develop. This is particularly true if the decision turns sour and must still be lived with (a leader is chosen who proves inept, a tax is imposed to curtail inflation and unemployment increases, a surplus is guaranteed and a deficit is incurred). Similarly, people are prone to make the best of a bad thing when the decision is out of their control. Thus, even though it can be proved that most school grades lack objectivity and validity, students and parents alike will still defend them. Outwardly they may go through a long and involved defense of the system noting the value of the grading process, while inwardly the push for keeping the system lies in the more clouded area of, "I've suffered and so should you." This rationalization has nothing to do with the questions surrounding the legitimacy of grades. Still, such internal justifications tend to reduce dissonance and forge the major source of support for maintaining the status quo. Therefore, one needs to justify one's expenditure of time, energy, and hope in a decision that is a matter more of image than principle and which will shape future behaviors and accompanying decisions (Festinger and Aronson, 1968; Festinger and Carlsmith, 1959).

Advantages and Disadvantages of Group Problem Solving

By this time, it's clear that any problem solving implies the possibility of change and the exploration of alternatives to the status quo. Because of this point of potential discomfort, using a group can compound the difficulties of problem solving beyond those resulting from the volatility created by a roomful of different personalities with all of their individual needs, biases, and personal agendas. In addition:

1. The results of group decision making are often dismissed by those in positions of influence who are unwilling to give credence to a process few of them have experienced positively.
2. All too often, participants learn that what appears to be a fair, democratic process is in reality a charade in which decisions have already been made and group participation is provided as a means of placating those having to live with the ultimate decision.
3. Unless well designed, a group effort at problem solving can be a colossal waste of time, money, and effort.
4. Few group leaders are trained in the effective utilization of group members, which can result in a deterioration of both the process and task dimensions of group life.
5. As a result, instead of morale and team spirit improving because of a group approach, it may degenerate.
6. If the selection of group members is not related carefully to the task at hand, the technical and experiential components simply will not be available when needed.
7. It is common that members of a problem-solving group fail either to be briefed adequately prior to the meeting or fail to do the pre-meeting work that enables the group to establish a common point of beginning with clearly defined goals.
8. A few individuals can often take over a group and dominate its process or inhibit the participation of members whose contribution represents the reason for the group in the first place.

With so many possible pitfalls for groups, there appears to be justification for questioning their use. Assuming that groups are not always the best vehicle for problem solving, why and when should a group be utilized?

1. A group problem solving together will provide those participating a baseline of common understanding and information that cannot be replicated in a memo or less personal means. Such involvement inevitably results in a greater sympathy toward the complexities of the problem and provides the substance for the group's acceptance of the eventual solution. Increasingly, we are understanding how critical the acceptance of any solution is by those who must live with it. Thus, effective communication, understanding, and the eventual accepting of the solution are tied closely together.
2. As we've indicated previously, it is only natural and to be expected that individuals enter into a problem-solving situation with personal biases, ignorance, and misinformation. A group setting provides an environment that legitimizes a variety of viewpoints. Usually, provided that good information is translated in an

intelligent manner, a group, like any individual, will move toward the best ideas, assuming that a climate exists in which individuals are not compelled to defend their positions.
3. Given even a minimal level of trust and good will, a group is capable of producing greater quantity and variety of ideas than the average individual.
4. A good experience in a group can generate enthusiasm and can be contagious. The commitment toward eventual action can be born out of the teamwork, arguing, building of alternatives, and movement toward choice.
5. The give-and-take of open and free discussion among a group can spring new ideas into play that may have never been considered by an individual. Different from the formal presentation of new ideas in a group setting, open discussion enables group members to be irreverent, to question even the unquestionable, and to push against old absolutes. We are all aware of how rigid group norms can restrict discussion, and how those members considering new ways can be chided.
6. Problem solving is a multidimensional process. It is quite possible to involve large numbers of people (fifty, one hundred, five hundred, or more) during the problem identification, diagnostic (clarification) and ideational (generating alternatives) phases. As suggested earlier, people feel quite differently about the solution they have to live with if they have been given a fair opportunity to participate and are aware of all the factors underlying an issue.

Many problems can certainly be solved more easily, efficiently, and with less conflict and stress when done individually. On occasion, such individual problem solving can be justified because of the special expertise of the problem solver, because of limited time, or the nature of a crisis that might exist. However, when we realize that a solution, the decision itself, is in many ways only half the battle, then group participation makes increasing sense. In a society that cultivates the idea of participative democracy and raises its expectation, it behooves the sensitive leader to encourage the development of vehicles to allow this to happen. This is particularly true at a time when individuals often feel organizationally impotent and have diminished respect for leadership and bureaucracies. Individuals simply want to be heard, to have a piece of the action, and to have a sense that their organization reflects their ideas. It is more the knowledge that their ideas are considered than the demand that they had better be used. The passive hostility and dependency that results from nonparticipation is increasingly recognized as a price that affects everything from morale, to the quantity and quality of production, to down time and absenteeism. This is too heavy a price for many.

Open-Ended Versus Closed-Ended Problems

Before taking an in-depth look at a structured, rational problem-solving methodology, we shall explore a major reason problem solving often fails. The issue is one of attitude and perceived opportunity. Many of us, as individuals or as part of a group, enter a problem-solving situation with a predisposition toward the problem itself and the range of solutions open to us. Quite often, we establish what might be called premature boundary conditions for the problems that by their very nature restrict our ability to see creative alternatives. For years in school, we were drilled to discover the single correct answer previously established by the teacher. Thus, problem solving became a search for the answer the teacher desired. That simple answer point of view immediately removed the challenge of discovering the most creative alternative. Instead, the task became the name of "giving them what they want." Everyone has known the irrascible Jimmy Bradock, the kid who always asked the extra questions, who wanted to know the "why" behind everything. For most of us he was a predictable annoyance, especially to the teacher. It was Jimmy who would have the nerve to discover another route to a predetermined math problem, forcing the teacher to think beyond the normal boundaries to the solution. It is a wonder what drove Jimmy Bradock, what kept him going in the face of both student and teacher hostility. Today as a scientist and inventor he is still asking—rare indeed.

In addition to becoming dependent on what Rickards (1974, p. 10) calls the "defining authorities" for the boundaries of solution and the accompanying reluctance to challenge those boundaries, many people feel impotent within their institution and, without real power, feel it does little good to look for creative alternatives. More likely than not the individual suspects that additional effort will result in embarrassment or rejection. Thus, there is little perceived payoff for extending oneself.

The fact is that there are few problems that are closed and not open to a variety of creative solutions. It is the job of problem solving to help draw out the best alternatives and to break down artificial barriers, including those that psychologically bound the problem.

Groups themselves, when composed of a variety of individuals, often provide the different perspectives necessary to push boundaries away. Most of us have experienced the situation when our own close proximity to a problem reduces our ability to see the logical or creative solution. Marriages are particularly prone to this, with predictable patterns of behavior between both individuals establishing a very limited vision of what is possible. One purpose of therapy is to reduce boundaries and help us reframe problems (Watzlawick, 1978, p. 117).

In one situation, the management of a steel manufacturing company took the revolutionary first step of involving a group of union line workers in the problem solving of companywide problems. Their enthusiasm and different perspective ushered in a variety of solutions that had originally been artificially bounded by management because of their own narrow experience. Not only this, but the union workers developed a much greater respect for the complexities of other problems they had failed to understand, never having stood in the shoes of management. Their sympathy increased in relation to their new involvement and responsibility.

Thus, most problems are open-ended, except for such things as math equations, puzzles, and a variety of controlled scientific problems. The great majority are not so restricted or governed by theoretical tests. Rather, they tend to be more variable in scope, with relatively fluid boundaries controlled more by our own habits, inflexibility, limited experience, and personal agendas or vested interests. Productivity on a production line, a communications problem, the need for a new invention, marketing strategies, selling, public relations, or interpersonal conflicts are often approached from a limited perspective, thus resulting in limited solutions. Again, effective problem solving, whether structured and quite rational or more intuitive, is to a large degree dependent upon our ability to open ourselves to all of the possibilities.

The Right/Left Hemisphere View of the World[1]

In current scientific language, many of the behaviors conceived as successful in problem solving would be classified as originating in the left hemisphere of the mind, that part of the brain where most logical and rational thought takes place. It is here that intellectual ideas are translated in a linear fashion into words and eventually articulated through speech. But, in recent years scientists have dis-

[1]This discussion is drawn in part from the thoughtful studies and application of research by such writers as: Richard Bandler and John Grinder, 1979; Henry Mintzberg, 1976; Robert Ornstein, 1975; Paul Watzlawick, 1978; and Benjamin Young, 1979. They are exploring previously uncharted territory in the related areas of thinking, problem solving, and change.

covered that the right hemisphere of the brain is responsible for another type of thinking crucial to problem solving, although less valued for its own sake. Here, the brain processes emotional cues, nonverbal behaviors, and visual clues that add significantly to our baseline of knowledge and understanding of a problem. We are talking about the intuitive domain, the area of the irrational, illogical, and often spontaneous reactions to an event that often move us to act without much apparent thought. The question is not whether we use right hemisphere thinking, but how much and in what ways does it interface with that of the left? Some believe that, like anything else that has not been valued, rewarded, or supported in its cultivation, it remains an area of tremendous unutilized potential. One school of thought is that although we spend much of our time acting as if we are rational, objective, and in control of our problem-solving faculties, we are in fact driven to solutions all too often by subtle influences of the right hemisphere by feelings, emotions, and nonverbal information we pick up and translate as if by osmosis. We then justify the eventual decision based on rational, scientific arguments, giving no credence to the often critical right hemisphere influences. It is not uncommon for some leaders to make an intuitive decision and then scurry around finding the logical reasons to support it. An extreme but true example follows:

The eight executives were called together for a critical, two-day problem-solving meeting designed to provide the most logical and systematic approach possible. The eventual decision resting on the shoulders of this group and its leader was whether or not to build a nuclear power plant at a cost of over one billion dollars. The group represented legal, accounting, and engineering expertise and was led through a complex and thorough problem-solving process whose structure was designed to weed out the irrational and to focus on the data-based realities. Order and objectivity rather than emotions and personal bias were stressed.

Thus, they dug in, looking carefully at constituency needs, costs, factors influencing efficiency, environmental factors, financial considerations, management problems, training, lag time, and maintenance issues. Each question was systematically attacked and conclusions drawn. After two days of careful, highly controlled and, for the most part, unemotional assessment, the members of the group gave the president their recommendation, which was unanimous. It was decided that it would be more economical in the long run to utilize coal as a generating alternative to nuclear power. A major consideration was the social ramifications underlying the nuclear plan.

The president took their recommendations with thanks and went home for the weekend to make some last-minute deliberations and draft a formal proposal based on the recommendation.

On Monday, the statement was passed among the top executives. In their total amazement, the proposal supported the decision for nuclear power. The recommendation incorporated a number of vague generalities and justifications based on their need to be involved with innovation, to be leaders in the field. All of the rational problem solving had come down to a basic, gut-level feeling on the part of the president. Not only this, but the group now perceived that he had made his decision months before and that the formal problem-solving session was a final search for justification and legitimacy. Thus, it was an affirmation of his personal, intuitive decision that perpetrated the charade—a speaking of left hemisphere rationality for right hemisphere feelings and knowledge.

Increasingly, management is referred to as an art at the upper levels of business, while at lower levels, rationality and the scientific method are still stressed. There appears to be a growing recognition that any complex decision goes far beyond rational data, far beyond what can be formally organized and tabulated, and that the underlying, nonquantifiable issues such as morale relationships, individual egos, power, competition, and what is generally called "politics" are of equal importance. These are the areas translated by the right hemisphere and forged into the decision-making process, seldom overtly recognized as valued. In a similar fashion, many group leaders wish and act as if groups are rational. They simply are not. Efforts to look at the process domain, at the maintenance side of group life, provide a legitimate avenue into some of the underlying issues of the right hemisphere that need to be translated into a usable language for the group. The more this occurs, the more understanding will occur and the more alternatives for constructive action will be open to the group itself.

RATIONAL PROBLEM SOLVING: A LEFT-HEMISPHERE FOCUS

Most of us have cut our teeth on the scientific method. From our earliest years in school, teachers have been trying to organize disorganized thought patterns into what might be conceived as a rational pattern of thinking. Hypothesis testing, gathering empirical data, defining, clarifying, establishing criteria, measuring, evaluating—

most of us have squirmed and floundered as we have tried to apply the scientific point of view so carefully laid before us. The question that is difficult to answer is that for anything so logical, so straightforward and so necessary, why do we have such a tendency not to use it, to forget about it, to avoid it at all costs? Perhaps it is tied in our minds too closely to unappetizing teachers or courses, based on irrelevant examples. Or perhaps, like American history, it was recycled, repeated, and compulsively presented to us so often that we somehow became allergic to it. But, we ask you to take one more look, to view rational problem solving as an ultimately helpful means of organizing thoughts rather than as the imposition of an inconvenient necessity.

The fact is that the scientific method used to be looked at as an end in itself. The key was to try and isolate a problem, look at it impartially and objectively and then develop certain prescriptive solutions. This controlled, laboratory view of problem solving was translated into the world of everyday problem solving, and somehow it was found to come up short more often than not. Increasingly, we discovered (Gouran, 1972, pp. 23–27) that even the simplest organizational or group problem resulted in a variety of consequences ranging far beyond what appeared to be the original scope of the problem. Instead of something concrete and easily definable, the water of the scientific method would continue to be muddied by unpredictable variables such as personalities, organizational structures, protocol, informal norms, historical realities, fear and issues of territoriality. These and any of a hundred other seemingly insignificant factors could blow our rational decisions out of the proverbial water. Thus, from what seemed to be a rather simple and straightforward process, rational problem solving has grown, in some instances, into a nightmare of complexity—something so torturous to use that few people have the time, energy, compulsion, or know-how to do it right. Our purpose here is to provide you first with a general overview of rational problem solving, which integrates a small part of systems theory, a dash of communications theory, a bit of group dynamics, and a tad of organizational development. We will then provide a more comprehensive model that elaborates the more general perspective and its potential for use in a group setting.

The Six Stages of Rational Problem Solving

For the most part, the steps one moves through in solving a problem are quite simple. First, there is the identification and clarification of the issue, a developing of alternatives, a selection of one or more of these, and, finally, an implementation phase followed by an evaluation of the outcomes. It is a wonder, then, that a process so straightforward and so lacking in complexity can result in so many problems and pitfalls. The issue, of course, is that people initiate the process and set their own traps and barriers. Groups are no less susceptible to these problems than are individuals, and it would seem profitable to spend some time looking at them. (See the exercises at the end of this chapter for examples of how to implement these six stages of problem solving.)

Stage 1: Problem Identification The recognition that a problem exists can happen either by chance or as a result of systematic inquiry. More often than not it seems that problems arise naturally and announce their presence through increasing tension and conflict or, perhaps, inefficiency. Conditions will worsen if the presence of such tensions is not confronted, or if they are denied or covered over so that accompanying frustrations become a breeding ground for other problems. This is too often the case in groups where a little internal festering is somehow preferred to dealing directly with the issues as they arise. In some cases, there is simply no mechanism available to help bring the problems into the open. Something as simple as a suggestion box (if there is evidence that it is being used) can be a direct line to sources of individual, group, or organizational problems. Once recognized, it is then important to discover the degree to which they are shared by others, as well as the level of urgency. Occasional questionnaires or small group discussions can be helpful in drawing problems into the open before they become destructive. Such problem sensing of both task and emotional issues can help keep communication channels open. Other problems will arise, however, if the group or individuals are encouraged to identify specific problems, which are then avoided or minimized by those in positions of influence.

Stage 2: The Diagnostic Phase Once the symptoms have been recognized and brought to the attention of others, several steps seem to follow quite naturally. First, the problem must be clarified and relationships identified. Too often the symptoms are little more than a generalized recognition of discomfort or stress and tell little of the underlying factors creating the disturbance. At this point it must be discovered how much the problem is shared by others as well as its degree of urgency. A second step in the diagnostic process is to gather supporting evidence as to the nature of the problem. Third, with this new information, the problem should be restated in terms of a condition that exists and which, to some extent, needs to be changed.

Quite often problems are stated simplistically in relation to an "either-or" situation or in terms of "good" or "bad," which immediately polarizes the potential problem solvers into win-or-lose camps. If a condition can be shown to exist that is less than optimal, then the problem of the eventual decision-making group becomes one of identifying the factors that keep the condition from being optimal. Energy can then be directed toward isolating specific causal factors, such as a single person dominating the discussion, lack of time, or the need for clear goals. Thus, arguments become limited to the relative strength of such factors and not to whether they exist. This approach encourages compromise and multiple solutions. Finally, having gathered as much data as possible concerning the problem, stated it as a condition to be changed, and isolated the various causal factors, a determination must be made regarding how capable the group is to solve the problem. This involves looking squarely at the group's own power to influence the prevailing condition, what kinds of resources are going to be necessary, and how much impact their efforts will have on others. Nothing is more frustrating and deflating than for a group to design a clever scheme for solving a problem only to realize that it lacks the resources to carry out the plan. Therefore, before developing solutions, the group must test its own reality situation. The most important finding may be that because of certain limits (time, money, personnel, access to power) the problem should be stated more realistically, others should be drawn into the problem-solving process, or the issue should be directed to another group that does have the potential for solving the problem.

One final point should be made concerning this early stage of problem solving. For change to occur, those involved must see the problem as "their own." It cannot be imposed upon them. Thus, the diagnostic process is vital for involving those who will eventually be responsible for implementing the solutions. This has important implications concerning who takes part in the diagnostic process and which people are kept closely informed as to what is happening. Developing solutions will prove to be nothing more than an academic exercise if those to be affected have not even come to the point of admitting that a problem exists. It does little good for a doctor to prescribe specific therapeutic treatment for a patient if the patient believes himself or herself to be well. It is crucial that an individual or a group comes to accept the fact that there is a problem. This suggests implicitly a process in which the determination is not imposed, but evolves among all concerned.

Stage 3: Generating Alternatives Groups and individuals seek quick and easy solutions. It is one reason why the problem-solving process so often breaks down. There is a tendency to combine two distinct phases that can have important implications for the quality of the potential decision. Once a problem is identified, the usual reaction is to jump toward a logical solution. As we fasten onto what we perceive as a logical and ultimately resourceful solution, we automatically screen out numerous other possibilities, some of which may (difficult as it is for us to believe) be more appropriate. We commit ourselves to one idea and are then compelled to defend it. This may be particularly true in a group where some of us have a need to convince others of our wisdom and skill. Formulating solutions before ideas have been thoroughly explored not only reduces the potential quality of the eventual solution but also tends to inhibit open communication. It has the same effect as stating a problem in either-or terms. It forces individuals into a premature position of evaluation and places all members in defensive postures. Thus it is a major pitfall to evaluate solutions at a time when the intent should be merely to explore every potential solution that is possible. Done effectively, this process can reduce the tendency for groups to polarize around answers that are comfortable, and it may also help them to look toward new approaches.

After the ideas have been generated and explored in relation to specific causal factors (isolated during the diagnostic stage), then there should be a general screening process to integrate and synthesize the solutions into a smaller number. Again, the effort here is not to select a "best" solution, since the problem is likely to be multifaceted with a number of possible alternatives. Before any final decision is reached, a period of weighing and testing of the alternatives should be initiated. If time and resources allow, an effort should be made to gather data about the various solutions reached up to that point. This could range from establishing a pilot study to seeking the opinions of other individuals, such as experts.

Stage 4: Selecting Solutions With the new data and time to think about the alternatives, it is ideal to consider the consequences of each alternative in relation to the problem condition. Many times a group, anxious to get under way, will fail to explore the unanticipated consequences and focus only on the obvious benefits to be gained. Thus, it is at this stage that each potential solution should be carefully evaluated in terms of its possible limitations as well as strengths. The discussion should lead to decision by consensus, in which all members are willing to support a particular plan. Although this does not assume complete agreement on the part of all participants involved, it suggests at least a temporary accord during a period when the decision can be fairly evaluated. There are, of course, times when decision by consensus is impossible, but when effective implementation of the decision is based partly upon support of those involved, consensus has important advantages.

Stage 5: Implementation Many participants of decision-making groups, after being successful in developing a useful decision, have watched helplessly as the ideas so carefully designed and agreed upon are never implemented. Part of the problem often can be traced to the early stage of the process and the failure to involve or at least keep informed: (*a*) those with power to kill the idea and (*b*) those who would eventually be influenced by the final decision. Equally important is the failure to build accountability into the action or implementation phase. Too often, interest is not developed in the decision-making group. Accountability must be carefully cultivated so that individuals feel responsible for the outcome and are answerable to the others involved.

Stage 6: Evaluation and Adjustment One reason people are resistant to new ideas is that they believe that once change occurs, it will be just as impervious to change as was the previous idea. By building in a mechanism of evaluation as well as the flexibility to make adjustments once the data are analyzed, the entire problem-solving process remains flexible and open to new alternatives. Most important, it gives those who are being influenced by the decision the recourse to alternative procedures and a feeling of some potency in the process. Also, the notion of accountability is tied directly into the evaluation-adjustment procedure. Thus evaluation becomes more than a superficial exercise and tends to be used as an integral part of an ongoing problem-solving process.

The basic approach to problem solving outlined here is very reasonable, a natural falling out of steps that make sense. Rational problem solving is designed to keep us making sense, to protect ourselves from irrational biases and from limiting customs and habits, to expand our vision, to look at the consequences of our choices, and to make sure that ultimate decisions result in constructive action. Whether we work in the context of a group or as individuals, the same principles prevail. For those of us who spend our lives attempting to facilitate the problem-solving process, the key is asking the tough questions, not allowing our tendency to avoid pressures of reality (for example, limited time, habit, bias, limited experience) to take over and corrupt our positive efforts. Good intentions in problem solving don't count at all. A satisfactory outcome is the only measure. The orderly, stepwise process that follows is designed to draw the participant(s) along in a systematic manner. It takes time and should not be done in a hurry. Quick-and-dirty problem solving inevitably leads to quick-and-dirty solutions that simply mirror preconceived notions we wish to sell. Similarly, in our haste we fail to assess all of the viable alternatives or the consequences of those alternatives. It is rare that we allow ourselves the privilege of time to do the job right.

It should be noted as an aside, one reason we fail to put the proper time and energy into problem solving is that problem solving forces us to admit that there is a problem in the first place, that we are less than perfect, and that change is very likely to be called for. These are difficult pills to swallow. Change is usually uncomfortable at best and painful at worst. Who needs it? With that in mind, let's take an excursion into rational problem solving.

READER ACTIVITY

We all carry around an enormous set of problems. The world is simply too complex for them not to be a part of our lives. There are always problems within our relationships; or with how we utilize our time; or with a special project; or with not liking our job, or our boss, or both; or perhaps being dissatisfied with our life style. Most of these problems are open-ended from the point of view of the potential for change. Whether we personally are open to some of the possible solutions is quite another question. Our goal here is to open a problem you may have to thorough understanding and then to as many alternatives as you can discover. Whether you decide that change is not worth the price, or choose a conservative or even radical alternative is up to you.

Decide on three problems that somehow play an important part in your life—problems that have been with you a while and you would like to resolve. In this instance, a problem is defined as a state or condition in your life that you believe probably requires changing. Now, think of a friend you respect and trust and who knows you rather well. Ask the person to take an evening with you to problem solve one of the conditions you have indicated. Promise to do the same for that individual if that person believes the process is worthwhile. The purpose of the friend's involvement is to provide you with a different perspective, to help turn over stones that you might not see, to push you further, and to ask tough questions you might not be willing to ask. The friend will represent another side of reality.

If, on the other hand, you are part of a small group with a common problem, the process will also work. Since a group represents more ideas, concerns, and points of view, a bit more time is required to do justice to the problem-solving process.

Now, follow the problem-solving process outlined below a step at a time. Don't hurry. Probe each explicit question and then ask yourself other questions that naturally come to mind. Your goal is a better way. But, a warning. For the process to work, it cannot just be another activity. Problem solving is serious business, because it should influence you and your life. One reason we fail to take such activities seriously is that many of us have spent time doing just this with inconsequential problems. If there is nothing important you wish to explore and enjoy the stimulation of the search, then move on and read through the process with the thought that it may prove valuable at a later time.

A Model for Rational Problem Solving

Step 1. *Make a general description of the problem condition as you see it.* What is it that seems to be at the crux of the problem, how does it influence you? Where is the rub? Talk the problem over in general terms, trying to outline the parameters.

Step 2. *Describe what the defined condition would be like in an ideal but reachable state.* Here we are trying to establish a sense of the changes that would have to occur by looking hypothetically at, for example, how production operations in a factory might differ, how the attitudes of people might change in certain working relationships, how discipline in a classroom might change, how a group might solve problems differently. Again, it is important to talk the ideal condition over and obtain a feel for it. This in itself will often help sharpen the focus of the real problem.

Step 3. *Identify the specific discrepancies that exist between the present view of reality (Step 1) and the ideal state (Step 2).* The problem(s) should begin to take on a different shape as a result of this analysis.

Step 4. *Analyze the nature of the condition more thoroughly.* Do this by asking a series of critical questions.

1. Does there appear to be more than one problem existing, each of which warrants individual attention? Although the relationships between certain problems must be recognized, the more concretely a problem can be defined, the less difficult will be the task of problem resolution.
2. What benefits does the present condition hold for the individual, group, or organization that is defining it as a problem? One reason

problems don't just disappear is that there are very real satisfactions that will have to be given up. Consider for just a moment the benefits the smoker who "wants" to quit must give up, the benefits the obese individual will have to sacrifice, the benefits that accrue to the person with the volatile temper who would "really like to stop blowing up all the time," the benefits that come to the "talker" even though the individual realizes his or her talking alienates some people, the benefits to a group that constantly complains about starting a meeting late or about members arriving late. Until a group or an individual is willing to look such benefits squarely in the face, the chance of significant change occurring will be slight. If these benefits are not replaced, it might be crazy to change even though we'd be the last to admit it.

3. What are the blockages that have been thrown up in the face of previous attempts at change? Underlying a blockage may be a hidden benefit that subtly supports the existence of the status quo.
4. Finally, what are the present solutions that are currently being attempted albeit unsuccessfully? By taking a hard look at our unsuccessful efforts, we often gain a more clear understanding of the problem itself. For example, one supervisor would procrastinate in getting back corrected work reviews to her subordinate, who often needed the information but would "simply make do." The subordinate hated conflict and couldn't confront his boss. Not only this, but he couldn't stand the feeling of being a nag. Thus, instead of mentioning his need, the subordinate's inclination was to withdraw from the problem and never face up to the difficulties being created. The boss, on the other hand, was being subtly reinforced (not being accountable meant not having to do the work). Thus, part of the problem was being created by the subordinate's attitude to the problem itself.

Step 5. Now in light of all the new information about the problem condition, redefine it as clearly and succinctly as possible. Again, it is not negative to discover that there are several problems. But, for our purposes it is necessary to isolate the one that is most important to solve and that might have greatest impact on other existing conditions. By selecting the problem that can be solved and that might have a positive ripple effect, one can assume that the time will be put to good use. Several examples of clear, succinct problem conditions follow. Note that a problem condition simply describes a state that needs changing. There is no implication of good or bad and no implied solution.

1. The present level of shared participation in our meetings.
2. My ability to state my personal opinions in a group.

3. The present level of productivity by Team A.
4. The present level of absenteeism in this department.
5. My ability to assert myself in conflict situations in which I wish to make my point.

Step 6. Without considering the implications of a particular solution, generate as many alternatives as possible. Potential solutions might result from reflection back on any of the previous steps. The key in this stage is not to worry about implementation or consequences, but simply to develop real, concrete choices that presently are not available to you.

Step 7. Screen the various alternatives by changing them into specific objectives which by their nature suggest direction, quantity, and where and when they will occur. Also make an effort to determine which of the resulting objectives will have the greatest impact with the least cost to you or the organization, and which, for whatever reason, seem impractical. This is a preliminary screening step with more occurring later. Taking the first condition used as an example in Step 5—the present level of shared participation in meetings—a number of objectives might include:

1. To provide each participant with ten chips at the beginning of each meeting. They must give up a chip each time they speak. Once the chips are gone there is no opportunity to speak again unless more chips are negotiated from other members.
2. To establish the rule that during periods of discussion an individual can make a second point only after each other person in the group has been given the opportunity to speak.
3. To have a person appointed prior to each meeting the role of "participant observer" and to point out through various means (for example, a chart of how often each person speaks) how effectively the group is communicating.

An objective for the fourth condition—the present level of absenteeism—might perhaps be to use the last six months as a baseline for absenteeism and to provide a paid day off every two weeks when team members have, as a group, averaged 30 percent below that baseline for a month. Recipients of the paid day off would be rotated throughout the team over time.

The most effective objectives are those that are specific enough to be measured in some manner. At this point we are less concerned about practicality than we are with clarity and specificity. Once clarity has been assured, then other considerations relating to the value of a particular objective can be discussed.

Although you may not necessarily agree with the objective for attacking absenteeism, it is clear, specific in intent, provides information as to what would occur, when, to whom, and under what conditions—and that is our aim during Step 7.

Step 8. Consider the consequences—the price to be paid—the impact on the individual or the organization or group if each of the selected objectives were to be implemented. Then decide whether to alter the objective, either to improve its effectiveness or to reduce the negative consequences that will result. This hard-nosed step of looking at consequences is often overlooked because of the enthusiasm and blush of success that often surrounds the generation of solutions. As we showed previously, there are numerous attempts at change and many reasons for their failure. Thus, it may be decided that the passing out of chips as individuals talked would initially be seen as fun, but later would be resented as a game and tossed off as impractical. The idea, however, to have a process observer with a legitimate role of keeping participation high and communication open seems not only appropriate but feasible. Even this idea has its limitations, however, because most observers must be trained, given time in the group, and supported.

Thus, this step is down-to-earth and ultimately practical. Its purpose is to make an objective workable or to discard it. Questions need to be raised, depending on the particular condition being changed and the nature of the objective, about such issues as:

motivating people to accept a particular idea

insuring the necessary skills to facilitate success

overcoming cost factors

educating people to the value of a new idea

exploring issues of timing and the pace of implementation of a new idea

overcoming previous failures as well as the we've-done-it-before-syndrome

These and any number of other factors could render a potentially good idea, as translated into an objective, ineffective. This step, then, explores not only consequences but the strategies necessary to overcome potential resistances and to polish the objective into something that will work.

Step 9. Monitor and develop appropriate support systems to insure the stabilizing of most change efforts. The distance between the full cup and the lip can be long indeed. So too, the easiest part of much of problem solving is in generating alternatives. Getting those alternatives into action often proves to be impossible. Looking at consequences and building support strategies in Step 8 will prove helpful. But, equally important will be establishing the means of effective accountability. The reason the good intentions of a New Year's resolution seldom have any payoff is that there is no accountability built into the process, not to mention the problem of consequences that we aren't about to consider at the time we commit ourselves. Increasingly, we find that for change to work, groups and individuals need to experience the line of tension created between commitment or promise to act and some point in the future when results are assessed. Used in this light, monitoring can be utilized as a means of support and development rather than as a means of punishment and control. Monitoring suggests adjusting and adapting a process to insure success as much as anything. Thus, it should occur early enough to be motivating and helpful, before mistakes are made that cannot be corrected and prior to the setting in of guilt.

Support systems are vital to most monitoring processes. For example, legitimizing the talking about the process of change with people who have been through it before (AA, Weight Watchers, new students, ex-convicts), or with people who are undergoing similar experiences in the present can be gratifying, reassuring, and profitable, because it is helpful to learn from others' experience. In addition, the making public of one's own objectives tends to have a positive impact as one's commitment to act is supported by others who now expect action to occur and will reinforce it.

Step 10. Evaluate problem-solving efforts to decide what steps should be taken next. There are several ways in which a relatively simple evaluation can occur. First, at a designated time in the future, assess the degree to which the discrepancies between the present situation and the ideal have increased or decreased from the period of original assessment. A second approach is to take the objectives established in Step 7 and compare them to specific outcomes.

It is at this point that further problem solving can occur. It is ill-advised to consider problem solving, as many do, a one-shot operation. Most problem solving is a jerky, inconsistent process that results in success and failure and, one hopes, an overall sense of accomplishment. Taken a step at a time with the willingness to continue looking at the process of change, there is a good chance that the natural mistakes that inevitably occur and the natural resistance that accompany virtually any efforts at change will disappear and we will be left with the weight of a credit balance.

Some of you may be wondering how a simple problem became so complicated. It is we who are complex, along with the multitude of factors that impinge on us and make simple decisions seem terribly difficult. The problem-solving process in which you have been wandering about is, in fact, rather simple compared to some (Easton, 1976; Kepner and Tregoe, 1968). The problem is that it is rare for individuals to have the time, patience, endurance, or courage to expend the kind of energy required to tackle some of these creative and technically sound approaches to problem solving and decision making. Furthermore, the issue is complicated by raising the question of who should be present during the problem solving. Technically, sound and highly structured problem solving can cost thousands of hours of time; and with many problems, it is difficult to measure whether a complex or more simplified process is just as useful. Perhaps it is most important to help people internalize a tough-minded view of problem solving that allows them to utilize an appropriate question or problem-solving vehicle without requiring adherance to a cumbersome and time-consuming procedure except in special situations.[2]

Ordering Information for Rational Problem Solving

As you can see from the problem-solving model that we have just presented, rational problem solving is keyed to the digging out and systematic ordering of all of the factors that might impinge upon a problem. It is a process that forces the participants to at least focus on critical issues of cause and consequence. Thus, any logical way that the ordering of information can be facilitated should also prove helpful to the problem solver. Something as simple as a 2×2 grid can not only provide a basis for generating certain types of information, but can also focus our attention on relationships critical to the particular problem we are investigating. In the previous example of absenteeism, we might decide that additional information was required to facilitate the diagnostic phase of the problem solving. Asking various questions of all line workers across three shifts and across three working teams on the first shift, we can develop a simple grid to help order the data in the following manner.

[2]For an alternative strategy for individual or group problem solving refer to the activity at the end of this chapter utilizing Force Field Analysis (Lewin, 1948, p. 51).

Present Level of Absenteeism: Contributing Factors

Contributing Factors	1st Shift	2nd Shift	3rd Shift	Male	Female	Older	Younger
Low pay							
Physically tired							
Environmental stress							
Boredom							
Dead-end job							
Dislike of boss							
Too much pay							

Clearly, the grid can aid the organization of thinking and help in both the building and testing of hypotheses. It also lends increasing objectivity to what can easily be a process of whim or personal bias. Placing a numerical value between 1 and 10 in each box would allow comparison across groups.

Sorting, classifying, and building priorities through an orderly process of structuring available information is not difficult to do, given the willingness to take the time to do it. Without this effort, both the individual and the group tend to fall back to selective perception, personal bias, and conjecture. The trouble is that such a rationally sound approach will tend to make it more difficult to respond as we often do—according to what *feels* right. In some ways, it can make the decision-making process much less comfortable and more difficult.

Let's take another problem and explore an alternate means of bringing order and hoped-for rationality to the problem-solving process. In this instance, the problem is the present level of supervisory effectiveness between first-line supervisors and the foreman who report to them. One means of gaining a better understanding of the nature of the problem is to break it down categorically and to determine just how healthy each of the parts is and what factors influence each of them. Following is a list of possible categories drawn from this problem statement that might influence the quality of supervision within a particular manufacturing organization.

1. *Role clarity* (lack of clear definition, general rather than specific objectives, overlapping roles among peers, responsibility without authority)
2. *Performance review* (too casual, irregular, not constructive, not specific, not developmental, tends to be focused on the negative)
3. *Time utilization* (work is too sporadic, not consistent, often subordinates feel underutilized, too much time devoted to non-job-related tasks)
4. *Skill development* (task related skills developed through observation)
5. *Career opportunities* (few career path opportunities for supervisors, fewer still for foremen)
6. *Interpersonal relations* (supervisor's role seen as formal [last name basis], tends to be top down, emphasis on talk when there is a problem, negative)
7. *Interest—motivation* (routine, limited rewards, limited involvement, foremen seen as policemen, few "extra-work" activities)

The list of seven areas thought to be crucial to a supervisory relationship was generated in a structured discussion of supervisors who had felt tension developing over time with their foremen. The group was then simply asked to probe each of the catgories for areas of tension they felt the foremen might consider most important. There was no effort in this process of problem clarification to worry about whether the individual categories were totally independent of one another or whether they even made sense. The key was to begin organizing the general problem statements into more specific areas that might lend themselves to work.

The next step was to look for patterns across the various categories. From this effort at categorizing and clarification, there emerged three problems that the supervisors thought warranted their attention. They also became aware that representative foremen should be involved with them at least in the diagnostic phase and perhaps in the generation of alternatives. By asking them to respond as the foremen might, the supervisors would be forced to take a step back from their own involvement and could legitimately begin to put themselves in the shoes of their subordinates. The three problem areas that evolved were:

1. Performance review
   a. need for role clarity
   b. personnel utilization
   c. regularity
2. Career development
   a. developmental focus
   b. job enrichment
3. Motivation
   a. issues of authority
   b. management style
   c. nonmonetary rewards

This attempt to reorganize the problem as originally stated resulted in the participants' having a much better opportunity to develop effective solutions because implied solutions almost fell out of their new problem statements. In one sense, they had developed three clear hypotheses that could be easily verified by collecting data from their subordinates by means of anonymous questionnaires or even interviews. The simple clarification activity put them much closer to some real solutions than they had been previously.

Any person of average intelligence is capable of effective problem solving. Even though we presented a rather formal view of one problem-solving approach, the key is not to become too inflexible. Your job is to help yourself and the group to:

develop a clear, succinct, and workable definition of the problem.

take the necessary time required to do the job.

make sure the right resources are available at the various stages of problem solving.

not jump prematurely to solutions, but to grasp all the critical factors clearly.

generate a number of alternatives before developing final strategies for solution.

look carefully at the consequences of each potential solution.

be sure there is necessary follow-up and monitoring at the implementation phase.

Basically, if these issues are considered, the chances of success will be much higher. There are few rules of how actually to organize the information available, catagorize the information, set priorities, weigh alternatives, or consider consequences. It is for problem solvers to develop the necessary procedures as they feel appropriate. Problem solving can and should be creative, developmental, and fluid within the constraints of some relatively hard questions.

INTUITIVE PROBLEM SOLVING: A RIGHT-HEMISPHERE FOCUS

The only problem with an orderly, systematic, linear approach to problem solving, which has been our bent to this point, is that it encourages

restrictive rather than expansive thinking.

old, tried-and-true approaches rather than original and creative ones.

logical rather than illogical thought.

rigidity rather than flexibility.

Many of us would like to believe that the product of a rational approach will be the most reasonable, appropriate, and qualitatively best response possible. But, anyone who has experienced serious problem solving realizes the truth that many of the most creative decisions of greatest excellence result from some unexpected thought, from an aside said in jest, from a moment when defenses were down, or at a point of exhaustion, frustration, or exasperation that had never been programmed or anticipated. Finding sudden insight or wisdom under a stone left previously unturned or never even considered leaves the thoughtful individual joyful and perhaps a bit humble. We are humble in the understanding of just how many good answers never see the light of day because of our own inability to tap into the pool of alternatives lying in our right hemispheres, which are somehow blocked from reaching the left and our cognitive awareness.

Thus, a key to effective problem solving is not just providing order and a tough-minded approach to viewing causes and resulting consequences. It also involves bringing to the surface as many of the solutions that exist as possible. This means overcoming our personal predispositions, defenses, and habits so we really do have as many choices as possible. So it is that serious problem solvers are forever looking for ways of becoming "unstuck," of taking a new and different look, of redefining the problem in a manner that may provide a new perspective and freedom to alternatives not yet accessible to them.

Becoming "Unstuck"

Opening the floodgate to new ideas may occur in many ways, some of which are created through the turn of a question or by posing one that gives permission to look outside of the psychological boundaries we often impose subtly on a problem. One approach is simply to step back and take the time to redefine the problem as stated in a number of ways, using totally different words. Legitimizing new words and pushing ourselves to other definitions often uncover a useful approach to the solution, because many answers fall neatly out of a problem statement itself.

Another approach to uprooting one's mind-set is to discuss the problem in terms of analogies, thus forcing ourselves to think about a problem in terms of its similarity to other nonrelated situations or objects. Utilizing this approach is based on the assumption that if two things are similar in some respects, it is very likely that they are similar in others. A stubborn person standing in the way of progress, seemingly intransigent, may be likened to a boulder or a rooted tree stump. Pausing to consider ways of moving the boulder or removing the stump may open the discussion. Thus, pushing the boulder in the opposite direction from forward progress may free it and then allow rapid forward movement to be attained. Similarly, an individual may be holding onto a point of view simply because individuals have refused to recognize its legitimacy. Giving due credit is often difficult to do in the best of arguments, but may be the key to allowing the individual to give up the position. Saving face seems a long way from the pushing of a boulder, but may be similar indeed.

Another related approach is to have problem solvers think metaphorically. Thus, talking about old age and retirement conjures up fear and resistance in many individuals. For some it suggests the end, giving up, reduction of one's sense of potency, and dependency. But, if a discussion by people about to retire could be framed as a discussion of how to enjoy the "evening of life" or the freedom years, then resistances might be reduced. A twist, a turn of a word can open new meaning and new feelings and alter attitudes toward a particular reality.

A bit more perverse route is to have individuals take apparently very different situations and ideas and probe for similarities. Such an exercise again forces a relook at the givens, at the reality of a problem, which in turn opens the possibility of new insight.

Stopping a discussion and injecting one of the following open-ended statements may be all that is needed to discover new entrances into the problem:

This situation or problem is just like. . .

A different way to describe this is. . .

The only time anything like this happened before was. . .

This feels like a. . .

This situation reminds me of. . .

READER ACTIVITY

Consider any recent open-ended problem you or your group has had or still is concerned about. Now take at least two of the methods suggested and reconsider the problem as stated. Try to use different words to describe it, perhaps developing several definitions of the same problem. Utilize a metaphor or develop an analogy and try to uncover every possible similarity between the problem and that to which you are making the comparison. You will be surprised at the number of insights that result from being forced to stretch your thinking in this simple approach to problem assessment.

Brainstorming

The first real break away from strictly rational, linear, and highly controlled approaches to problem solving came nearly a half century ago when Alex Osborn introduced the concept of brainstorming. He discovered that by establishing a few simple rules and utilizing a limited amount of time in a different manner, he could dramatically alter the atmosphere in a problem-solving session and in his estimation create more and often better ideas than might otherwise occur. Needless to say, such innovation is always the source of controversy and the real value and significance of brainstorming has been a subject of heated debate for years. For our purposes, there is no question that this procedure opened the gate to more creative approaches to problem solving. Whether, for example, brainstorming groups are actually more productive than individuals working alone is a question that needs discussion. But, first, let's discuss the concept itself further.

Brainstorming is a tool designed to help individuals share their ideas without the interruption of discussion. By allowing the participants to associate freely and present any idea that comes to mind, it is believed that more ideas will be generated and that their quality will be better than if the same individuals worked independently. However, the climate previously established is not altered just by setting a few rules, for example, no evaluation or no discussion. For many people, brainstorming is a strange sort of experience, and it can create an initial sense of discomfort (Hammond and Goldman, 1961; Vroom et al., 1969; Collaros and Anderson, 1969). Consider an example.

Some time ago, a specialist in small-group behavior was asked to discuss brainstorming as a technique to approximately one hundred army officers at a college for career officers seeking promotions. He was met at the door of the auditorium by a colonel who informed him that he would feel right at home since things at the college were conducted in a very informal and casual atmosphere. The consultant wondered what he meant by that. The colonel said, "Oh, everyone here is on a first-name basis, ties aren't required, and we really have some wide-open discussions." The ultramodern lecture hall was arranged in three tiers. On the ground level were the students (mostly majors and colonels on the way up). A second tier held visiting dignitaries, nonpermanent staff, and those holding the rank of general. Finally, a third level contained the permanent college staff who passed judgment on the merits of the various students. It was in this "casual" and "informal" atmosphere that the lecture took place.

At one point the consultant wanted to loosen the group up a bit and involve them in the brainstorming process. He gave the officers the same warm-up example that he had given groups of high school and college students, sisters and priests, and a variety of other groups. The officers were asked to think of as many unusual uses as they could for a certain ladies' undergarment. They were given one minute to generate as many answers as possible—the wilder the idea, the better. When they were told to begin, all you could hear was a restless shuffling of feet by those on the ground level. Finally, one tough-looking major, risking his potential two-star rank, shouted, "basketball knee guards." The immediate laughter (and subsequent release of tension) was enormous, and slowly but surely the group squeezed out its self-conscious replies. At the end of a minute, which seemed like a year, there were eleven replies—not very many for an informal group accustomed to some "wide-open discussions." It should be noted that not a single idea was offered by the upper two tiers as they peered down at the students. Things, however, did loosen up a bit when the group was told that the record of thirty-five responses was held by a group of nuns for whom "promotion" was no real problem.

Thus, to a relaxed group familiar with the process, brainstorming may be a stimulating and useful approach to generating ideas. To a restricted, self-conscious group, however, it could actually prove a hindrance, since it forces members into new patterns of behavior and breaks certain norms that usually protect the participants (Bouchard, 1969; Bergum and Lehr, 1963). Conversely, the use of brainstorming can break open a stuffy and inhibited group, if used at the right time. Much depends on the facilitator's ability to read the behavioral cues of the group effectively and the group's familiarity with the ground rules of brainstorming, which are listed below.

1. Criticism or evaluation of an idea is not allowed. Ideas are simply placed before the group as rapidly as possible without any discussion, clarification, or comment.
2. The session, which can last anywhere from a minute to perhaps fifteen, is to be freewheeling and open to all ideas. The wilder and more fantastic or absurd the better, and the more chance there is to begin to draw from the pool of ideas that might provide a new way of thinking or a breakthrough.
3. Quantity is very important; thus, all ideas should be expressed and not screened by the individual—opportunity for this will come later in an integration and synthesis phase of the process.
4. Everyone participating should be encouraged to build freely onto the ideas of others so that thoughts are expanded and new combinations of ideas result.
5. When moving around the group, it is often helpful to limit members to one idea at a time so that less vocal or less aggressive individuals feel encouraged to get their ideas out.

But, what is the real value of brainstorming? Are there ways in which a group might profit from its use outside of simply generating a large quantity of ideas in problem solution? The fact is that we are living in a time when people are demanding to be heard and involved. It is hardly a question of whether or not a group is the most productive means of solving a problem. People are using group decision-making procedures for an ever widening variety of problems. The question becomes, How is it possible to facilitate the work of these groups? Brainstorming, given the proper exposure and a relatively nonjudgmental climate, has much to offer a decision-making group, particularly during the diagnostic and the generating-of-alternatives stages of problem solving. For example:

1. It reduces dependency upon a single authority figure.
2. It encourages an open sharing of ideas.
3. It stimulates greater participation among the group.
4. It increases individual safety in a highly competitive group.
5. It provides for a maximum of output in a short period of time.
6. It helps to insure a nonevaluative climate, at least in the ideation phase of the meeting.
7. It provides the participants with immediate visibility for the ideas that are generated (assuming they are posted).
8. It develops some degree of accountability for the ideas among the

group because they have been generated internally and not imposed from outside.
9. It tends to be enjoyable and self-stimulating.

Thus, the process is self-reinforcing; it draws the participants into new avenues of thought and into a new pattern of communication. How efficient the method is—and it can be efficient—is a factor of secondary importance to its potential for facilitating shared problem solving.

READER ACTIVITY

After all of the discussion concerning brainstorming, it seems like a good idea to provide you the opportunity to try it on for size. It seems appropriate to introduce you by providing an experience that we might use as a necessary warm-up activity for a group involved in serious problem solving. Like anything else, brainstorming requires the participants to be in the right mood if the benefits of the process are to be gained. Thus, the following could be used with a group of fifty or with as few as two or three people. It helps if a large sheet of newsprint is available to write down the ideas as they are generated.

We live in a highly critical society where competition and a win-lose atmosphere is often the rule and not the exception. It is not uncommon that getting ahead is done by putting down someone else so that we may appear better. Brainstorming is a means of reducing this inclination. Still, it requires practice to overcome the inclination to be negative and overly evaluative. With this in mind, get ready to brainstorm. Read the following brief story:

> A small wholesaler in the hinterlands of Mexico had called his buyer in Vera Cruz and asked him to obtain an order of pipe cleaners from the United States. Senor Gonzales, the buyer, agreed. He also agreed to advance Senor Gomez (the wholesaler) 5,000 pesos to finance the deal. A month later, just as the ship was arriving in Vera Cruz, Senor Gonzales received a disastrous phone call from Senor Gomez. Apparently the warehouse and outlet store had burned down and there simply was no more business. Gonzales was suddenly faced with somehow selling 200,000 pipe cleaners.

You and your group have exactly three minutes to generate as many creative alternatives as possible. Don't think, don't hold back, anything goes.

After three minutes, your list may include as many as twenty, thirty, or even forty items. Depending on how far you wish to go, the next obvious step would be to take the five or six best ideas and spend time creatively developing them further. Usually the screening process is based on criteria that are developed by the group that incorporate parameters important to the problem solvers.

Other Methods of Generating Ideas

Brainstorming is meant to reduce inhibitions, to encourage new ideas, to legitimize the unthinkable, and to push the participants past the bounds of their normally restrictive thinking. Over the years, a variety of methods related to brainstorming that can be used with small groups have been developed (Gordon, 1961; Prince, 1970; Phillips, 1948; Rickards, 1974).

Nominal Groups The process of creating a variety of alternatives can be done outside of a group context by simply having individuals develop their own separate lists of ideas or solutions without recourse to the ideas of others. Thus, individuals each generate their own ideas without discussion. They then bring these to the group and pool them. Such a group, in which the ideas of members working independently are pooled, is called a nominal group. Through a systematic integration of these ideas, common themes are extracted, discussed, expanded upon, and eventually critiqued.

Trigger Groups An attempt to build on the strength of nominal groups (no fear of group competition, domination of a few individuals, or the constraints of time as in traditional brainstorming), this approach has each member read his or her individually developed ideas to the total group. The group gives its total attention to each person. Thus, each member of a group may be asked to consider the ten best or ten wildest ideas in relation to a particular problem. The group's task in a series of five- or ten-minute periods is to take each idea and clarify, expand, build on, or in some manner trigger new ideas that will develop further the thought. Each individual has his or her ideas exposed to the constructive assistance of the group, and during the first round, no attempt is made to criticize. Members feel heard, their ideas are received positively, and a real effort is made to see where each idea may have value. After enough of the ideas have been explored in this manner, criteria are developed by the group to determine which ideas might have the greatest value. Then a discussion screening the variety of ideas ensues that will inevitably reduce the suggestions to a single one or perhaps two or three.

One can easily imagine a number of creative variations on this theme. For example, a group of seven or eight people is asked to develop perhaps ten ideas around a particular theme in a brainstorming fashion. The ideas are then discussed openly in terms of the strengths of various items. A second group, which has been watching the activity quietly from outside the group, is then asked after ten or fifteen minutes of discussion to switch places with the first group. They are requested to develop ten new ideas, considering the strengths discussed by the first group. They then discuss their ideas, again focusing on strengths. Finally, after fifteen minutes or so of discussion, clusters of four are formed by two members from each group. These clusters (approximately four or five) are then asked to develop two or three of the best ideas that seem to incorporate as many of the strengths suggested in both of the large groups. These finely tuned ideas are then presented back to the entire group perhaps an hour later. A representative group of the whole could then determine which idea(s) or combination of ideas best suit their requirements. The advantages of such a design include:

the generating of ideas in a nonthreatening atmosphere

a relatively nonjudgmental screening process

the opportunity to build on one set of ideas after listening to benefits, thus, reaching beyond ordinary limits or boundaries

the development of competition in the best sense of the word as individuals in the second group of eight try to generate new and even better ideas, and the groups of four attempt to develop the best idea or combination of ideas knowing that three or four other groups are doing the same thing

full participation by a large number of individuals

the utilization of individual resources, both in the development of ideas and in the important critiquing and selecting phase

the efficient utilization of time itself

Round Robin Groups Another creative means of generating ideas involves a small group of perhaps five individuals (simultaneous groups of five can be working on the same problem at the same time). Five problems that need solving and are recognized as open-ended are selected by the group. Generally, these are operational problems, those which influence people who work together, are hindering a particular task, or are seen as within the purview of the group working on them. They should tend to be problems that do not depend on some outside authority for determination. Each individual writes the problem they have been given on the top of three five-by-eight-inch cards. They then proceed to write a different idea or solution on each card. After perhaps five minutes, each individual passes his or her three cards onto the next individual, who then writes a new solution or idea on each of the three cards they receive. Their ideas can be original or simply be constructive additions to what is already there. After each individual has had the opportunity to respond to all of the problems, each original problem is summarized by one person, integrating the ideas from each of the three cards onto newsprint. These are then presented back to the whole group of five, whose job is then to discuss the pros and cons of the various ideas and determine whether a creative and operationally viable idea emerges. Obviously, such an approach assumes that the group has the necessary time (from one to two days) to work effectively on the various ideas. But, with less time available, a variation of the design would be to develop only one or two ideas with a slightly different manner of passing the various cards through the group.

The Wildest Idea Strange as it seems, one reason creativity is minimized in a group of creative individuals is that permission is not given to be "wild and crazy." By telling individuals in either a brainstorming group or a nominal group to generate the wildest ideas they can imagine in relation to a particular problem, thoughts will emerge that never would be considered in the normal course of events, when most people are usually worried about their image or what is appropriate. Thus, when a group is bogged down, ideas are not coming, and frustration is mounting in a problem-solving effort, the simple request to drop all pretense and let go for five minutes with the most outrageous ideas possible will inject fun, energy, and new interest into the group. In addition, it is very likely that among the "wild and crazy" ideas lie the seeds to some creative new approaches to solution.

Synectics

William Gordon (1961) worked for years with methods for expanding the vision and creativity of people in problem-solving situations. Dissatisfied with the constructive and yet limiting approaches of brainstorming and nominal groups, he experimented with a variety of methods that would release some of the restricted capacity we have for creative problem solving. Gordon saw our ability to speculate as the key to removing normal resistances and the stereotypical and predictable traps we often fall into while solving problems, and from this assumption Gordon developed his *synectics* theory:

> The word *synectics,* from the Greek, means the joining together of different and apparently irrelevant elements. Synectics the-

> ory applies to the integration of diverse individuals into a problem-stating, problem-solving group. It is an operational theory for the conscious use of preconscious psychological mechanisms present in man's creative activity (Gordon, 1961, p. 1).

Most of us recognize that we are often victims of habit, ritualized routines that slowly drown our potential for creative thought. Gordon's goal became one of taking people on excursions away from predictable and preconceived approaches to problem solving into the world of fantasy, metaphor, and analogy. By doing so, he opened portals into creative problem solving using methods meant to unlock the potential of more traditional, stepwise, structured, orderly, and controlled approaches to rational problem solving. Most often a synectics approach is utilized as a complementary means of adding energy and creativity to more structured, left-hemispheric methods.

A colleague of Gordon's, George Prince (1970) added to these original ideas and began to apply many of the synectics approaches into designs for group problem solving. Although many of their problem-solving strategies demand skilled facilitators or leaders with a keen understanding of the synectics point of view, many of their ideas can be adapted by individuals or groups who simply need a little help in breaking out of unproductive patterns of thinking. Again, the aim is to allow people access to submerged ideas, to utilize the unexpected, to vitalize a group, and to provide freedom to our thoughts in the context of some useful structure.

A Typical Synectics Meeting Most meetings that utilize this approach are best served when the group is from five to perhaps eight, and the group members have some measure of expertise in the subject to be explored. A person experienced in synectics is selected as the guide or leader for the problem solving. One of the members of the group assumes the role of client and it is his or her problem that will be the focus of attention for the entire group. It is unusual when one person in such a group doesn't have either more responsibility or legitimate authority than others present, even though others may have considerable interest and skill in the problem that is selected.

At this point, the client states the problem for the group, defines what he or she would like to accomplish by the end of the meeting and how the group might help him or her. The client also provides the other members with whatever background and historical information that he or she believes is necessary. He or she outlines what efforts have been attempted up to this point to solve the problem. At this time, and during the entire several-hour problem-solving session, the primary function of the group is to seek clarification and add insight. However, critical to this early phase of the problem solving is the work of the group members to help establish realistic and workable goals that can be accomplished during the session. Often, for example, as the result of tough questioning and redefining the problem, the group discovers that the problem as stated is not the problem at all but merely a symptom of a greater source of underlying tension. Or, they may find that what was perceived as a single issue represents several discrete problems, each of which deserves attention. Thus, as in any other affective problem-solving approach, it is essential to isolate the real problem as specifically as possible because it is upon this factor alone that success will often hinge.

For example, a member of an executive team recently asked his colleagues to discuss his problems with a key staff member who had been with the organization twenty years and had become a disruptive force to the department by doing things "his way" and being insensitive to his own subordinates. His behavior had alienated other staff, some of whom had quit, and had created a deplorable state of morale. The executive laid out the problem for the group, whose job became one of clarification. By the end of thirty minutes (the usual length for such an orientation period) it became clear that although the subordinate was acting inappropriately, the real problem was that the individual had been left alone to do his job with virtually no effective supervision over the years. Not only this, but he had been praised again and again for a job well done. The need for retraining and more effective supervision were identified as the critical issues. In addition, it was necessary to focus on a third goal, which was the current situation of low morale which would have to be dealt with, because retraining the manager and developing effective supervision would both take considerable time.

For this group of executives the three agreed-upon goals of the session became:

1. altering the present level of supervision (or lack therof) in the department. Implicit is the need to retrain the executive in this area.
2. the need for retraining and personal development for the subordinate supervisor.
3. improving the present state of morale, which needs immediate attention.

In the next logical step in the problem solving, the group, including the individual designated as client, listed in a quasi-brainstorming fashion possible solutions, without discussion or criticism, for one of the goals. Wild, impractical, and unorthodox ideas were encouraged as well as more appropriate and traditional solutions.

Each suggestion was then discussed to discover anything and everything that was positive and possible. If something appeared impractical, it was still milked for any aspect that made sense or seemed doable and constructive. For example, when the group was working on the second problem—the need for retraining the subordinate supervisor—one suggestion was to send him off to the South Seas for a year's worth of relaxation and recovery—relaxation for him and recovery for the department. Looked at in terms of serious strengths, it became evident that morale (problem #3) would be helped if the antagonist was removed for a period of time, and the kinds of new behavior required of the individual could not be hoped for with anything other than intensive training outside of his present environment, where he was reactive and defensive. The discussion led to several other ideas of a more practical and useful nature.

After each item is discussed in relation to strengths, a tough critiquing period occurs. It is at this point that questions of consequences are raised as are any issues that might block the use of the idea or any of its parts. It should be noted that such a meeting may be well into its second hour before a negative word is spoken. By this time, it is usual that a constructive and positive climate has developed that is laced with humor and obvious good will. The client usually feels helped early into the session and, without having to defend his previous actions, is able to respond to the group in a nondefensive manner. The process is so analytical that other issues and subgoals inevitably are raised, and if they are seen as important to the goal under scrutiny, the facilitator will suggest that a list of solutions relating to it be generated. There is no rigid format; rather, the format is responsive to the unfolding goals. There is, however, a strict adherence to the procedure of focusing only on strengths and waiting for criticism.

After strengths and weaknesses or consequences have been carefully analyzed, a final phase of the problem solving for a particular goal—that of *integration*—takes place. The most valuable suggestions are considered in terms of a single constructive plan. Action steps for implementation are seen as critical. It is the forcing of diverse ideas together, the weaving of a tapestry of ideas that proves just as creative and interesting to the participants as anything else in the problem solving. Such a product is measured against the goals of the session. Done like this, it is easy to recognize how the benefits of a synectics session could be of tremendous value for a group's development in addition to the final product. Support, trust, openness, creativity, and humor are the threads of the weave and result in a sense of cohesion that problem solvers and participants rarely experience in the all-too-common crisis-reactive approaches to problem solving and group interaction.

But, we have not yet discussed the element of synectics which, although only a small part of the total process, has been the aspect that has gained greatest notoriety. Previously, we mentioned the *excursion* as a method for untracking problem solvers, moving them away from predictable, noncreative responses to an issue. Basically, it represents an intervention into a problem-solving group that is stuck and is designed to free their vision, to lead them into the hidden solutions lying in the right hemisphere. Quite simply, it is a means of "fooling us," of pushing us away from our predispositions and canned, predictable solutions and allowing us access to new and better ideas.

There is no right way to conduct an excursion. It does, however, demand a sensitivity and alertness on the part of the facilitator. Synectics, as outlined thus far, requires no great insight or skill on the part of the leader. It is a well-ordered approach to problem solving in a group context that stresses a positive approach to problem analysis and encourages creative alternatives. The excursion, on the other hand, demands timing, a firm grasp of the problem, and sensitivity to the group—its needs and readiness as well as the problem solvers' own capabilities. The idea itself is simple enough; it is execution that can be demanding. An example is required.

In the example of the executive who identified three goals to problem solve, the first two resulted in useful solutions without much difficulty by simply following the process of clarifying each goal, developing alternatives, and seeking the positive aspects of each solution. Finally, each idea was critiqued, considering the means of removing negative aspects and forming an integrated plan of attack. But, in attacking the third goal, concerning the present state of morale that had deteriorated because of the supervisor's behavior, the group became bogged down, unable to generate anything but the most routine solutions for a situation that they felt was serious and could disintegrate rapidly. Sensing their "stuckness," the facilitator suggested that they take an excursion that might open them to fresh solutions. First, he selected the critical word *morale* and had the group list qualities assumed in morale. The group suggested:

1. cohesiveness
2. enjoying being together
3. good spirits
4. support
5. feeling productive
6. feeling worthwhile to the organization (rewarded)
7. a certain sense of being interdependent—depending on each other

Briefly studying the list, the facilitator was looking to find an idea that might open up the group. He landed on the word *spirits*. He asked the participants to help him take two brief excursions. The first was to list all the qualities of spirits as an alcoholic beverage, including its influence on the user. The group listed the following:

1. volatile
2. intoxicating
3. loss of balance
4. uninhibited
5. depressant
6. fun
7. out of control
8. drunk

The facilitator then had the group, with his help, conjure up a supernatural spirit who had been watching over the department under discussion. The group was asked to develop a series of statements from the characteristics written about alcoholic spirits that the group thought represented the feelings of the supernatural spirit about the state of morale in the department or what might be done about it. Stressing the spirit's dissatisfaction with the present situation, the facilitator wrote the following statements from the group. The spirit thinks:

1. The members of the department are dispirited and inhibited, afraid to express themselves.
2. The atmosphere in the department is volatile, people are ready to explode with frustration because of not being heard for so long.
3. The department could use a good drunk, they need to throw off their anger and have a good time—have fun.
4. The atmosphere is so depressing that there needs to be something dramatic done to change things, to create a new balance.
5. Those in the department need to be intoxicated, provided with a sense of hope.

The facilitator then asked the group to take five minutes with each of the five spirit statements and develop a series of ideas for altering the problem of low morale within the department. Thus, following a rather normal synectics format, the first item was used as a stimulus for solutions, positive comments were allowed, then each was critiqued with the effort made to resolve problems with any single idea. Finally, a concrete proposal(s) for each of the five spirit problems was developed. The group responded to the creativity of the situation, energy and ideas flowed easily, and the group was successfully unblocked. Two of the ideas generated from this process that had not even been considered previously included:

1. It was agreed that so many bad feelings existed that the department had to have the opportunity to vent these. Not only this, but, instead of a format for griping, individuals had to be provided the means to solve identified problems constructively. If the manager supervisor was to remain in the job, he had to be open to taking the flack that would come. But upper management would stand behind him and support his new approach to working with his people. In order to break routine, to reveal the seriousness of their intentions, and to provide an environment conducive to play as well as work, the problem identification and work sessions were to occur at an off-site recreational area over a three-day period.
2. As a result of their problem solving, those present became conscious of the fact that supervision in the total organization was basically punitive when it occurred, taking place only when something went wrong, or, in more cases than not, it didn't exist at all. The excursion helped them be open to certain realities about people's attitudes that they had been unwilling to face. A task force was then established to evaluate the supervisory process for the total organization and to recommend the training necessary to humanize the process so that what had happened would not occur again.

Basically, then, the excursion utilized in this situation did the following: First, it forced the group to create a new view of morale, expanding the definition, and while so doing, increased the potential dimensions for problem solving in the group. Second, the problem solvers were "spirited away" from their own "stuckness" by the use of the two spirit excursions. These introduced new interest while, at the same time, they disoriented the group from predispositions and a certain rigidity of thought that had evolved. Also, the group was naturally tiring after two successful efforts. Finally, while the solutions were not tremendously creative, they had been blocked from the group's view of alternatives; the excursions provided them new access as well as understanding. One could imagine a never ending source of possible excursions through the use of analogies, metaphors, and a variety of other devices intended to disrupt our tendency to think in patterned, restricted, and predictable ways. For Gordon (1961, p. 141), a founder of this process, the key is to make the familiar suddenly unfamiliar, to disorient and at the same time to excite, to challenge the mind with planned irrelevancies. He desired to preoccupy the rigidly occupied minds of the problem solvers and through such momentary excursions, create access to untapped attitudes, feelings, and ideas hidden in their fantasies.

Prerequisites of a Successful Synectics Group In their study of problem solving with groups, Gordon and Prince and their colleagues discovered a number of necessary elements that must be present. Just as a synectics excursion must evolve out of the needs of a group and cannot be canned, so too it is obvious by this time in this book, there are no easy formulas for helping a group work more effectively together. Following are some critical areas for attention in any problem-solving approach but felt to be of particular importance by those using synectics in a group context.[3]

1. It is necessary to maximize participation and help individuals feel that their ideas are valued, which will increase feelings of trust, openness, and willingness to risk. Destructive competition and the developing of common win-lose attitudes are the aspects of traditional problem solving that must go. Thus, a prerequisite of any effective synectics group is attention to the need to develop effective group process, with members aware of their impact on the group, and a willingness of the group to monitor its own behavior.
2. A critical problem to any problem-solving group is the willingness and ability of group members to listen. People are so busy selling their own ideas, proving themselves, hearing only what in others supports them, or reacting to personality rather than words that it is a wonder that we hear as much as we do. There are so many intrusions into our listening that something needs to be done that legitimizes "hearing" others and that protects us from our own inclinations to "yell and sell." Active listening (see Chapter One) is one of the central themes of synectics. Basically it is being aware of both the verbal and nonverbal messages being communicated as well as the feelings that often carry the real information. By rephrasing, paraphrasing, or in some manner feeding back what we hear another saying, speakers tend to be less defended because they realize they have been heard. Individuals who hear that they have been heard have less of an inclination to say it one more time. Thus, if active listening can be built into the problem-solving process and individuals know that it is part of the game rules, an immediate change in climate can be detected. When we feel understood, it matters less whether the idea itself is eventually adopted because we personally feel accepted.
3. A rather simple pattern that occurs in many groups is created out of our desire to protect ourselves and to put responsibility or, on occasion, blame onto another individual. The pattern is one of asking questions. You might respond to this "wondrous" revelation by shouting back at the page, "of course, problem solving is all asking questions, probing reality, testing the value of an idea." True, but the

[3] See Rickard's discussion (1974, p. 71–73).

trap is that behind nearly every question that an individual asks is a statement, often an implied answer. The question, then, often incorporates a message we are trying to give with minimal risk, putting the individual who responds on the spot and removing us from the possibility of having to be responsible for the implied statement. Thus, by asking questions, we often shift attention and a sense of blame or guilt onto the person being asked. Questions tend to corner the individual, leaving him or her feeling trapped and vulnerable and having to justify or rationalize his or her response with much more vigor than might otherwise be necessary.

Individuals using synectics attempt to reduce the number of questions being asked and instead encourage participants to make statements that provide new meaning or clarification to a problem or to another idea. In addition, statements act to provide information about one's own point of view, indicate alternatives, or request additional information. Statements are direct and they tend to establish ownership immediately. This leads to a sense of integrity in the group itself and again helps to build a climate of trust.

4. One of the crucial differences between a synectics-type meeting and others is that individuals with special influence or power are requested not to run the meeting. It is assumed that their ideas are crucial but not their influence. All too often, the leader is also the boss, with his or her own agenda, whose style and status can create a sense of intimidation, resistance, fear, dependency, or even anger that can corrupt the open problem-solving approach that is being attempted. How often is the boss's idea somehow magically accepted with suprisingly little discussion? Such leaders are so often biased toward predetermined outcomes that people begin to acquiesce at the first proverbial whiff of his or her idea. Participants will take great pains not to place themselves in a win-lose situation with their boss. Thus, by selecting a group leader who can lead the problem solving from a position of neutrality and have a more objective view of the process, the chance of success goes up. Of course, such a leader should be acceptable to the group, not easily manipulated, and with skills in such areas as active listening, goal setting and group maintenance.

5. Perhaps one of the most insidious factors that undermines productive group problem solving stems from the natural inclination of a group and its members to be overly critical of each other. Most of us have cut our teeth on a view of group participation that encourages criticism. The problem is that such criticism is often born out of a desire to minimize someone else's success and has little to do with a desire for group success. Implicit in this is often the assumption that if we can make someone's ideas appear inadequate, perhaps our

own will grow in stature, not realizing that the game is a two-way street. A simple rule developed in synectics meetings is one that focuses on the strengths of an idea, is validating of the giver, and reduces the inclination to defend. In addition, by focusing on the positive aspects of an idea it is often found that part of a solution can be used, thus reducing the inclination to throw the baby out with the bath. Even when the idea is clearly inadequate, the first effort is directed at finding ways of improving the idea so that it may be workable. It is the sound belief of this method that the more good choices a group has, the better will be the quality of the final product.

6. A synectics viewpoint encourages effective group process. To this point, many of the suggestions would be useful even in more traditional, well-ordered, rational approaches to problem solving. But, they are absolutely critical for synectics participants because at the center of their approach is the excursion. There are many ways for an excursion to occur, but all have the common theme of pulling the problem solver away from premature solutions, from patterned, expected, or predictable thought. The excursion is designed to relax the group, to build a sense of purpose with the added dimension of fun so that participants will be inclined to move beyond the "tried and true," away from any preconception of the right way. The excursion is a trip away from reality and the constraints of expected thought into a realm where unrelated, untested, and creative ideas are valued. Clearly for this to occur, for participants to allow themselves the freedom to try on totally different, often outrageous and absurd ideas, requires a climate of trust and openness where there is no fear of being judged and recriminations don't occur.

WHO SHOULD DECIDE— THE LEADER OR THE GROUP?

The bulk of this chapter has been involved with how groups and individuals can become more effective problem solvers. Problem solving is nothing more than the process used for developing alternative forms of action that resolve a source of tension, an uncertainty, or difficulty. Without giving some thought to *how* we problem solve, we have seen our inclination to fool ourselves, narrow our vision, not look at relevant consequences, and sabotage ourselves without even knowing it. Although we have alluded to the actual selection of various choices after certain creative deliberations, we have not talked about the implications of decision making itself.

The Leader's Role

As we have suggested previously, many decisions fall naturally out of a thorough problem analysis in which clear goals have been established, alternatives developed and systematically weighed, and potential consequences measured. Yet, there is a way of thinking about decision making that is essential for both the leader and the participant. The critical question that forever plays on the mind of the leader is "Should I make this decision? Can I risk leaving it to the group?" Leaders get themselves into trouble when they are not willing to define the parameters of their power, when they refuse to let their constituencies know their domain of influence. By not defining their range of authority, they are capable of arbitrarily making literally any decision they wish, or they can benevolently turn it over to the group if so inclined. Their failure to give real definition to how decisions will be made creates a climate of uncertainty, suspicion, and dependency. Many leaders are simply fearful that by defining their real areas of decision-making power they will leave themselves vulnerable to the irrationality and perhaps irresponsibility of the group for which they will ultimately be accountable. They don't realize that groups tend to be thoughtful and rational (often too much so) and, if provided the necessary time and affective problem-solving procedures, will often contribute significantly to effective problem solving and realize the best decision.

Thus, effective leaders are willing to look carefully at their areas of influence and let their subordinates know categorically those decisions they will always make and those for which other groups or individuals will be responsible. At the beginning of any problem-solving activity, the group should reach an understanding as to how any decision will be made. If the group's ideas are advisory, this should be made very clear with an appropriate rationale. If the decision for eventual action is to be in the hands of the group, then the particular decision-making method should be understood and discussed. This decision on making a decision occurs prior to the problem solving because people often lose their rationality as vested interests are threatened. Thus, people will naturally protect themselves regardless of their good will toward the group. Prior to problem solving, a group is usually more willing to go with a fair decision-making method such as a two-thirds majority before they experience the fear that they may actually have to give up their own idea and go along with one that is less acceptable to them. The integrity of the problem solving is often saved by this simple rule. Let's briefly discuss a few simple approaches to making decisions.

Simple Majority Rule

A group should only be willing to accept this approach when the decision is of relatively little consequence and they need a rather quick response. We all know that there are many ways to block a decision so that implementation never occurs. In theory, decisions are made to be implemented. But, if 45 percent of a group disagree with a decision that has significance for them and the life of the group or organization, then the quick-and-dirty majority rule will be perceived as a means of control and manipulation by the majority. Even if the decision is implemented, the large minority may spend its time deviously attempting to disrupt the decision or find the means of overthrowing it. Finally, majority rule is an easy way of shutting off discussion and the thoughtful views of the minority. This can leave the group wounded and result in future insensitivity and psychological "paybacks." Thus, the consequences of how a decision is made may have important repercussions later.

Two-thirds Majority Rule

Obviously, such a method should be used for decisions of greater consequence. Psychologically, if individuals have had the opportunity to discuss various alternatives thoroughly and they agree that a two-thirds vote is "fair" prior to the discussion, our experience is that members find it easier to accept the eventual decision. Somehow, the feelings of manipulation that often accompany the simple majority occur less frequently when members know that 34 percent of the group can influence the total group.

Consensus

This is a terribly misused and abused approach to problem solving and decision making. There are many implicit assumptions made when using it that rarely occur in many working groups. These include:

1. That a level of trust exists in the group that allows honesty, directness, and candor.
2. That the group is aware of its own process and can deal with its own behavior openly so that individuals cannot dominate or manipulate the group, so that ideas are actively solicited, so that members listen and support each other as individuals even when disagreeing with each other's ideas.
3. That the group is not leader dominated.

4. That there is time available to consider opinions, alternatives, and consequences so that time itself does not become a coercive element in the process.
5. That members of the group are privy to necessary information prior to the meeting so that they are familiar with critical issues and can respond intelligently.

It is the rare group that enters a decision-making situation with these conditions. Groups who do experience them find consensus to be an invigorating and often efficient approach. Those who do not will often find the process to be painful, aggravating and nonproductive. The major reason why consensus fails is that people do not understand it conceptually. It does not represent a method that demands agreement by the total group. It simply requires that individuals must be willing to go along with the group's predominant view and will carry out the implications of the decision in good faith. People may disagree with the view of the great majority and they are encouraged to hold onto their position until they are, in fact, willing to live positively with the decision being recommended. It takes trust to argue for one's position in the face of group pressure and just as much trust to back off one's own position and go with the group.

Delegated Decisions

The more decisions that can be delegated to representative bodies or even to individuals, the more efficient most groups will be, especially those with over seven or eight individuals. Delegated decisions depend on parties being willing to seek out the pulse of the group, testing ideas thoroughly before moving ahead with a particular idea. Again, this depends on a basic trust by members that decisions for the group are not based on the vested interests of individuals. Clearly, decisions of a controversial nature should be provided a problem-solving forum that allows a maximum participation. Essential to delegated decisions is the presence of procedures for critical review and accountability so that members of any delegated task force realize how the effectiveness of their efforts are to be measured.

Double Vote

Many organizations would like to involve relatively large numbers of people in decision making on a wide range of issues that influence their lives. This rarely occurs because leaders do not believe that individuals will be as rational as they will be or that an emotional speech or bandwagon effect won't influence the outcome in some irrational and undesirable manner. The following method, although imperfect, has proven highly successful in minimizing these legitimate concerns. Let's imagine that as a result of a thoughtful problem-solving process, a large number of alternatives are being considered for a department of one hundred people. The ideas have been drawn from committee and task-force recommendations that have involved a large number of the one hundred individuals at one time or another. Being considered are ideas ranging from alternatives for using the parking lot, to methods for participative management, to recommendations, some of which require choosing between two alternatives. The management has agreed that all the ideas are acceptable if the group desires them. Ideas not selected may be explored further at a later date, but present policy will continue if a new idea is not selected. The method for decision making is as follows:

1. Each alternative that is proposed is written as a brief paragraph stating as specifically as possible what changes will occur, when and how they are to be accomplished, as well as how they will be monitored.
2. The task force or committee responsible for the recommendation is noted so that further clarification and discussion can be gained during the coming week.
3. The following week a ballot is distributed. Ideally, this would occur at a large meeting(s) where time would be provided for further clarification but not debate, because it is assumed that controversial issues have been discussed at length during the preceding week. Because the voting process is based on 100 percent of the distributed ballots, a meeting with those present voting is often preferred.
4. Individuals are requested to vote for each alternative they find acceptable.
5. Any alternative that receives a two-thirds vote is placed on a second ballot. A report containing all of the results is distributed the next day. The following week, members are encouraged to lobby and discuss their views prior to a second vote. We find that the first vote acts as a reality test and energizes the participants to new levels of interest.
6. The second vote, held a week after the first, relates only to those items that received at least two-thirds of the vote on the first ballot. Those alternatives receiving two-thirds vote on the second ballot are then accepted to be implemented. A representative group is selected from the total population to help in the implementation phase and to monitor the progress of each recommendation for several months.

The benefits of such a complex and time-consuming process are many. First, participation, discussion, and influence on the system are being exercised by individuals who must live with their own decisions. Second, by knowing that ideas not accepted in the voting can be raised at a later date, members of the organization are actually encouraged to seek constructive change and the support for their ideas. As a result, they feel potent and interested in the life of the organization. The major drawback is that the leader of such an organization must have a clear view of which areas of influence are shared, and this must be communicated specifically so that false expectations are not raised. Finally, the leader must be willing to go with the recommendations even though he or she does not necessarily agree totally with them. Limits as to policy issues and the expenditure of funds can be explored prior to alternatives reaching the ballot.

In many ways, effective decision making requires creativity and judgment as much as does the formal problem-solving process. There is no question that decision making should not be taken for granted, and that it can be designed in a variety of ways depending on the realities of each situation and the goals of the leader. Issues of participation, acceptance, and overall morale are all influenced by how the leader decides to decide. As participants, it is important always to understand what is happening, why, and whether the process seems equitable. Often, leaders have never considered the implications of their own decision-making behavior and would be open to alternatives once educated to the consequences of their own actions.

Following are a number of questions frequently asked in relation to the use of groups in problem solving and decision making.

QUESTIONS FREQUENTLY ASKED ABOUT GROUP PROBLEM SOLVING AND DECISION MAKING

Do groups appear more effective in problem solving than individuals, especially considering person hours invested? There are good reasons for using groups in some problem-solving endeavors. However, few of these reasons involve efficiency. Increasingly, research is pointing toward the conclusion that individuals and nominal groups are equal to or more effective than natural groups (assuming no training) when undertaking problem-solving activities (Campbell, 1968; Rotter and Portergal, 1969). A nominal group is one in which the ideas of members working independently are then pooled. A natural group is where members would work cooperatively at the same task. Some working groups are slowed down and reduced to a level of performance equal to the slowest member (McCurdy and Lambert, 1952), while others become polarized as a result of the group discussions (Moscovici and Zavalloni, 1969). Furthermore, even in groups designed to facilitate the open sharing of ideas, differences in status and perceived authority can inhibit productivity (Vroom et al., 1969; Voytas, 1967; Collaros and Anderson, 1969). Yet it is true that in a number of rather specific instances, it does seem that a group effort can be justified over that of nominal groups or individuals. For example, when a task involves the integrating of a number of perceptual and intellectual skills, it has been found that group members tend to supplement one another as resources (Napier, 1967). Also, when a major goal of the group is to create commitment to certain goals or to actually influence opinions, the involvement of individuals appears essential (Kelley and Thibaut, 1969; Lewin, 1948). However, when it comes to simply producing ideas in quantity or even quality, the evidence (although in some cases mixed) does not support the faith shown in recent years for working in groups.

Does training help the problem-solving capabilities of a group? Work groups have existed as long as there have been problems, and there seems to be a rather casual assumption that the process is natural and even simple. But, as previously shown, using group resources effectively requires great skill on the part of the facilitator as well as skill and understanding on the part of the members. One reason why problem-solving groups tend to fare poorly when compared to the work output of individuals or nominal groups is that they are invariably untrained and, to make matters worse, they are usually "stranger" groups. The result is that the group members not only have to coordinate their work efforts, but also they are caught in the midst of tensions common to any developing group (see "The Stages of Group Development," Chapter Eight). Virtually no research exists in which the quantitative and qualitative products of trained and well-practiced groups are compared with those of individuals or nominal groups. There is evidence, however, that laboratory training sessions, in which individuals are given the opportunity to learn group skills through the systematic observation of their own performance on a variety of tasks, have impressive transfer value to other group situations (Hall and Williams, 1970; Stuls, 1969; Tolela, 1967).

In one interesting experiment, requiring the solution of a specific task-oriented problem, groups were involved in an interdependent, multistage problem-solving process. Trained groups revealed greater improvement, had higher quality products, and used the knowledge of members more effectively than untrained groups. In fact, it was shown that groups of institutionalized, neuropsychiatric patients scored significantly better than untrained managerial groups that were assumed to have greater knowledge of procedures and problem-solving operations (Hall and Williams, 1970). Although such research is limited because of the type of training undertaken and because of the problems involved, the implications are clear. Effective training can maximize the benefits that are possible to achieve within the framework of problem-solving groups.

What are the strengths and limitations of the democratic approach to decision making in groups? For most working groups it seems that the key to decision making is found in a rather loose concept of the democratic process and the rule of the majority. It provides governing "by the people," reduces the threat of tyranny from within the group, and insures that at least half the members will be in support of a particular issue. Nevertheless, this approach to decision making has a number of severe limitations when applied to a group that must live by its own decisions. For example:

1. Under the pressure of a vote, individual decisions are often made for the wrong reasons. This is partly the result of different levels of knowledge and understanding present in the group and partly because of extraneous pressures (friendships, propaganda, payment of past favors, and so forth). Thus, issues are often lost sight of in favor of other variables such as voting for "the person."
2. During the discussions leading to a vote, it is assumed that people will have an opportunity to express their opinions and to influence the group, but this is seldom the case. Many individuals simply do not have the skills to influence their own destinies in groups. It is the rare group where silence is not taken as consent, where the shy person is drawn into the discussion, and where the intent is to consider all ideas and not just to project one's own. Therefore, the basis upon which a vote is taken is often faulty or, at least, premature.
3. The will of the majority can be used effectively as a means of reducing tension (strong differences of opinion) and the time needed to discuss a problem. A vote can be a means of getting on to other business. This, of course, fails to take into consideration whether or not the support for the decision is enough to insure effective implementation.
4. There is also the problem of power and despotism in a democratically run group. How often is the dissenting minority perceived as a

disrupting influence? How often is the minority opinion seen as a threat to the cohesion of the group? And how often are such pressures used to coerce the dissenters back in line? If the vote is used to override the opinion of this minority faction, the vote itself stands to further polarize the group and magnify the divisive lines upon which the vote is taken.

5. Similarly, rather than providing a solution, the vote may actually create more problems. Instead of resolving differences, the minority may spend its time proving the vote wrong and reasserting itself in the eyes of the group. Or, labeled as radicals or discontents, it may try to live up to the image and really become a disruptive force.

6. Finally, by encouraging a move toward quick decisions, there is a tendency to simplify problems in terms of either-or dichotomies, and there is a resulting failure to explore all the issues influencing the problem condition. A quick vote based on an inadequate exploration of issues will inevitably create difficulties. Members may have second thoughts and fail to support the vote in terms of behavior or rationalize their vote and become intransigent.

Therefore, when weighing these kinds of problems often linked with a simplistic notion of the democratic process, it might seem worthwhile to study other alternatives. The fact is that a democratic group in the real sense of the word requires enormous patience, understanding, and cooperation. It also is very time-consuming. Few groups are willing to face these realities and thus reduce the process to one of convenience rather than effectiveness.

How useful is Robert's Rules as a procedure within which to make decisions? Robert's Rules is based upon an assumption valuing the notion of debate (Robert, 1943). It is a complicated procedural method keyed to the majority vote and democratic process. It probably can be stated fairly that nearly everyone who has worked within a variety of groups has at one time or another been frustrated by the limitations of this system. Those who understand the complexities of the process can easily control the meeting, but few people know the rules for a quorum, tabling a motion, adjourning, or even amending a motion.

On the one hand, the moderator of a meeting can be in a position of considerable power. On the other hand, since chairpeople are not chosen necessarily for their understanding of Robert's Rules, it is fairly easy for them to lose control of the meeting to a few individuals who know the finer points. Another problem is that because the system is based on debate, there is a constant tendency toward polarization. True, the amending process does allow compromise, but usually these are political compromises, and the real issue can be pulled to pieces as factions based on broader ideological issues use the problem at hand to solidify their political position rather than seek the best solution. Furthermore, it is relatively easy for the majority to stifle discussion by pushing for an early vote or using some other defensive measure to change the focus of discussion. Finally, because the system is not based on a cooperative and interdependent approach to problem solving, there tends to be a great deal of politicking, bargaining, and bidding for power outside the meeting itself. In relatively small groups (under twenty-five or thirty), the method reduces open communication and the amount of participation. In larger groups, if the participants understand the system, it can prove useful in organizing discussion and stabilizing work procedures. Again, large numbers of participants present a limiting factor in the decision-making process, and accepting Robert's Rules must be done with the view that while it is gaining order, it is at the price of interdependence and, to some degree, cooperation.

Is decision by consensus a viable method for small-group decision making? Reaching a decision through consensus represents the ideal in terms of group participation, but it is by no means the most efficient or least tension-producing approach to decision making. It assumes that a decision will not be made without the approval of every member, but it does not mean that each member must agree totally with what is going to happen. It simply indicates that each member is willing to go along with the decision, at least for the time being. The process provides for full group participation and a willingness to compromise. Immature groups that lack skill in processing their own interpersonal behavior may find this a painful approach to problem solving. Unlike a system based on majority vote (basically a tension-reducing system), decision by consensus seeks out alternative viewpoints and then struggles to find a solution at the expense of no particular group or person. The value, of course, in using this sometimes slow and belabored process is that by the time a decision is reached, it does represent a group decision and therein lies an important component of support. At times, a provisional straw vote is used to test sources of differing opinion so that the full dimension of the problem can be explored. If it becomes coercive, the process breaks down. Usually, it requires time, familiarity within the group, and trust in the process before consensus becomes effective. Once this occurs, however, decisions can be made rapidly because there is a willingness to get to the core of the issue quickly, analyze the alternatives, and then compromise in finding the solution.

Why do institutional committees become so ineffective? Most committees are part of an inefficient hierarchical system that has developed over time with little built-in flexibility for change. Procedures become routine, more complex, and the interest of those participating wanes. Within the committees themselves, there are other barriers. For example:

1. Decision-making procedures are usually imposed and based on tradition rather than what is most useful.
2. People are often appointed to committees, and, even if they volunteer, they may be there for a variety of reasons (from interest in meeting important people to helping out a friend who is chairperson).
3. Often committees lack the power to implement the decisions they make and thus feel their own impotence.
4. Committees seldom see processing their own interpersonal behaviors as part of the job, especially if the group only meets once every three or four weeks.
5. The committee is not necessarily composed of the people best equipped to discuss the issues confronting the group.

The above-mentioned factors do not mean that most committees are not designed with a functional purpose in mind or that their participants are not well intentioned. They merely suggest that such groups often become self-defeating because of their membership, their decision-making procedures, and their lack of potency within the larger organization.

Are there useful alternatives to the committee system? One possible alternative is the use of a task force. Ideally, when a special problem arises within an organization, instead of pushing it off to an already overburdened committee, a task force is appointed or elected. This group is composed of representative individuals (or, in some cases, individuals with special skills) who are given the job of solving the problem. It is assumed that their recommendations will be taken most seriously and, in essence, they are given the power of the large group. Unlike a committee:

1. They often have more power.
2. Appointments are for a short term.
3. A definite measurable outcome will be the result.

4. The members may develop working procedures that best fit the nature of the task and are not limited by tradition of previous groups.
5. They must work through all phases of the problem-solving process including the diagnosis, actual implementation, and follow-up.
6. Because of the immediacy of the problem, there should be high motivation and involvement, especially since the product will be its own reward.

One potential problem with a task force is that given support and some feeling of potency, these groups generate recommendations that are much less conservative than might have been expected and that are much more thoroughly documented than usual. Unhappy is the executive who turns an issue over to a task force and then, instead of the problem losing importance and momentum, which often happens when problems are referred to committees, the task force provides the organization with clear methods for altering the situation. These methods may unveil other problems.

It is often true that task forces are also used for political purposes. Instead of expecting solutions, the aim is to look as if something is being done while, in reality, the objective is to mark time. An example of this situation occurs when prestigious people are appointed to the task force and, because of other commitments, find it impossible to do the kind of job necessary. The final product is a watered-down and poorly conceived attempt to look competent with a minimum commitment to action.

What influence does an individual who talks excessively or in a dominating manner tend to have on the group? There are most certainly talkers who are not heard and wield little influence on the life of a group. But, for the most part socially verbal talkers tend to have a high degree of influence over other members, regardless of their measured knowledge or ability (Hoffman, 1979, p. 378). The simple fact of talking more than others may be all that is necessary to increase one's impact on the group. It has even been shown that known talkers, when not dominating the conversation, will often influence the problem-solving effort through their support of someone else's ideas. Thus, regardless of the intent of the talker (whether devious, out of personal need, or out of interest) talking garners influence. This is a major reason that many problem-solving designs have built-in mechanisms for insuring greater freedom of participation and access of other members to the ideas of more retiring members. Brainstorming techniques and synectics excursions are but two approaches that support this view.

EXERCISE 1 Brainstorming: Useful in Conducting an Effective Large-Group Needs Assessment

Action

Brainstorming can be used as a central activity in a design for assessing the needs of a group where active participation is desired on the part of the group members. In this example, the directions are to be given to a group of twenty-four participants. Many alternative formats could easily be designed, but the following one has proven successful: After the problem has been stated clearly (one hopes it is an issue of relevance and concern to members), three large sheets of newsprint are placed next to each other in front of the group. Three participants are chosen as recorders, given markers, and asked to stand in front of one of the newsprints. The group is instructed that it will have between 3 and 5 minutes to list all the possible causes for this particular problem (on another occasion they might brainstorm solutions). The first recorder posts the first cause, the second recorder the second, and so on. After approximately 3 minutes, the result should be three sheets with an equal number of responses.

The large group of twenty-four is then broken down into three groups of eight, each taking a sheet with causes for the particular problem. Their task in a period of about 30 minutes is:

1. To clarify and expand any of the statements.
2. To integrate similar statements and to delete any that are irrelevant.
3. To develop from the items a list of causes that are most important to deal with immediately and which are within the power of the group (basic priorities).
4. This high-priority list of perhaps three causes is presented to the total group of twenty-four. In all, nine high-priority causal factors will have been identified. Some of these will be very nearly the same so that the real list will be about five items.

If the group agrees that something must be done with these five causal factors, it may prove useful to halve the groups of eight and have each group of four design specific action solutions to two of the problems. It is suggested that within 45 minutes or an hour the groups of four reconvene with their original group of eight and present their ideas to the other for a critique. This will take another 20 or 30 minutes. Again, if the possibility for integrating the ideas (solutions) exists, it should be done.

Finally, the crystallized ideas of the three groups are presented to the entire group. There should be about nine separate ideas presented. It is very important that the facilitator stress the need to minimize the time of the various presentations (no more than 5 minutes). The main purpose of this session is to give visibility to the various ideas and to bring some closure to the problem-solving process. The total time for this session will be between 2½ and 3 hours. It is an exhausting process and may not result in final decisions. What often helps is to have a representative of each of the groups of four act as a steering committee and, at a later time, report back specific recommendations that incorporate the various solutions offered.

Discussion of the Problem-solving Sequence

There seem to be a number of practices in this sequence that could be used under a variety of circumstances and with different kinds of problems.

1. It is important that the ideas being explored are the result of the group's effort. This is the first step in building accountability for the eventual solutions.
2. If the ideas are developed in a nonevaluative atmosphere, there will be less of the vested interests that tend to be present in any group and that surround any problem of importance.
3. The process forces a look at a variety of alternative approaches *after* important causal factors have been isolated. This builds a norm into the group for exploring new ideas and stimulates interest and involvement in the process itself.
4. The participants are held under strict time limits during their various work sessions. By being held accountable to other groups at the end of brief work periods, a continuous flow of ideas is assured and withdrawal because of disinterest or boredom is almost impossible. It seems to be very true that people will use the time made available to them.
5. Each product is a product of a number of people's ideas, and this reduces the possibility of one or two vociferous individuals taking over the group. Even in the presentation, it is important that the presenting groups do not try to sell their ideas, but simply reveal them.
6. By having a representative body make recommendations to the entire group based on *all* of their efforts, consensus is much more easily used as a final decision-making device. By this time the group should be ready to stand accountable for its own product. And, of

course, the decision is only as good as the group's willingness to implement it.

EXERCISE 2 Phillips 66: Discussion and Decisions in a Large Group

Objectives

To involve large numbers of people in discussion of topics relevant to them

To maximize the use of time as a factor in reducing argument

To insure greater accountability in large groups, which often tend to be impersonal

General Description

D. J. Phillips (1948) at first saw this method as a means of involving large numbers of people in the discussion of a particular issue. For example, after a presentation, debate, panel, and so forth, he would have the large group break into groups of six and develop a question that the group could agree was important to them in a period of about 6 minutes. The relatively small groups would have their interest focused, many individuals would have an opportunity to interact, and they could have some impact on the total group's discussion. It was impossible for a few people to dominate the discussion, and it assured a high rate of interest.

More recently, the method has been adapted to meet the needs of many groups. Members are given 6 minutes (it could just as easily be 10 or 15) to develop an agenda for the meeting. With all groups reporting, certain items are immediately perceived as having interest to many in the group. Or in other sessions, members are asked to offer a solution for a particular issue. Usually 6 minutes is long enough to define and clarify the issue, but it has been shown that in a relatively short period of time, an enormous number of good ideas can be generated and then refined at a later time. Brief reports on these findings can be an important stimulating factor in the large group.

Others, including Maier (1963), have used adaptations of this method for larger groups with as many as two or three hundred people. Groups of six are given a problem to solve or an issue to discuss, and then these groups are polled by the facilitator in terms of certain logical categories. Immediately the group gains a picture of how others feel and the range of ideas that exist in the group.

Of great importance in using this method is to make sure that the topics used for discussion are specific enough to allow an almost immediate discussion to get under way. Questions that are moralistic in tone will only frustrate people since there is no hope for any kind of resolution in a limited period of time. Similarly, when looking at a particular problem, the participants should not be limited to an either-or type of response. The enjoyment lies in the opportunity to be creative and to look beyond the commonplace response. The great value of having many people together doing the same thing is that one is assured of wide-ranging responses that may not develop in the more traditional committee work group, which is partly controlled by past experience and behavior.

EXERCISE 3 The Chairman Role: Dealing with Conflict in a Group

Objectives

To practice diagnosing and working with a framework of conflicting viewpoints in a problem-solving group

To generate alternative chairperson behaviors in situations where vested interests of a personal nature seem as important as the task that has been defined by the group

Setting

This activity involves a six-person role-playing situation in which various members of the group are coached in positions for and against a particular point of view. The chairperson is to help the group reach agreement on certain specific recommendations involving the disputed issue. The design is based on a situation in which there is an even number of teams taking part in the role-playing scene. For purposes of this discussion, it is assumed that there are four groups of six players each; each group is enacting the same role play and has its own chairperson.

Each group will take part in the initial role play simultaneously for approximately 30 or 40 minutes. The chairperson, who is working under a time pressure, attempts to facilitate the group in any way he or she can. After the allotted period of time, the individual groups analyze the sources of conflict and the ways in which the chairperson did and did not help the situation. Then they outline other procedures that might facilitate the chairperson's role in a similar situation. At this point, one member of the group is sent to another group, and the previous chairperson takes a member role. Thus, group 1 will have a chairperson from group 2, and group 2 will have a chairperson from group 1 (the same switch will occur in groups 3 and 4). At this time, group 1 reenacts the role play with the new chairperson, and group 2 observes. After about 15 minutes, group 2 reenacts the scene with its new chairperson. Both chairpeople try to include the suggestions raised in the discussion after the first role play. Groups 3 and 4 are doing the same thing. After the second reenactment, the two groups discuss their observations and draw a number of generalizations concerning the chairperson role in situations of conflict; these are then shared with the total group (all four groups). From 2 to 2½ hours should be allowed for the activity.

Action

The facilitator makes a general statement on the objectives of the activity and then proceeds to give a general description of the role-playing situations.

> You are the members of a junior high school faculty in a deprived area of the city. On returning to school on a Monday morning, it is discovered that vandals (apparently students from the school) have caused damage in a number of classrooms. Worse than this, however, the windows of the rooms of three teachers have been defaced with obscenities. Since these rooms overlook the playground and the parking lot, nearly all of the students and the faculty saw what was written. The school is in an uproar, and there is no end of talk about "why those particular teachers" and who could have done it. There is also some fear that such behavior could lead to much worse problems in the future.
>
> The principal is outraged and takes immediate steps by appointing a six-person faculty committee to investigate the incident. If possible, they are to discover who the culprits are and recommend punishments for the acts. The principal has given the committee 2 days to submit a report with their recommendations.

SPECIFIC ROLES OF THE SIX COMMITTEE MEMBERS
(Name Cards are needed)

> *Mr./Mrs. Simmons (Chairperson):* You have been appointed chairperson by the principal. You tend to have a liberal outlook and believe that there must be good reason behind the "selective madness" of the student vandals. Beyond your personal views, you are anxious to be an effective chairperson, and you realize how important it is for both students and

faculty that a fair decision is reached by the committee. As chairperson, you begin the first meeting by saying: "Our principal asked me to chair this committee. . . ."

Mr./Ms. Vincent: You have been appointed to the committee as a representative of modern elements in the school. You are well respected for being able to act as a harmonizer and integrate diverse points of view. You are not considered a person out for personal power or gain, and you have the trust of most of the faculty.You do have strong opinions and are not afraid to express them, but, on this issue, you have very mixed feelings and do not enter with a firm position in mind.

Mr./Ms. Paul: You are often perceived as antiadministration and antiestablishment. You tend to see students as pushed down by the system, and you think that, in many ways, the administration is guilty of the pent-up frustration many students feel. In the past your outspoken views have gotten you into trouble. In fact, you know that a promotion was rejected because of your strong position on one occasion. Actually you are attempting to mend fences and want to be perceived as less extreme so that possible promotion may come your way, and put you in a position whereby you can be more effective in changing those conditions that seem to inhibit the learning process. At this moment you feel very anxious abut your role in this meeting.

Mr./Ms. Peters: You take a strong stand in defense of whoever did this thing. It is your belief that only very angry and frustrated persons could have done it and that they are more in need of help than punishment. The environment is to blame, and you personally believe that the three maligned teachers contribute more than their share to that environment. Also, you know one of the culprits well, like him, and respect his family. This could be a scandal to them. For these and other reasons you wish to see what happened as a symptom of a larger problem that needs to be attacked with vigor.

Mr./Mrs. Johnson (Assistant Principal): As the administration's representative on this committee, you feel that something decisive must be done as a deterrent to further problems of this nature. You are more concerned with how you are perceived in an administrator's role than anything else and realize that your performance will be carefully observed by the principal, who has a strong vested interest in immediate action. You do not like friction and find personal confrontation difficult to take. This is a problem because the principal is a very confronting person and admires directness and toughness in people.

Mr./Ms. Smith: You personally feel that the children involved must be punished and taught a lesson. Not only does your security as a teacher depend on this, but severe punishment is the only possible vindication for the three teachers (two of whom are close friends). You have been in the school for a long time and feel resentment at the changes you have witnessed during the last five years. Not only have academic standards deteriorated, but discipline has almost vanished in the classroom and in the halls. In your mind, the only way to handle these students is through immediate and severe action.

Action Continued

The facilitator has the individuals playing the same roles in groups 1 and 2 and in groups 3 and 4 spend a few minutes practicing and shaping their roles so that they enter the discussion with some conviction and involvement.

Second Role-Play Session

After the discussion of the first role play, the chairperson, Mr./Ms. Simmons, takes the role of Mr./Ms. Vincent in each group. Now, the original Mr./Ms. Vincent goes from group 1 to group 2, and from group 2 to group 1, and takes over the chairperson role that has been vacated. He or she attempts to implement new approaches to the problem and the various conflicts that appear to exist.

Discussion

After the conclusions of groups 1 and 2 and 3 and 4 have been shared with the total group, the facilitator may find the following questions and observations helpful in bringing further issues to the surface or in summarizing the various issues.

Should the chairperson be neutral or take a substantive position?

Have the decision-making procedures been defined for the group? Who has made this decision?

What approaches has the chairperson used to reduce personal goals and develop a shared group goal?

Was there an opportunity for individuals to release personal feelings without polarizing the group further?

What efforts have been developed to insure more data for the committee in its deliberations?

Given the urgency and short time, how could the most efficient use of the group's resources be insured?

EXERCISE 4 Force Field Analysis:[4] From Diagnosis to Action

Objectives

To provide participants in the decision-making process a means of thoroughly diagnosing the factors causing the particular problem

To help those involved to look beyond the obvious and into new responses to the problem condition

To focus upon the possible repercussions of any decisions

Rationale

It is assumed that in most of the decisions we make we fail to have access to or, if available, to use all the information relating to a particular problem.

Basically, we fail on three counts in the problem-solving process. First, we often enter problem situations with some preconceived notion of what we would like the outcome to be. Thus, we fail to do a very thorough diagnosis of all the causal factors that created the tensions, and it is seldom that all the relevant data reach us. (It takes time and energy, and may lead to conclusions not desired.) Second, for similar reasons, we often limit our perception of all the possible alternatives for changing the existing condition. Finally, when making particular decisions, it is seldom that we look beyond immediate reactions and explore all the possible repercussions that could result.

The primary aim of this exercise is to give the participant access to more data and alternatives and a greater awareness of their possible implications in terms of later consequences.

Setting

Groups of about six or seven are ideal for this exercise, but the method can be effectively used by individuals or large groups in general problem-solving sessions. Often the method is quickly presented and the group fails to understand the reason behind the approach. Superficially it may appear to be nothing more than listing positive and negative forces that influence a decision. However, it represents a way of thinking about a problem and the changes that any decision resulting from the problem analysis will create (inten-

[4]This design is drawn from Lewin, 1948; 1951.

tional or unintentional). Enough time must be allowed for the process to develop, although it is difficult to suggest how long since this depends on the nature of the problem, the motivation of the group, and the actual time available. Some designs have allowed 2 days for the analysis and development of specific solutions. Less than 2 or 3 hours on organizational problems or those of a small group may prove to be sufficient. Newsprint, markers, and tape should be available so that the ideas developed can be easily seen and recorded as the group moves through various phases of the exercise toward solution.

Action

A group using the Force Field Analysis should be involved in a brief theory and practice session. For example, the facilitator may point out that groups, when looking at a problem that needs to be solved, tend to: (*a*) move too rapidly toward a solution, (*b*) begin to argue and polarize, and (*c*) fail to look at all the causal factors behind the problem.

The following method helps to get out the data and explore them in a rational nonjudgmental fashion. The first step is to view the problem as a condition that needs to be changed, and success will be determined by just how much this condition is altered. For example, smoking is a condition that exists to a certain degree in some people. If a person smokes a pack of cigarettes a day and thinks he or she should stop, then the problem becomes altering the particular condition (one pack a day). One reason individuals who attempt to stop smoking find it so difficult is they make it an all-or-nothing proposition, and it becomes a test of personal will. As in many other problem conditions, the person fails to look at the multitude of factors that are causing him or her to smoke. The self-will issue is only one and by no means the most important. Unless all the restraining forces inducing him or her to smoke are understood, a strategy for altering the condition will tend to be limited in its impact. The strategy must attack many of these factors. Some of these factors are depicted in the figure below.

In the mind of the individual, each of these factors has a different weight and importance in restraining him or her from reducing the amount of cigarettes. Also, these restraining forces may change in weight and in character. For example, a mother-in-law visiting the house may add another source of tension, another force that may actually increase the number of cigarettes smoked. Thus, with more weight pushing downward, the problem condition changes, and more cigarettes are smoked. See the figure on page 420.

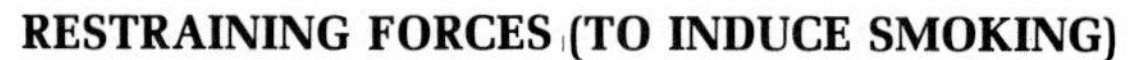

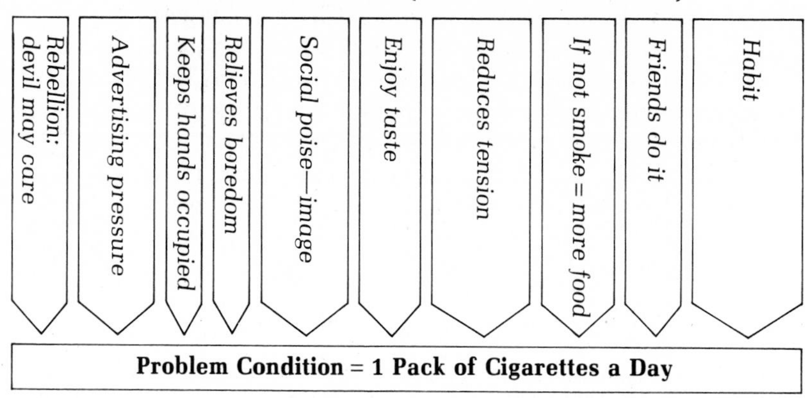

However, we have looked at only half the picture. There are also existing in this individual's life a multitude of forces driving or pushing him or her to give up smoking: health advertisements, family pressure, sore throat, his or her own test of will, cost and so forth. Each of these forces also has its own weight and is pushing against the restraining forces. They too can change in their importance (weight), as, for example, when the individual is visited by a relative who has had a lung removed because of lung cancer and it is discovered that he or she smoked two packs a day before the operation. For a while, anyway, this may add enough of a push to actually reduce his or her smoking output for a few days. It is the point at which the restraining forces and the driving forces impose the same theoretical amount of pressure that the level of smoking is determined. With more pressure from below (cancer reports, sick relative, and so forth) the problem conditions will change to less smoking. More tension at work, a visiting mother-in-law, fear of weight gain may alter the balance of forces and increase the amount of smoking. See the figure on page 421.

In one sense, the one-pack-a-day level represents a point of temporary equilibrium between the two competing forces. If a person is really interested in altering this level, or problem condition, then he or she may work at reducing the number of restraining forces, or he or she may work at increasing the number of driving forces that will then push the point of equilibrium (amount of smoking) to a new level representing fewer cigarettes a day. (Note: During the process of this brief theory session, it is helpful to have the participants add the forces in this example while the facilitator sketches them on newsprint or a blackboard. It is important that they grasp the feeling of weight or forces that are acting upon the person.)

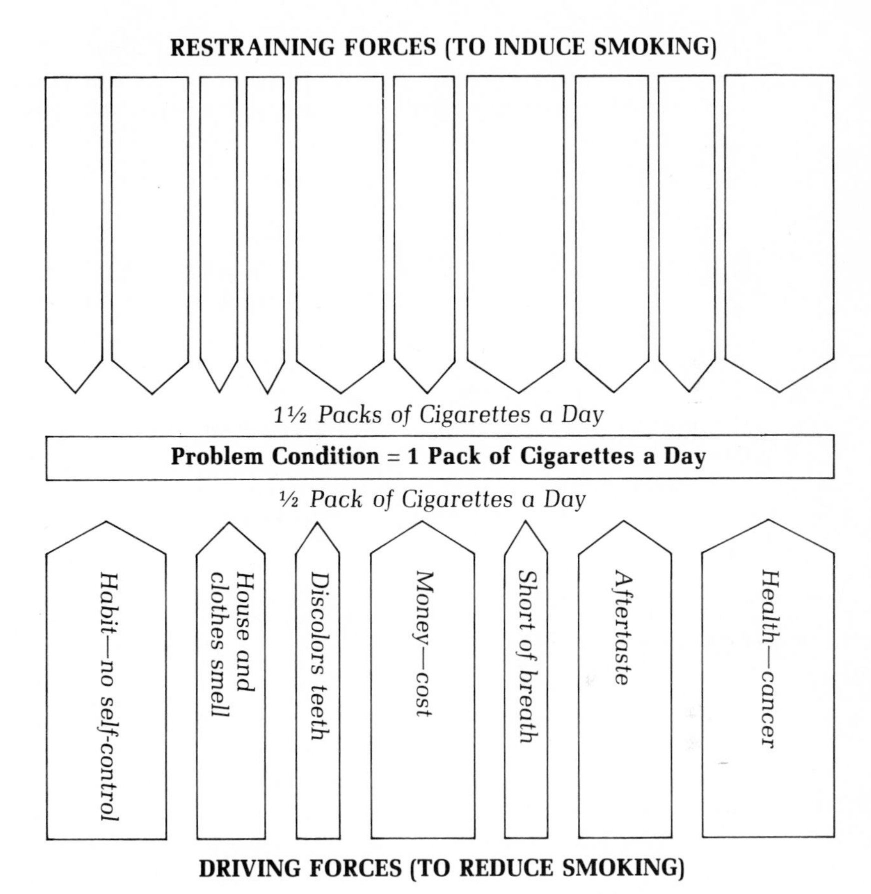

At this point, the facilitator may wish to ask which (if there is to be a choice) appears more useful: to try to remove the restraining forces, or to add driving forces in order to reduce the level of smoking. The majority of individuals will tend to feel that adding to the driving forces would be easiest and most appropriate. Given more time to think and discuss the question, there will be a movement to a position of removing the restraining forces. The fact is that for most people change is the result of force or coercion. The problem is that it seems to hold true in the social sciences almost as much as in the physical sciences: for every action there is a reaction. People, whether individually or in groups, react to force or pressure. They build defenses to offset such driving forces. For example, "If you keep smoking, you're going to cut five years off your life" (coercion, threat). A frequent response may be, "Well, I would rather die sooner and be happy than cut out all the little pleasures of life"

(rationalization, compensation). The response by cigarette companies to the surgeon general's famous cancer report was to increase cigarette advertising to a point that six months later total consumption had actually risen (by now, however, significant declines have taken place). The point is that none of these compensating reactions occurs when effort is directed at reducing the restraining forces, particularly those with the greatest force.

The use of driving forces can be an important factor in any movement toward change. The problems arise when unanticipated reactions create more problems than the newly introduced driving force is worth. It is from the driving forces, however, that creative new ideas tend to be generated. Thus, it may be anticipated that a new manager will bring many problems as people adjust to his or her new style and expectations, but, hopefully, these will be anticipated and offset with orientation sessions, social hours, personal discussions, and employee involvement in some of the accompanying changes.

Problem-solving Sequence Using Force Field Analysis

Having introduced the group briefly to some of the theoretical notions (it may be useful to take a few minutes to help the group work through an issue more relevant to them from a group perspective on the board), they are then given the opportunity to use the approach systematically in exploring a problem and developing specific solutions. Taking 3 hours as an example in this case, a possible problem-solving sequence might look like this:

Step 1 (10 min.) The problem is defined as a condition that exists with forces impinging upon it. The issue must be specific and one in which the group has an interest and some power to change. (Again, how the problem is arrived at is a very important step for the facilitator to consider.)

Step 2 (20 to 30 min.) The driving and restraining forces impinging upon this particular condition are brainstormed. It is best not to qualify or evaluate any responses during this phase, but simply list them on a large sheet of newsprint (the restraining forces are placed on the left of the sheet going down the paper and the driving forces on the right).

Step 3 (30 to 40 min.) The groups now concentrate their efforts on the restraining forces (particularly if time is limited). They do this by ordering the list they have brainstormed into priorities. Thus, they are, in a sense, weighting the forces and can more easily see those with which the group must come to grips. This process of establishing priorities is also a diagnostic process for the working group. It is best if the priorities can be agreed upon through consensus and if individuals try to be relatively flexible since a variety of solutions will probably be forthcoming.

Step 4 (10 to 15 min.) Having arrived at agreement on priorities and defined the restraining forces more clearly, it should be possible for the group to eliminate some because they simply do not have time, money, resources, or the power to do anything about them. This is an important period of reality testing since the group can now focus its energies on those restraining forces that they have the power to change.

Step 5 (40 to 60 min.) Having established those restraining forces that can be worked with by the group, the participants attempt to establish specific ways of reducing these restraining forces in a manner that will minimize possible offsetting reactions. It might be best to have subgroups of two or three people focus upon particular forces and, after they have developed a number of concrete solutions, report them back to the larger group. If time permits, suggestions for new driving forces may also be developed.

Step 6 (20 min.) Specific proposals including methods for implementation and group accountability are developed.

Step 7 (30 min.) The total group is presented with the specific proposals, and a discussion follows (the group will probably not be able to handle more than 30 or 40 minutes after working for 3 hours). At this time, suggestions should be entertained for insuring that the various proposals are integrated into a strategy for action. This step is of utmost importance since it insures follow-up and commitment of further involvement by the very people who have developed the proposals. A cross-group steering committee may be a useful mechanism.

Step 8 An evaluation would take place at some agreed-upon date in the future and would provide the large group the opportunity to appraise what impact the various suggestions have actually had.

Obviously, there is no magic in these eight steps, or in this particular time frame. They simply provide a structure in which to implement the Force Field as a tool in the problem-solving process. It focuses on the building of constructive alternatives for action and attempts to draw out as many factors as possible that may inhibit potential solutions.

EXERCISE 5 The Acid Floor Test

Objectives

To analyze decision making and problem solving under extreme time pressure

To develop supportive roles within a group

To observe individual behavior under stress—both facilitative and inhibiting

Setting

This task is more effective when it is done competitively. If there are enough materials available, as many as three or four teams of from six to ten players can participate at the same time. If materials are limited, competitiveness can be achieved by placing a time restriction on the exercise (each team works in isolation). For each group that participates it is necessary to have available eight empty no. 10 cans. It is also helpful to have observers assigned to each group to watch carefully the nature of the problem solving that takes place, how resources in the group are maximized, and how leadership develops within the context of the activity.

Action

The facilitator gives the following directions to the team(s):

> You are members of a gang that has taken a short cut across another gang's territory. You have been spotted by one of their lookouts and are aware that they are mobilizing a group of some thirty boys to come after your small group. You have decided to cut through an alley that leads to your own territory. Halfway down the alley you see that the other gang has spread acid across the alley. It stretches at least thirty feet or more ahead of you. You realize that it is too late to go back the way you came. Near the point where the acid begins the group sees a number of large cans. They see the cans as offering a way over the acid. But how? Can they figure how to get all of their gang over the acid before the other gang arrives?

There are some specific rules. No part of a person's body may touch the acid floor. If it does, he or she must return to the starting point and bring his or her cans with him or her. No person may sacrifice himself or herself and run through the acid setting up cans. When the last person crosses, he or she must bring the cans to the far side with him or her so that the pursuing gang cannot follow over the acid.

This event is timed.

Discussion

By teams, the observers describe how they saw the decision-making process develop. The group may be asked to comment on these observations and discuss what might have been done better. Why or why not were other alternatives developed? Was there dependence in the group on one or two people? Why? Was there consideration for all the members of the team? Was their performance (going across on the cans) ridiculed, or was there a supportive, all-for-one attitude?

EXERCISE 6 The Devil's Dilemma[5]: Combining Reality and Creativity in the Problem-solving Process

Objectives

To stimulate a creative problem-solving session

To help the participants face directly the kinds of realities that often keep good ideas from getting off the ground

To insure maximum participant involvement

Rationale

Problem-solving sessions all too often become so serious and tense that participants stop enjoying what should be, in many cases, an exciting and innovative process. On the other hand, many planners and problem solvers get carried away with the creative process and forget the kinds of here-and-now realities that prohibit the execution of many ideas. The think-tank approach to problem solving is usually too removed from the source of the problem, and critical factors are often overlooked. The intent of this exercise is to make sure that reality is tested again and again throughout the entire problem-solving process and that participants have an enjoyable time while working.

[5] The authors are indebted to Clark Abt, who developed the original design from which this exercise has been adapted.

Setting

This particular exercise can be conducted with a minimum of seven or eight individuals to as many as sixty or seventy (depending on some adaptations to the design presented below). An ideal number ranges between about sixteen (two working groups) to about twenty-four (three working groups). The use of three groups provides a most stimulating design possibility and allows for the effective use of intergroup competition and a great assortment of ideas. Below is an outline of a three-hour work session that is part of an all-day workshop. The reader will be able to alter the time schedule and numbers involved to suit his or her particular needs. In this case, it is assumed that the group of twenty-four participants has come together to work through a variety of problems. The morning session is being devoted to a single problem of importance to the entire group (how this problem is arrived at and defined may be another "problem").

Action

The members of the large group are divided randomly[6] into four working groups (names should be known or name tags provided). Actually, three of the groups will work at the task, and one group will be composed of "devils" (devil's advocates) who will help the various work groups test the reality of their ideas. The work session is scheduled to begin at 9:00 A.M. of which the first 30 minutes is an informal coffee session. The facilitator feels that by 9:30 all the participants will have arrived and be ready to get down to serious work.

9:00–9:30 Arrival of participants, coffee, informal talking.

9:15–9:30 Directions to those individuals (six) selected to be devils and assignment of groups to other members. The devils will play a key role in the success of this session and must clearly understand their function and its implications in the decision-making process. Two of these individuals are assigned to each group (it is also possible to have all six moving freely among groups, but this may prove confusing). As the discussion develops around the problem, issues may evolve that reduce the effectiveness of the group. The devil can bring these to the attention of the members. This cannot be done verbally; it must only be done by writing a

[6] If the issue represents a polarization of views in the group, it may be interesting and useful to build one group with members of similar opinions, another group with members of opposing views, and a third group composed of individuals with mixed opinions. Very often the solutions of the three groups are quite similar once all the realities have been considered, and once the groups begin to plan and stop being defensive over their particular positions.

note to the group and letting members respond to it among themselves. This keeps the devil in the role of an objective observer interested in the group process and asking questions he or she believes need to be raised. Obviously too many notes to the group could be disruptive and eventually ignored. Discretion calls for a limited use of this role. Examples of the kinds of interventions are many. For example, the observer may make observations about time, the fact that only a few people seem involved, that the group is stuck on a few ideas and has not explored other possibilities, or that they are straying from the topic into irrelevant areas. Or the devil may question a certain conclusion—in other words, actually take the role of someone outside the group who may be affected by a particular idea or decision. This gives the group a chance to test reality from an outside perspective. The following are a number of interventions that a devil might make during the planning session.

1. You are talking about solutions before you have defined the problem clearly.
2. You have spent a half hour talking about an issue that can be worked out later.
3. I represent an administrator. I question your estimation of cost; it may be twice your figure if you consider training costs.
4. Three people in the group have somehow not had much of a chance to get an idea in edgewise.
5. Time is getting away from the group. Perhaps it would help if you spent some time brainstorming particular solutions, without evaluating any of the ideas until later.
6. I represent a subordinate in the company. You are assuming that I will not resent spending additional time on this task. I am already overburdened, and I do not believe you have taken my feelings into consideration. Depending on how you present the idea to me may determine my level of acceptance or hostility to the proposition.

Thus, the devil must occasionally help paint a more complete picture for the group. His or her statements must be objective, directed at helping the group clarify an issue, or aimed at facilitating the communication and decision-making process among the participants. Again, moderation in the number of inputs will give them greater value and effectiveness.

9:30–9:45 The problem is clearly defined (again, it is assumed that the problem is of interest), and a schedule is presented. The schedule should be placed on newsprint in view of each of the working groups:

9:45–11:15	problem-solving session
11:15–11:30	break
11:30–12:30	presentations
12:30– 1:45	lunch break
1:45– 2:15	summary of morning session
2:15– 5:00	afternoon program

At this point, the role of the devils (other names for these participants may be preferred) must be clearly explained, and it should be stated that their purpose is to help the group with their own work process and also to test reality in terms of their specific proposals.

9:45–11:15 It is during this 90-minute session that the problem is discussed, analyzed, and proposals for its solution developed. By 11:15 each small group must have a plan it can present to the total group outlining intended action and the steps to be taken for their implementation.

11:15–11:30 This is scheduled as a break, but it will also enable groups that have not completed their task additional time. The facilitator should also meet with the devils for a few minutes to outline their role during the next hour. (See below.)

11:30–12:30 Group presentations will be made. The presentations themselves (with the help of newsprint) should take no more than about 5 minutes. During the remaining 15 minutes allocated to the first planning group, the total group of devils will be able to criticize the group's ideas. These individuals are instructed to respond to the ideas presented as if they were outsiders who may be influenced by the proposal. Thus, one of them can say, "But, as a person who is going to be influenced by this decision, I resent not having been involved in the planning and decision-making process."

Or another devil may say, "As an administrator, I see no place in your implementation phase for accountability. Who is going to see that your plan is actually implemented, and who will be responsible for its overall success or failure? The plan seems sound otherwise."

These comments should be heard without comment from other observers, although the group members making the presentation can reply to the role-played questions. Other members of the group who are observing may also frame questions in a similar manner. The point is to raise the kinds of questions that could result in the failure of the particular proposal. The facilitator will probably have to help keep the discussion within the proposed framework, otherwise the time will not be sufficient. If all the questions have been discussed, the presenting group may wish to accept suggestions that would strengthen their proposal. The three groups proceed through this process. Thus, each has time for a presentation, an opportunity to clarify certain points, and a chance to hear the kinds of criticism that might well be generated once the proposal was implemented.

12:30–1:45 It is suggested that a representative of each of the three problem-solving groups and two of the devils have lunch together. With the previous discussion in mind, it is their task to integrate the various ideas into a plan that can be presented back to the total group after lunch. There should be considerable overlap among the proposals so as to ensure the possibility for integrating the three proposals. This is not easy, and the facilitator may wish to help the group in the organization of their task.

1:45–2:15 Often it is difficult after lunch and a morning of hard work to motivate a group into further work. This is not usually the case when the lunch-planning group presents its integrated proposal for action to the other members. During its presentation, the large group may raise some serious questions that require more work. If they cannot be resolved on the spot, it may be possible to allow this work group to reconvene and consider alternatives that will satisfy the criticism, while the other group members continue with the afternoon program. Hopefully, this will not be necessary. Early closure brings a real sense of accomplishment and new vigor to the afternoon program.

EXERCISE 7 Developing Feedback: An Important Step in Open Communication

Rationale

Any system, whether it be an individual human being, a small group, or an organization, must have some source of information about itself (we are notoriously poor judges of our own performance). This process of receiving information about one's self is referred to as feedback. An examination is one source of feedback for a student even though more and more it is merely used as means of classifying individuals and often loses its value as a meaningful feedback instrument. In the course of a group meeting, thousands of verbal and nonverbal communications take place, thousands of internal responses occur within individuals, and a great deal occurs that is never recorded in the minutes. Yet, the next meeting of the same individuals may be dramatically influenced by just these events. Unless a feedback system is established within any working group, an enormous amount of time and energy may be poorly used as individuals attempt to cope with the many procedural and human problems that can hinder a group. The diagnostic process described in the previous exercise is, in fact, a form of feedback. However, if it is to really benefit the group over time, it must be legitimized as an important part of the work process. Feedback ideally is a welcomed source of information for the group as a whole or for individual members. If misused, it can be a destructive, inhibiting, and degenerative force; if skillfully integrated into the group, it can facilitate more effective communication and better working relationships. The following exercise represents one means of introducing the concept of feedback to a group so it can explore whether or not it is ready to be used more systematically.

Setting

There are many ways that information can be developed to provide a group with insights on its operating procedures and the consequences of its behaviors. Such information, however, will not be useful if it is imposed or if there is no opportunity for the group to do anything about it. The group, like the individual, must be ready to hear about itself and desire to act on what it hears. If the facilitator has evidence that the group is ready to deal more directly with its own process, the following exercise may be useful. One means of discovering readiness is simply to raise as a point of discussion whether or not the group would like to take the time to explore how it operates (that it will take too much time is an easy defensive response and suggests a rather low priority for the process efforts).

Action

The group is divided into subgroups of about three people, and each is asked to develop a set of rules that might be discussed and applied to any feedback efforts. For example, how can the group learn about its operation? How effective and ineffective is it? How does it facilitate and hinder the working process? Perhaps there are rules the group can agree on that will make the feedback more easily received and thus more easily given.

After 15 or 20 minutes, the results are posted by the facilitator on newsprint, with each group requested to give the most important recommendations first. The posted list represents a loosely grouped statement of rules and recommendations that have some priority to the various subgroups. To this list the facilitator may wish to add a few suggestions and integrate those already listed into a brief theory session. The key is that if people are going to hear about themselves, it must be done in a manner that creates the fewest possible defenses and insures a climate of acceptance. Some of the suggestions that may be forthcoming from the session are:

1. The information about the group is descriptive and not colored by value-laden adjectives of a good-bad nature.
2. The examples of group behavior being discussed should be specific and clarified through examples.
3. Whenever possible, the information should be given sooner rather than later. The longer the time between the behavior and its discussion, the less value.

4. The information must be confined to matters that are within the power of the group to do something about.
5. The group must seek the feedback. If it is not solicited, it will be met with resistance and probably have little positive impact.
6. A person and a group can only internalize a certain amount of information about itself at any one time. It is important not to overload the system with more than it can handle.
7. If a person presents his or her perception of how the group behaved in a certain instance, it is important to discover whether or not others share his or her view.

Following this, it may be useful to focus upon the areas of group activity about which information might be collected in a somewhat objective manner. A 5-minute brainstorming session might result in some of the following:

Communication patterns in the group (who talks and how much)
How problems are actually solved and decisions made
Problems that seem to keep the group from moving
The goals of the group (implicit and explicit)
The level of involvement and interest among the participants
Whether there is shared leadership in the group
The physical structure of the group
The roles taken by group members (constructive and destructive)
The degree to which the group is open or closed

Thus, in a matter of about 40 minutes, the group can establish some suggestions that it feels necessary for the presentation of feedback to be effective as well as useful. Now, following these guidelines, the final 20 minutes is designed so that the original groups of three reconvene and agree on one piece of feedback that should be presented to the larger group in a manner that will benefit the whole group in its future work together. It may be in the form of a comment on the group behavior that indicates the need for new behavior. However, it should be left for the large group to interpret the feedback and decide on any relevant action. The idea, at this particular point, is to get the group practicing the process they have been discussing. This gives them a good starting point for the next session.

Follow-up

If the group has been able to accept the notion of feedback with some equanimity, the way will be open for the facilitator to suggest (or, if time, have the group develop) a number of ways feedback within a group may be carried out. For example:

1. Appointing an observer of the group's process who reports his or her systematic observations at the end of a meeting. A discussion of the implications would follow (it may take no more than 15 minutes).
2. Having the group fill out reaction forms to the meeting at the end of a session. These may focus upon various aspects of the group process.
3. Taping the meeting and then replaying a portion of the tape and briefly discussing its implications.
4. After building confidence in the feedback process and trust in one another, members look at the roles and specific behaviors of individual members to see how they influence the group. This level of feedback may take a long time to evolve, and it is best for it to develop naturally out of a deeper concern and involvement on the part of the members. In other words, a group is not just suddenly ready for individual feedback. It is simply a natural extension of the slowly developing willingness of the group to process its own behaviors.

EXERCISE 8 Post-Session Feedback

Rationale

Most task groups hardly have time to complete their business commitments, let alone spend much time exploring the process of the group. The comment is often heard, "If we get into that subject, we'll be here all night." This fear of personal overinvolvement outside the actual working agenda often can shut down all efforts to develop more effective working relationships. It is not necessary for a group to spend an inordinate amount of time in its process efforts, nor does it necessitate that members become overly personal. However, time must be allowed for the group to improve its own working relationships, or tensions and problems will subtly build up and eventually reduce effectiveness. Following are two simple suggestions for helping to keep the process level of group work legitimized.

1. After the working session, 10 minutes is set aside for a discussion, in pairs, of the question: "What are one or two ways that this group could improve its working procedures or relationships the next time it meets?" It is important that the participants focus on specific, constructive suggestions. If, for example, one member dominated the discussion and created hostility among many of the other partici-

pants, it might be helpful to establish a temporary mechanism for insuring that more individuals have an opportunity to share their views and also to help move the group past individual roadblocks. In a group where the level of trust and acceptance is high, it would not be inappropriate to share with an individual the problems created by his or her particular behavior. But, as suggested previously, the climate must be such that the individual desires the information and those giving it are skillful enough not to appear punitive or judgmental.

The group may also find the post-meeting session an avenue for suggesting a structural format in the meetng, for example, the role of the chairperson. Similarly, the way the agenda is being formed may influence the feeling of individuals about participating, and a change in how this is accomplished may affect other aspects of the meeting.

The issues briefly raised by the individuals in the paired groups are then shared briefly with the total group. They are first shared without discussion in order to establish how much agreement there is among the different members. Then suggestions are taken to remedy the situation in time for the next meeting. The total process should take no more than 20 or 30 minutes. After the group has developed acceptance of the feedback idea, the first step of breaking into twos or threes will be unnecessary, and observations can be shared by the whole group. This would cut the process time down to 10 or 15 minutes, although, as groups become more open and communicative, there is a tendency to broaden the scope of the feedback process. It is possible that this can become a problem because many members may find feedback a fascinating and personally satisfying experience. Groups have been known to spend more time discussing their process than the task. Obviously, it is a sign of a mature group if it is able to use feedback in a constructive fashion rather than a means of meeting individual emotional needs that extend far beyond the purpose of the group.

2. If certain members of the work group are involved in establishing the format of a particular meeting (often this is a rotating responsibility in which agenda building and building procedures for a meeting change hands regularly), the following feedback procedure may prove useful: At the end of a meeting, a reaction sheet is passed out to participants, who answer explicit questions on operation of the meeting and improvements. The group responsible for the next meeting analyzes these responses (thus, it takes only about 5 minutes of the group's time), and they make plans to incorporate various changes in the format of the next meeting that they feel respond to the concerns and suggestions given. These new procedures or innovations are then evaluated in the reaction sheet developed for that meeting. In this

way, there is a constant willingness to look at how the group is working together and an opportunity to develop new ideas. Theoretically, each person eventually has a chance to improve the meeting. It may be that a certain format develops that is basically satisfactory to the members; this too will be found through the regular use of reaction sheets.

BIBLIOGRAPHY

Adams, J. K. and P. A. Adams. "Realism of confidence judgments." *Psychological Review*, 68 (1967), 33–45.

Argyris, C. and D. A. Schon. *Theory in Practice*. San Francisco: Jossey-Bass, 1974.

Bandler, R. and J. Grinder. *Frogs into Princes: Neuro-Linguistics Programming*. Moab, Utah: Real People Press, 1979.

Bergum, B. O. and D. J. Lehr. "The effects of authoritarianism on vigilance performance." *Journal of Applied Psychology*, 47 (1963), 75–77.

Bouchard, T. J., Jr. "Personality, problem-solving procedure and performance in small groups." *Journal of Applied Psychology*, 53, No. 1, Part 2 (1969), 1–29.

Bouchard, T. J. "Training, motivation and personality as determinants of the effectiveness of brainstorming groups and individuals." *Journal of Applied Psychology*, 56 (1972), 324–331.

Brehm, J. "Post decision changes in desirability of alternatives." *Journal of Abnormal and Social Psychology*, 52 (1956), 384–389.

Campbell, J. P. "Individual versus group problem solving in an industrial sample." *Journal of Applied Psychology*, 52 (1968), 205–210.

Collaros, R. A. and L. Anderson. "Effects of perceived expertness upon creativity of members of brainstorming groups." *Journal of Applied Psychology*, 53, No. 2, Part 1 (1969), 159–164.

Crombag, H. R. "Cooperation and competition in means-interdependent triads: A replication." *Journal of Personality and Social Psychology*, 4 (1966), 692–695.

Dalkey, N. C. *The Delphi Method: An Experimental Study of Group Opinion*. Santa Monica, Calif.: The Rand Corp., 1969.

Davis, J. H. "Group decision and social interaction: A theory of social decision schemes." *Psychological Review*, 80 (1973), 97–125.

Davis, J. H., P. R. Laughlin, and S. S. Komorite. "The social psychology of small groups: Cooperative and mixed-motive interaction." *Annual Review of Psychology,* 27 (1976), 501–541.

Delbecq, A., A. H. Van De Ven, and D. H. Gustafson. *Group Techniques Program Planning.* Glen View, Ill.: Scott Foresman, 1975.

Deutsch, M. "An experimental study of the effects of cooperation and competition on group success." *Human Relations,* 2 (1949), 199–231.

Deutsch, M. "A theory of cooperation and competition." *Human Relations,* 2 (1949b), 129–152.

Deutsch, M. "Socially relevant science: reflections on some studies of interpersonal conflict." *American Psychologist,* 24 (1969), 1076–1092.

Easton, A. *Decision Making: A Short Course in Problem Solving.* New York: John Wiley and Sons, 1976.

Edwards, W. "Behavioral decision theory." In *Decision Making.* Ed. W. Edwards and A. Tversky. Baltimore: Penguin Books, 1967a, pp., 65–95.

Edwards, W. "The theory of decision making." In *Decision Making.* Ed. W. Edwards and A. Tversky. Baltimore: Penguin Books, 1967b, pp. 18–40.

Festinger, L. "Informal social communication." *Psychology Review,* 57 (1950), 271–292.

Festinger, L. *Theory of Cognitive Dissonance.* Evanston, Ill.: Row, Peterson, 1957.

Festinger, L. and E. Aronson. "Arousal and reduction of dissonance in social contexts." In *Group Dynamics Research and Theory.* Ed. D. Cartwright and A. Zander. New York: Harper and Row, 1968, p. 125.

Festinger, L. and J. Carlsmith. "Cognitive consequences of forced choice alternatives as a function of their number and qualitative similarity." *Journal of Abnormal and Social Psychology,* 58 (1959), 203–210.

Fiedler, F. E. and W. A. T. Meuwese. "Leader's contribution to task performance in cohesive and uncohesive groups." *Journal of Abnormal and Social Psychology* (1963), 83–87.

Gordon, W. J. *Synectics.* New York: Collier Books, 1961.

Gore, W. J. *Administrative Decision Making: A Heuristic Model.* New York: John Wiley and Sons, 1964.

Gouran, D. S. *Discussions: The Process of Group Decision Making.* New York: Harper & Row, 1972.

Hackman, R. J. and N. Vidman. "Effects of size and task type on group performance and members' reactions." *Sociometry*, 33 (1970), 37–55.

Hall, J. and M. S. Williams. "Group dynamics training and improved decision making." *The Journal of Applied Behavioral Science*, 6 (1970), 39–68.

Hammond, L. and M. Goldman. "Competition and non-competition and its relationship to individuals' non-productivity." *Sociometry*, 24 (1961), 46–60.

Hare, A. P. *Handbook of Small Group Research*. 2nd ed. New York: Free Press, 1976.

Heider, F. *The Psychology of Interpersonal Relations*. New York: John Wiley and Sons, 1958, pp. 75–82.

Hoffman, L. R. "Applying experimental research on group problem solving to organizations." *Journal of Applied Behavioral Science*, 1593 (1979), 375–391.

Hoffman, L. R. and M. M. Clark. "Participation and influence in problem solving groups." In *The Group Problem Solving Process: Studies of a Valence Model*. Ed. L. R. Hoffman. New York: Praeger, 1979.

Kelley, H. H. and J. Thibaut. "Group problem solving." In *The Handbook of Social Psychology*. 2nd ed. Vol. 4. *Group Psychology and Phenomena of Interaction*. Ed. G. Lindzey and E. Aronson. Reading, Mass.: Addison-Wesley, 1969, p. 1.

Kepner, H. and B. Tregoe. *The Rational Manager*. New York: McGraw-Hill, 1968.

Lewin, K. *Principles of Topological Psychology*. New York: McGraw-Hill, 1936, p. 25.

Lewin, K. *Resolving Social Conflicts*. New York: Harper and Row, 1948.

Lewin, K. *Field Theory in Social Sciences*. New York: Harper and Row, 1951.

Maier, N. R. F. *Problem-Solving Discussions and Conferences*. New York: McGraw-Hill, 1963, pp. 193–195.

McClintock, C. G. and S. P. McNeil. "Reward and score feedback as determinants of cooperative and competitive behavior." *Journal of Personality and Social Psychology*, 4 (1966), 606–613.

McCurdy, H. G. and W. E. Lambert. "The efficiency of small human groups in the solution of problems requiring genuine cooperation." *Journal of Personality*, 20 (1952), 478–494.

Mintzberg, H. "Planning on the left side and managing on the right." *Harvard Business Review* (July–August 1976), 49–58.

Moscovici, S. and M. Zavalloni. "The group as a polarizer of attitudes." *Journal of Personality and Social Psychology,* 12 (1969), 125–135.

Napier, H. "Individual versus group learning: note on task variable." *Psychological Reports,* 23 (1967), 757–758.

Newell, A. and H. Simon. *Human Problem Solving.* Englewood Cliffs, N.J.: Prentice-Hall, 1972.

NTL Institute for Applied Behavioral Science. *Reading Book: Laboratories in Human Relations Training,* Washington, D.C.: NTL, 1969, p. 33.

Ornstein, R. *The Psychology of Consciousness.* San Francisco: W. H. Freeman, 1975.

Phillips, D. J. "Report on discussion 66." *Adult Education Journal,* 7 (1948), 81.

Prince, M. *The Practice of Creativity.* New York: Collier Books, 1970.

Pruit, D. G. and M. J. Kinnel. "Twenty years of experimental gaming: critique, synthesis and suggestions for the future." *Annual Review of Psychology,* 28 (1977), 363–392.

Rickards, T. *Problem Solving through Creative Analysis.* London: Halsted Press, 1974.

Robert, H. M. *Robert's rules of order.* Chicago: Scott, Foresman, 1943.

Rotter, G. S. and S. M. Portergal. "Group and individual effects in problem solving." *Journal of Applied Psychology,* 53 (1969), 338–342.

Stuls, M. H. "Experience and prior probability in a complex decision task." *Journal of Applied Psychology,* 53, No. 2, part 1 (1969), 112–118.

Tolela, M. *Effects of T-group Training and Cognitive Learning of Small Group Effectiveness.* Unpublished doctoral dissertation, University of Denver, 1967.

Voytas, R. M. *Some Effects of Various Combinations of Group and Individual Participation in Creative Productivity.* Unpublished doctoral dissertation, University of Maryland, 1967.

Vroom, V. H., L. D. Grant, and T. S. Cotton. "The consequences of social interaction in group problem solving." *Journal of Organizational Behavior and Human Performance,* 4 (1969), 79–95.

Watson, G. "Resistance to change." In *Concepts of Social Change.* Ed. G. Watson. Washington, D.C.: NTL Institute of Applied Behavioral Science, 1967, pp. 10–25.

Watzlawick, P. *The Language of Change.* New York: Basic Books, 1978.

Young, B. I., Jr. "A whole brain approach to training and development." *Training and Development Journal* (October 1979), 44–50.

Eight

The Evolution of Working Groups: Understanding and Prediction

The complex weave of relationships that comprise any small group eventually become rooted in roles, norms, problem-solving procedures, membership criteria, communication patterns, and a number of other conveniently labeled intellectual concepts. These concepts provide us with a means of establishing some order in viewing what might otherwise be a confusing labyrinth of factors that impinge upon the group and determine its success. However, what makes group behavior difficult to predict, even when one has a firm grasp of these concepts, is the never ending stream of personal needs that individual participants are seeking to meet and that can invariably influence the entire group process. In the following pages we look more closely at some of these needs that appear to be shared within most groups. Then, with an understanding of these needs in mind, we discuss the developmental characteristics of many working groups. Finally, we examine distinctive qualities of successful groups as a means of integrating many of the facts learned to this point.

THE TASK AND EMOTIONAL ASPECTS OF GROUPS: A SOURCE OF UNRESOLVABLE TENSIONS

Societies, institutions, and small groups are the breeding grounds for tensions generated from contradictory forces. The forces are real, and their nature must be comprehended if one is to understand the sources of many stresses that originate as a result of human interaction. Several examples will help describe the nature of these forces (Cooley, 1909; Parsons, 1951).

The large mental health hospital is built on a model of informality, individual care, and attention. Its brochure describes the atmosphere of love and consideration that is cultivated in the hospital wards. There, the patients are called "residents," and many of the features so common to institutional living have been purposely removed. Still, one has merely to walk down the corridors, go into the staff lunchrooms, or observe the interaction among the staff to feel the strain that exists. The informality so clearly outlined on paper fails to permeate the roles and expectations of the staff. Psychiatrists have little to say to psychologists, nurses to social workers, social workers to teachers, and no one seems to talk with the day-care workers who spend endless hours with the ward patients. The administration, composed mostly of ex-psychiatrists, attempts to maintain the open doors of informality in the tightly run staff hierarchy, but the tensions remain, and those who suffer most are the patients.

The trouble is that the family-oriented treatment center is built on a power-based status hierarchy and a set of clearly established behavioral norms that more than offset the good intentions. Staff roles are rigid and narrowly defined, respect is based on years of learning rather than experience and performance with patients, and staff relations are based primarily on role stereotypes rather than knowledge of one another.

Another example of countervailing forces follows:

The young religious novice enters the convent in an order that specializes in medical services. Her goal is to be a missionary and to work as a hospital administrator in one of the developing countries. Her personal training focuses on two areas; one is the technicalities of hospital administration (accounting, deployment of services, supervision), and the other involves her own personal growth as a sister dedicated to virtues like love, charity, and faith. Her relations with the other novices and sisters are warm and affectionate, and she finds it difficult to leave for her first assignment in a small Ghanaian hospital. The transition proves to be overwhelming. From the nurturing atmosphere of the convent where acceptance was immediate and unqualified, where gentleness and consideration for others was rewarded, she enters her new environment. Here, there is not enough time, every day is a new crisis, decisions are immediate and based less on human feelings than on expediency and efficiency. Her value to the hospital has little to do with Christian virtues. It depends how effectively she can keep the hospital out of the red, how efficiently she can keep the illiterate workers working, and how well she is able to marshal the limited resources available. Failure means transfer, perhaps into an even less desirable situation and very likely into another area of work.

Thus, from an atmosphere of trust, openness, and acceptance the young sister finds herself in a highly competitive, crisis-oriented situation in which she is vulnerable and where virtues of work far outweigh the personal dimensions so important in her previous life. There can be no right or wrong labels attached to these two situations. Each is real and must be understood in its own context. What is important here is to look carefully at the source of tension that results and its implications for the goals inherent in the particular situation.

In both the example of the convent and the mental health hospital there developed a large discrepancy between expectations and reality, between what was desirable and the conditions that prevailed. What stands out is the constant struggle between work efficiency and personal needs, between measured success in terms of task roles and success in terms of emotions. The dichotomy, of course, will be less apparent where the dual set of expectations is not present. In the army the rules, regulations, and codes of behavior are clearly detailed. A career soldier is fully cognizant of these and can accept the depersonalized nature of many of the relationships that exist. Similarly, a young person applying for a job with a Wall Street bank will find a complex set of rules (both explicit and implicit) that governs his or her behavior, clothes, hair style, language, accepted level of feelings to be displayed, and relations with other workers in the bank. If the individual accepts these rules and knows fully well what is expected of him or her, he or she will probably experience little tension, since the personal-emotional factor will have been screened from most of his or her work involvement.

In most groups and organizations, however, no such clear dichotomy exists, and this is exactly where the stress often originates. For example, in most groups, people wish to be accepted for themselves and not because of academic degrees, superficial knowledge, or other artificial standards; but most groups tend to define success in terms of some visible achievement. Material wealth (how much do you earn?), status (executive assistant to whom?), power (how many people under you?), or tenure (you mean you have twenty-five years with this department?) are much easier to grasp than the hazy variables that form the basis of most personal relationships. Thus, most groups and organizations are simply not established in such a manner that personal acceptance is not conditional and based upon some implicit or explicit achievement criteria.

Work-oriented task groups tend to be high in control, depend on material rewards for motivation, stress accuracy, organize their use of time, and minimize the range of free expression and autonomy allowed. Similarly, such conditions encourage individual competition rather than interdependence, and conformity rather than individuality. Tension occurs when informality creeps in, when personal relations begin to get in the way of efficiency, and when regulations are altered to meet the peculiar needs of particular individuals. The fabric of army life would break down if exceptions were made to the rule. The same could be said for big business, the organized church, and, to some degree, large school systems.

For example, it is difficult to imagine the following situations occurring:

On the day of a major battle, a young private remarks to his field sergeant, "I hope you won't mind, but I probably won't be going to the front today. This headache is killing me." The sergeant looks on with great sympathy and says, "That's all right Joe, we all have days like that; why don't you just rest and take it easy today and save yourself for tomorrow?"

"Miss Jones, you mean you didn't get that rush report in like I asked?" "No, I'm sorry," she replied, "but Johnny surprised me last night, and we went out for the most heavenly lobster thermidor you can imagine." "Well," he said, "I know how you feel about John, and there will be other contracts. After all, we're only young once."

"Well, Mr. Gibbs, how are you today, and did you like my sermon?" The minister waited expectantly for the usual monitored reply. "To tell the truth, I really felt you were talking down to us, and if there is one thing I don't appreciate, it's being lectured to in a condescending fashion. Also, you tended to stray from the point by bringing in humorous asides which, although interesting, tended to distract me from the issue."

Criticism to a minister, sympathy from a field sergeant, acceptance of gross inefficiency from one's boss are difficult to imagine within the context of expected role behaviors and institutional demands. But multiplied a thousand times these expectations condition our behaviors in nearly all groups in which people are involved. They are tied to the strands of a Puritan ethic and to years of involvement in schools, businesses, and churches where acceptance is linked to one's output, dependability, efficiency, and conformity.

Implications for Small Groups

When entering a group, we look for familiar hooks on which to hang our hats, signs that make the unpredictable predictable, sure ways of being accepted. Even in a social group one will begin with credentials, strengths, or skills in order to establish a tone of respectability. When working on a task, meeting a deadline, or in some way remaining highly task-oriented, there tends to be an order and safety in working relations; but if the curtain of formality is drawn away, one can almost feel the strain of another set of forces pushing for greater intimacy and the personalizing of behavior as well as greater authenticity. It implies increasing one's vulnerability and willingness for risk in a group where trust is typically built around performance. Indeed, it is the rare group that can effectively combine social-emotional interests with those necessary for getting the job done. It is not so much that they cannot be combined, but that the combination necessarily increases the complexity of the existing relationships and the risks for the participants. They simply may not be worth the trouble and, in fact, efficiency may decline and overall problems increase.

READER ACTIVITY

We seldom take the time to sit back and consider the subtle, rarely discussed sources of stress and tension that occur within an organization simply because the values and priorities of the organization differ from our personal needs, self-interests, and values. These differences are natural and, to some degree, we must accommodate them. Take a moment and respond to the following questions in relation to an organization (and group) in which you are a participant. This could be a church, social or professional club, school, place of work, or even your family.

1. What gives members status? Are there certain attitudes, behaviors or accomplishments that are particularly important to the organization that seem much less important to you as an individual?
2. If a new person were to join your group in the organization, what would the individual have to do to gain the most immediate acceptance? How do you feel about this? Are there behaviors that you believe are important but are given almost no weight by the organization?
3. In a typical day, what are the kinds of emotions people will tend to show? Are there emotions or feelings you wish would be acceptable but are not?
4. Are there certain rules (stated or unstated) within the organization that are absolutes which you feel are too inflexible, or that tend to dehumanize its members? In what areas is there no room for individual differences where there should be?

At this point the question is not how personal or impersonal, formal or informal, efficient of inefficient a group is. The issue is that the individual interested in understanding groups must be aware that underlying much overt, symptomatic behavior will be causal factors reaching far beyond the particular incident. In many cases the nature of the group or organization will simply not allow the legitimate expression of even basic social-emotional needs. If the resulting pressures and frustrations do not find release outside the group or if informal avenues are not created within, it is likely that tensions will be released indirectly. Usually this occurs by creating interpersonal conflicts around the task issue at hand. Thus, the task itself becomes the outlet for nontask-related tension release. Strange as it may seem, the very presence of restrictions to minimize extraneous issues from undermining the group's work creates new areas of stress that may be even more insidious and difficult to deal with. Exactly how the weave of interpersonal forces influences a group in its development can be understood by following what appears to be a typical group from its inception through various stages of its growth.

THE STAGES OF GROUP DEVELOPMENT

People working with groups often fall into a trap because groups appear to evolve through a series of discrete and visible stages in much the same way that children appear to move through rather specific periods of physical and emotional development. Although in both cases there do appear to be rather clear patterns of development, it would be unwise to take them too seriously. Each group, like each individual, is unique and must be understood in terms of exceptions, and in terms of when, how, and why changes in development occur. True understanding results in grasping the subtle variations in time and place and in recognizing the nature of the precipitating factors and how they contribute to the uniqueness of each group or individual.

Nevertheless, it appears that group development is on many occasions more predictable than individual behavior. The fact is that social systems are generally conservative, seeking to maintain themselves in relative comfort by establishing protocol that insures predictability. In addition, because there tend to be so many vested interests (held by both individuals and groups) within complex organizations, change becomes very difficult and often occurs only as a result of extreme pressure.

The following description of the events that may occur as a group develops, and the accompanying driving forces, is a composite of many views.[1] It is presented not as a model for all groups, but as an example of events that may take place and needs that may exist. By becoming more keenly aware of the changes in group behavior and raising appropriate questions in relation to them, the facilitator should be in a better position to respond effectively to the group's present needs. The following case assumes an ongoing group, one with a reason for being, relatively little personal information about fellow members, differing perceptions of task, and methods for reaching the goals. Finally, it assumes a group that is starting out, although much that is suggested will have relevance for groups at various stages of development.

The Beginning

People have expectations of what will occur in a group even before they attend. They flavor their first perceptions with these expectations and their personal needs. They bring with them their individual histories and experiences in previous groups. It is these factors that provide the glasses through which the group is perceived.

[1] The composite is drawn from the views of many people working with a wide range of experiences and types of groups. They include: W. Bennis, W. Bion, R. B. Cattell, V. Cernius, A. M. Cohen, E. Erikson, J. Gibb, G. C. Homans, D. C. Lundgren, E. A. Mabry, R. D. Mann, R. D. Mills, T. Mills, F. Redl, W. Schutz, H. Shepherd, R. D. Smith, C. Theodorsen, H. Thelen, and B. W. Tuckman (see References at the end of this chapter).

First, it is necessary to survive, to protect themselves, and to be relatively secure in an unknown situation. For some this means acting out, almost immediately attempting to gain a measure of control through a strong offense. But for most it appears to be a time for waiting, for observing what lies ahead, for sorting out potential dangers, and acting with discretion. Thus, there is a period of data gathering and processing through the screen of our own previous experiences, biases, and stereotypes. Like a child on the first day of school, we tend to:

Feel inadequate, but are afraid to show it.

Feel tentative, but often need to appear fairly certain.

Be watchful.

Lack a feeling of potency or sense of control over our environment.

Act superficially and reveal only what is appropriate.

Scan the environment for clues of what is proper: clothes, tone of voice, vocabulary, who speaks to whom.

Be nice, certainly not hostile.

Try to place other participants in pigeon-holes so that they become comfortable to us, able to be coped with in our own minds.

Worry about who we should try to be in this group.

Desire structure and order to reduce our own pressure to perform.

Wonder what price it will take to be "in" and whether the rewards are worth the effort.

Find it difficult to listen and look beyond our own immediate needs.

Wait for the leader to establish goals, roles, and who has responsibility (even if we resent it being done).

So, it is a time for testing, a time of inhibition guided by the rules of other places and experiences. It is not a time for heroics, but for first impressions. Often, it is an environment based more on suspicion than trust, partly because of our initial discomfort and partly because we simply do not know. However, because people always want something better and seek to reduce tensions, more often than not they will risk involvement, and at least a minimal sharing. Our needs to be liked and accepted tend to light the way, and though seldom satisfied, there usually are indications of better things to come. On either side of this position there are, of course, the groups that from the first minute are so tightly controlled that one's very breath and individuality are lost. Then, too, there are those equally rare groups where a sense of openness and security prevails immediately. However, it appears that most groups are a mix of hope and trepidation, where our own needs and the mixed views of others provide the ingredients for an initial climate of doubt and hesitation.

Movement Toward Confrontation

It is not until the initial probing into the boundaries of appropriate behaviors has taken place that facades are dropped and individuals establish personal roles and reveal more characteristic behaviors. Much of the new movement in the group relates to the patterns of power and leadership that are being established. The initial period of unfamiliarity often leads to an acquiescence to authority and a seeking of structure that allows members to move with a certain ease in the strange environment. However, for many this soon results in a desire for more influence in what is happening and focuses attention upon those with power. How to be liked and accepted by those with influence becomes of central importance to some, and others begin to seek personal recognition and their own spheres of influence. Suddenly the leader becomes not only a source of dependency and admiration but also an object of criticism whose inadequacies become a source of discussion. How things are to be done, how decisions are made and by whom, as well as issues of freedom and control all become preeminent. Whether there is a leader focus or a member focus, influence and so-called territoriality among the participants is central. It is the assertive seeking of one's place in the group that bares behavior formerly hidden, and, thus, it is a period of new behavioral dimensions for various members and a period in which stereotypes are often disconfirmed. This springing forth of new behaviors creates suspicion and mistrust in some and forms the basis for new alliances within the group. It is bound to cause tensions and conflict. Not uncommon are such statements as: "I wouldn't have suspected that of John"; "I knew there was more to him than that soft voice and smile"; "I didn't know he was capable of being so angry."

Within this more assertive environment, members begin to take more definite stands, and issues become polarized. Instead of an argument being looked at in terms of data and facts, it also becomes a testing ground for personal influence and prestige. Tenacity may be as important as rationality in winning, and for some it is winning or losing and not the issue itself that is important. The tentativeness is gone, hostility is legitimized, and in many ways the group is much more real than it was in the beginning. Alliances within the group are redrawn based more on experience and behavior than on expectations and wishful thinking. Along with the increased amounts of anger being shown there is also probably more laughter as well and a generally wider range of affective behaviors.

During the first phase it is often difficult to concentrate one's energies on task issues as long as one's own role and secure position in the group have not been established. Now, however, the task becomes a means of exercising other spheres of influence and expertise. Underlying issues facing the group may involve such things as status, prestige, and power. With increased signs of rigidity among the participants and an unwillingness to compromise, less assertive members tend to withdraw as others in the group now bring personality issues into what had been previously content or task issues. If the group is able to face its own destructive tendencies, there will very likely be a confrontation and an effort to get people together and back on the track.

Compromise and Harmony

A confrontation over work and personal issues will usually occur when individuals who are more willing to compromise recognize how self-defeating the present course of events seems to be. Acting as intermediates, they reopen issues and help to get individuals talking again. It may also result when some of the more aggressive members realize that their own personal aims are not being met as a result of their present course of action. They begin to see a more amicable climate as essential to any further movement or growth on the part of the group.

The result is a countermovement to shut off the growing hostility, to reopen communication, and to draw the group together into a more smoothly working body. It is often this effort that ushers in a period of goodwill and harmony during which time there is a reassessment of how people have or have not been working together and how conditions for work might be facilitated. The dissensions are eased, deviations in member behaviors appear to be more readily accepted, and self-expression is encouraged. Greater familiarity with each other and the experience of some success result in a willingness to accept people as people in light of both strengths and limitations. Compromise is no longer seen as equivalent to losing face. Collaboration is more readily sought and competitiveness is played down, if not rejected, by the members. The group tends to exude a new confidence and begins to actually see itself as an integrated unit that can be facilitative when it wishes to be. There is a genuine effort to look at issues, discover appropriate resources, and avoid the personalizing of issues that occurred earlier.

After the nearly destructive series of events and the mistrust previously generated, members are careful not to step on one another's toes, to avoid signs of hostility, and to make sure everyone is heard. Real honesty and openness are encouraged on the one hand, but, on the other hand, there is a subtle pressure to not raise any problems that might break down the harmony that has been so difficult to obtain. Thus, everyone is given "air time" and encouraged to voice his or her opinions. There is a tendency to let people talk (even extraneously) rather than cut them off. Joking and laughter are common, and personal irritations, unless couched in veiled sarcasm, tend to go unnoticed. An increasing discrepancy develops between feelings and behavior, but even so the group may increasingly talk of its openness and ability to work together. Yet the denial of personal issues tends to increase tensions that remain unexpressed. With this submergence of issues there is less participant involvement, stimulation, and overall interest.

Thus, although fences have been repaired and wounds covered, it has been done at a cost of some of the group's integrity and efficiency. Instead of the anger and overt blocking that had occurred previously, issues are overdiscussed, and it is very difficult to make decisions. Resistance appears to be more covert. Instead of leading to greater productivity, this harmony often leads to even less efficiency. Eventually there is the realization that the behaviors within the group are actually inhibiting authenticity and directness. One reason it takes so long to reach decisions is that covert resistance and passivity block progress. The initial elation shared during the beginning of the period gives way to disillusionment and increasing tension. The group's efforts toward harmony simply have not succeeded as they might.

Reassessment: Union of Emotional and Task Components

Having worked under a period of relative structure and under conditions of less control with neither resulting in a satisfactory climate for work, the group seeks a new alternative. One obvious solution is to impose greater operational restrictions to insure a more rational approach to decision making. Such a thrust would streamline work procedures and redirect the group toward the task with greater efficiency. It would not, however, face the source of many of the problems created within the group. As with many life problems, this approach only attacks the symptoms and eases the pain of the current situation, but it may be enough to insure a smoother decision-making process.

If, however, the group decides to delve more deeply into the problems at hand, into causal factors, then considerably more time, energy, and involvement will be the cost. It requires a sizable risk on the part of the members because many issues that have long been submerged will be forced to the surface. Member roles, decision-making procedures, and leadership and communication patterns are likely to come under close scrutiny as are the personal behaviors that facilitated or inhibited the group. Thus, this becomes a period of reflection on goals and performance, means and ends. There is usually a recognition of how vulnerable the group is to the personal needs, suspicions, and fears that can determine how successful the group is in reaching these goals.

If the group chooses this latter course of action, it must build a mechanism that allows it to appraise its own ongoing operations and to alter its pattern of working behaviors when it is obvious that current methods are not proving effective. The group must face the question of how honest it can be and just what level of personal intimacy must be reached before the group can accomplish its goals in the most effective manner. It makes sense that a third-grade classroom will demand greater intimacy than a Marine Corps platoon. Thus, the period of harmony and compromise, although not allowing all the issues that needed to emerge, did prove to be of great importance in the development of the group. Because it was a period of reduced competition, greater informality, and increased familiarity among members, it provided a needed foundation upon which to build. What had been missing was a means of legitimizing the feelings that were not positive, the communication of feelings and ideas that might create conflict or force the group to consider alternative approaches.

Often at this same time there is a realization that as the functions of the group become increasingly complex and there is a need for more resources, greater interdependence is necessary. Greater participation through the division of labor becomes essential, and with it accountability and personal responsibility are spread throughout the group. With greater freedom to communicate and methods of feedback built into the group's operations, necessary tasks are increasingly undertaken by those with particular skills and interests, leadership is shared, and participant involvement is generally increased. The notion of accountability is crucial here, and it suggests that individuals know what is expected of them, that their expectations are shared by the group, and that they are to some degree measurable.

There is, of course, the possibility of a temporary period of intense conflict resulting as tensions and stresses previously withheld are brought out. If the group can overcome the fear of such conflict and realize that it can be put to effective use without being destructive to individuals, then there will be less reluctance to deal more openly with such issues in the future. Thus, during this period the group realizes that if it is to survive, it must increase shared responsibility as well as personal accountability. This in turn will increase trust and insure more individual risk-taking as well as a willingness to devote the necessary time to resolve working issues of both a substantive and personal nature.

Resolution and Recycling

Effective working groups are not necessarily harmonious and free of tensions and conflict. There seem to be periods of conflict resolution and harmony and even times in which the group tends to regress into a pattern of indecisiveness and floundering. As a group matures it should find itself resolving conflicts more quickly and with a minimal expenditure of energy. And, like any mature person, it should be increasingly able to recognize its own limitations and build effectively around them. It has, however, been found that if the group is suddenly faced with a crisis, a series of critical deadlines, a number of new participants, or even a controversial new idea, it may usher in a period of readjustment and a reappearance of old and not necessarily helpful behaviors. For example, special interest subgroups might develop and inhibit the efforts of the entire group, or the group could enter into a period when lines of communication break down, feelings and emotions are denied, and tensions begin to build. Such tensions, quite often, will be released toward other members, thus creating further points of stress.

It is not a sign of group immaturity that such tensions develop, but the degree of maturity is revealed in how effectively the group is able to cope with these very natural problems. A new, influential person being added to the group will be a threat to some member's security, for another he or she will represent a potential ally, and for a third a competitive rival for leadership in the group. Because personal needs are involved as well as various levels of the group's working relationships, conflict is inevitable. Too many groups deny the various reasons for the increased tensions and proceed as if nothing has happened. This denial begins to decay further existing levels of trust. A mature group will stop the deteriorating cycle of events by openly exploring the possible causal factors and then providing at least temporary solutions during the period of adjustment. By confronting the issues that tend to debilitate the group, a norm of positive and constructive problem solving will be reinforced, and this will reduce the length and intensity of the regressive cycle.

Groups, like people can become immobilized at certain levels or stages of their development. Some individuals can never break away from dependency on their parents, and it carries into nearly every other relationship they have. Others fail to resolve their own needs to fight authority. Still others cannot tolerate conflict and the fear of being rejected. In a similar manner, some groups never move beyond a particular stage of development because of various unresolved issues among the members or external factors out of their control. Whatever the reason, such groups will find it difficult to reach an adequate level of functioning. For example, many groups develop a norm that inhibits the expression of anger. As a result, many emotions are bottled up, and the group remains at an artificial level of interaction where harmony inhibits the development of authenticity. Similarly, compromise may become a mechanism for escaping the true resolution of issues. Instead of tensions being reduced, they are increased by the passive behaviors used to cope with individual feelings of anger and aggression.

Other groups have other problems which, unless dealt with directly, will influence nearly every aspect of their work. If members do not feel involved in the decision-making process and must constantly respond to edicts from various outside sources, they will build characteristic response patterns to counteract the new edicts. Because this group is treated as immature and not responsible, it will tend to respond in an immature fashion. If it lacks the feeling of potency, it will not express itself through power. Rather it will respond through inaction or denial. Members may spend a great deal of time angrily discussing the issue among themselves, but never actively and overtly taking a position against the particular source of power behind the edict. Such in-group catharsis may reduce tensions momentarily, but in the long run it will tend to lead to only more feelings of frustration, impotency, and perhaps guilt. One would hardly expect such a group to move smoothly through the various stages of development and on to resolution of its own problems.

A final point needs to be made. Groups do not have to be perfect to get a job done or even to do it well. It is sad to say that in the short run, coercion, intimidation, anger, the reality of a deadline, or a variety of other motivating forces can result in a satisfactory product. If the particular group is not expected to have an ongoing relationship, the tensions and frustrations from such an unsatisfactory process will become merely another piece of information for the informal system of the organization that will influence overall system morale and attitudes. Individuals will simply become increasingly distrusting and probably passively resistant toward other demands made upon them. Over the long run, a continual pattern of isolated insensitivity will eventually affect product quality and relationships on virtually any task. Worse, if the group (task force, committee) has an ongoing relationship beyond the immediate task and there is no process built in to defuse feelings of abuse, one can predict an immediate degeneration of group effectiveness in covert ways that will influence productivity and overall efficiency.

OTHER VIEWS OF GROUP DEVELOPMENT

Forming, Storming, Norming, Performing, and Adjourning

Based on a twenty-year review of the literature relating to group development in therapy, natural, self-study, and laboratory groups, Tuckman and Jensen (1977, pp. 419–427) conclude that task groups, like all others, go through five basic stages of development that are rather predictable.

A first stage of *forming* incorporates all the discomfort found in any new situation where one's ego is involved in new relationships. This initial period of caution is followed by a period of predictable *storming* as individuals react to the demands of what has to be done, question authority, and feel increasingly comfortable to be themselves. A third stage is defined as *norming* in which the rules of behavior appropriate and necessary for the group to accomplish the task are spelled out both explicitly and implicitly, and a greater degree of order begins to prevail. Next comes a period of *performing* in which people are able to focus their energies on the task, having worked through issues of membership, orientation, leadership, and roles. The group is now free to develop working alternatives to the problems confronting it, and a climate of support tends to remain from the norming stage. Finally, with the task nearing completion, the group moves into what is called the *adjourning* period, in which closure to the task and a changing of relationships is anticipated.

Although some believe it is most difficult to consider the development of a group without stressing the interdependency of task and maintenance or process functions, Tuckman finds it helpful to view each of the stages of forming, storming, norming, performing, and adjourning from two points of view. First is that of interpersonal relationships. Thus, the group will move through predictable stages of testing and dependency (forming), tension and conflict (storming), building cohesion (norming), and finally establishing functional role relationships (performing) before the group adjourns. Each of these substages focuses on the problems inherent in developing relationships among members.

At the same time, the group is struggling with the problems of the task. In light of this, the initial stage focuses on task definition, boundaries, and the exchange of functional information (forming), followed by a natural emotional response to the task (storming), a period of sharing interpretations and perspectives (norming), before a stage of emergent solutions is reached (performing).

One can begin to see just how complex the development process can become and how it is quite possible to have problems in the interpersonal realm block progress in the accomplishment of the task activity. In some ways it becomes an academic question whether the major stages are broken down, because whatever progress is made is totally dependent on the interrelationship of task activities and the mix occurring in the interpersonal realm. Nevertheless, Tuckman's and Jensen's model provides a useful means for sorting the critical ingredients.

A Thematic Approach to Group Development

A different view of group development is taken by Cohen and Smith (1976) who, although also stressing the importance of separating social emotional issues of group development from those of task, do not see developmental stages as patterned and predictable with one following the other. Rather, they believe a working group develops around five themes that appear to occur periodically. It is their belief that every group must deal actively with these themes when they arise in the group and that not to do so, not to resolve the tensions inherent in each, will thwart the development of the group and restrict its effectiveness.

The themes are *anxiety, power, norms, interpersonal relationships, and personal growth.* Any of these themes (or more than one) can dominate the group at any point in time, and it is up to the leader and members to explore the issues underlying the theme, deal with them effectively, and move on. It is up to the leader to continue to view the group in relation to these themes and to intervene appropriately, depending on which of the various thematic issues are blocking the group. Thus, anxiety in the group will predictably result if roles are not clear, the task is not well defined, and methods of evaluation of performance are hazy. Similarly, issues concerning leadership and authority will arise in the group with particular concern about power—its distribution and control—and will inevitably result in tensions that will reduce group effectiveness and channel energies away from the task.

The development of restrictive norms or ineffective interpersonal relationships can occur at any time in the development of the life of a group. In a like manner, issues surrounding the personal theme focus on self-awareness, personal growth, and actual behavior change, and lead to possible feelings of tension, conflict, and dissatisfaction.

Although these themes were developed for what the authors call *growth* or *self-study* groups, the thematic approach seems to be one that is helpful for viewing the development of task groups as well. The model points to issues individuals face upon entering any group or organization. Because groups often are not static (new members, changing goals, changing leaders), these issues will periodically influence the feelings and concerns of individuals as well as the entire group.

The focus on themes does not suggest that generalizations cannot be made concerning patterns of group development. In fact, if one examines the reasons why a group is not developing according to predictable patterns, it is quite possible that one of these themes will raise its head. Nor is it our belief that these five themes are all-inclusive. Cohen and Smith suggest that these themes are always present to some degree and that a constant shift in balance occurs among them. Such shifts can be quite natural, expected, and desirable in the growth of the group, but all must be dealt with actively so that issues underlying a theme don't become blockages to the group's necessary development.

The Cybernetic Hierarchy

A third way of looking at the development of a group focuses on four ingredients any group needs to survive (Hare, 1976, pp. 14–16). The order and degree of these may differ according to the purpose of the particular group, but it is assumed that all must be present. First, the group must develop a sense of common identity with values and purposes that are consistent and agreed upon. Second, the group membership must be such that it is capable of generating the skills and resources necessary to meet the goals that have been established. Third, there must be rules and procedures developed that coordinate the activities of the various members and allow a feeling of interdependence and task effectiveness to evolve. Finally, the group must generate the leadership to facilitate the process of execution, to provide a sense of accountability, and to control the overall functioning of the group.

There is a tendency for the group to attack and resolve these needs in an order that reflects what is called the *cybernetic hierarchy*. This theory suggests that areas requiring information tend to control and take precedent over areas of high energy. Thus, people need to know what to do and how to do it before they can generate the energy to do it. In problem-solving groups, it is reasonable that establishing values, goals, and an identity will be followed by defining rules of operation and establishing roles and boundaries of influence. Leadership activity would then put to use the skills and energies of the members. Sometimes, however, it is more important to recognize that all ingredients are needed, because the absence of any of the ingredients will result in confusion, tension, and hostility. Sometimes a strong and skillful leader can mobilize resources around a pre-established goal and insure the development of necessary rules or conditions for effective execution. On other occasions, strong leadership is less important than a clearly defined task that the group itself can monitor. Real trouble comes with the absence of any of the central ingredients.

FACILITATING GROUP SUCCESS

Whether you are a member of a group or in a position of leadership responsibility, it is your responsibility to help the group meet its own potential. By being aware of factors that influence the successful development of a group, you will be in a better position to be effective in your own contribution. Each of the three methods for viewing the development of a task group can arm you with questions necessary to identify what is happening, what is needed, and how to help the group. Thus, it is essential to be aware of how the group is developing along theoretical lines, to know where deviations are occurring, to realize whether certain tensions can normally be expected, and to understand what might facilitate the group through a particular stage.

Similarly, it is important to consider the implications of possible critical incidents influencing the development of the group. This information, as well as insight into essential group needs, will provide anyone with the first essential tools for understanding. It will provoke useful questions for gathering data and testing personal hypotheses. At this point, as a member and/or leader, you will be aware of alternatives and begin to sense your own ability to influence the group. Individuals who feel impotent or victimized as members of a small group seldom have a framework for looking and conceptualizing the group as a developing entity. Obviously, it requires some skill to translate living theory into constructive action, but this is the first critical step.

READER ACTIVITY

Like individuals who become routinized in their life style and quite predictable, so too, groups can become just as predictable. Instead of moving through certain developmental stages, members of a group will find themselves caught in one and not able to move beyond it.

Can you think of a group of which you are a member that is stuck in a particular developmental phase and for some reason is not able to move ahead? What is the cost to the group of being immobilized?

In contrast, are you aware of a group that is not stuck—that is active and developmental? What makes it different from the first group? What are the factors that allow it to grow and develop? Is there any chance that it may become blocked like the first?

What are you doing as a member of the first group to help sustain its stuckness? What can you and others do to move it to a more productive level of activity?

The Right Conditions

To talk of an ideal or perfect group is to walk on very thin ice. Nevertheless, it may be worth the risk since it will allow an exploration of certain dimensions of group behavior that seem to result in a productive work climate and where both emotional and task needs are given necessary consideration. It will, at the very least, provide a list of pertinent questions to be used in helping to diagnose a group and the internal working relationships that develop. Special circumstances and particular demands require groups with specific orientations and skills, but there do appear to be certain conditions that facilitate the involvement of people in groups. These conditions include the following:

1. The group has a shared sense of purpose, and its members are aware of common goals. Their participation should be voluntary if at all possible.
2. Roles in the group are varied and differentiated according to both interest and performance. In other words, there is a concerted attempt to discover appropriate resources, depending upon the particular need at a given time.
3. Communication channels are open, and there is a specific effort on the part of members to listen and clarify what has been said. An interest is shown in what others have to say or feel. This also implies that the vocabulary and any special jargon is familiar to the group.

4. Similarly, dissent is freely expressed, and silence is not taken as consent; opposing opinions are sought as part of the clarification process. In this manner, inputs are seen as shared ideas, and they are not evaluated in terms of the presenter. On the other hand, members assume responsibility for themselves and their own ideas and are willing to stand accountable to the group for these ideas.
5. In this light, there is an acceptance of different participant styles, although the group is not dominated or controlled by the personality characteristics of a few members. Thus, the whole group, rather than a few powerful individuals, is in charge of its own destiny, and there is a sense of shared leadership based on changing needs and varying according to particular situations.
6. Because not to decide is to decide, the group makes decisions. Most of these decisions, however, are seen as provisional and are to be evaluated by the group with particular members being held accountable for assignments accepted. It is this assurance of reappraisal and accountability that enables the members to accept a consensual method of decision making in many instances.
7. Nevertheless, the decision-making process is flexible enough so that a number of decision-making procedures may be used, depending on the peculiar nature of the problem under consideration. Underlying this process is an awareness that an apparent decision (regardless of method) is not a real decision until it has been initiated, and members of the group are made responsible for its implementation.
8. Failure does not immobilize the group or its interest in experimentation. Thus, innovation is encouraged and support given for the implementation of reasonable ideas. However, success does not shut the door to further evaluation and exploration. This assumes that the group is willing to look carefully at its own productivity and, equally important, how it works as a group to accomplish its agreed-upon goals.
9. Thus, the group has developed the skills and interest necessary to diagnose those problems that minimize its effectiveness. This includes the collecting and processing of a variety of data that provide the group as a whole and individual members with information relating to their own behaviors and their impact upon others in the group.
10. It is important that impersonal problems are seen as issues that influence the entire group and do not become "member" problems that can easily result in an evaluative and even punitive climate. This also assumes that the group is fully involved in establishing its own system of rewards and is able to identify and maximize its strong points.

11. The group is responsive to its own changing needs and goals, able to create new functions and roles, work in subgroups or as a whole and, if necessary, assimilate new members with a minimum of disruption. This takes for granted a well-balanced interdependency while maintaining necessary flexibility.

Continual Adaptation to Stress and Tension

If a person wishes to understand another individual quickly, it is often helpful to ask, "What are his is her needs and the sources of tension he or she is trying to reduce?" Similarly, groups represent the composite needs of a number of different people and the attempt to reduce the level of existing tensions. Usually it appears that if there is a choice between facing such tensions and the conflicts that often accompany them, the tendency will be avoidance. However, by minimizing the disruption and agitation of the moment through such mechanisms as denial and avoidance, underlying sources of tensions tend to multiply, and tremors of dissatisfaction erupt into even greater turbulence later. It is the inclination of a group to maintain the status quo that severely hinders the development of a dynamic and flexible working group. Disproportionate energy is often spent on issues of little relevance and minimal controversy while emotional concerns focus on issues that may never be formally placed on the agenda. An example of this follows:

During a recent rather volatile spring, racial conflicts flared in a number of schools of a large eastern city. Many of the secondary schools waited for the incident that might ignite demonstrations, strikes, or even riots. It was during this time that a special faculty meeting was called at a special school for academically superior students. Because of their status within the school district and the reputation of having the best teachers and most talented students, an "it's their problem" sort of smugness had settled among this faculty group. Racial issues were seldom even the topic of conversation.

A brief incident of racial conflict several months before, however, had stimulated an interested group of teachers, students, and community members into action. Without publicity, they decided to discover whether, in fact, they did have serious problems within the school and, if they did, what constructive steps might be taken to alleviate them. Thus, it was at this particular meeting that these data were presented. Questionnaires, interviews, and discussion groups had been conducted in an effort to tap the representative views of students, faculty, and the community. As might be expected, the study uncovered areas of tension that had been denied or glossed over lightly by the faculty and revealed a potentially dangerous situation developing between a particular student group and certain faculty members. The proposal was met with immediate denial and anger and was voted down with almost no discussion. The only response to these concerns and the supporting information was an irate negation of the report and of the individuals who had developed it. The idea that such a literate, sophisticated, and learned body could be guilty of having even the seeds of racism or prejudice was inconceivable to the group.

Clearly, the single best indicator of the depth of this problem was the highly symptomatic denial of the issue, a refusal to even explore the problem at a time when schools were literally burning around them. The lack of positive response might have been the result of the faculty's own guilt, fear of imperfection, or discovery of their intellectual dishonesty. Whatever the cause, the denial of the present reality almost precipitated the chaos the faculty was so desperately trying to avoid. The hope was that the momentary overt expression of discontent expressed in the report would soon pass and a semblance of the previous peace and tranquility would again appear.

Similarly, it is not uncommon for a group or organization to handle a period of conflict or severe crisis by developing a commission to study the problem or by devising an experimental study to explore the parameters of the problem. What usually occurs is a reduction in the immediate level of tension, an avoidance of a direct confrontation, and a redirection of participant energy. Often what then occurs is a diluting of the issue over time as it loses immediacy and becomes entangled in the bureaucratic ritual of meetings and reports. Even with good intentions, such task forces frequently prove ineffective because the recommendations for action will be presented at a point in time far removed from the crisis and when there is less pressure for change.

It appears that a group will often be willing to sustain a gnawing source of tension and conflict over time rather than to dig the issue out and place it in the open. Dealing with the problem directly will inevitably rupture the status quo. Thus, to really understand a group, an individual must look beyond superficial behaviors and into the underlying stresses that exist but are not being dealt with. More often than not these tension points lie behind those concepts and values being most vociferously defended within the group and which, if altered, would force a change in the imaginary point of equilibrium around which expectations are built and habits are forged.

Abating the Abuse of Power in Group Settings

It is difficult to pinpoint historically the change. Perhaps it was after World War II, with the new assertion of individual rights. Perhaps it began then and blossomed fully during the era of flower children when, for the first time, the United States found its bedrock values questioned. The work ethic? Why? Marriage, sexual morality, material wealth and the role of the traditional family all were fair game. Then, even more outrageous was the questioning of our national leadership, an unwillingness for many to follow blindly, without question, into the heart of an Asian war that seemed to strike against every humanitarian value we held. But, to question the national leadership, the President, the head of the "family"—wasn't that going too far? The rise in high-level government corruption and the lies surrounding Cambodia poured openly into the cauldron of American debate in the Ellsberg Papers, and, finally, the ultimate catastrophe of the Watergate conspiracy seemed to put the finishing touches on the average individual's disillusionment with leadership in this country.

From an acceptance and even awe of leadership evolved a perception of it as a source of control, manipulation, mistrust, and opportunism. The democratic ideals of open communication, collaboration, interdependence, and shared decision making were seen as corrupted beyond repair by many. It is little wonder that power and control are issues of increasing importance in the life of any group. Not only is leadership seen as lacking in integrity, but, all too often, the insensitivities of the bureaucratic organization and its unresponsiveness to individual problems create both overt and covert resistances from the beginning of almost any group. The leader offers a prime target, a perfect escape valve for other frustrations.

To make matters more complicated, leaders are taught to play the democratic, participative game. But much leadership still carries the stamp of autocracy, because autocracy is the model on which most powerful leaders cut their teeth. As a leader, there is a no-win position developing in which we expect leaders to be ineffective at best, and corrupt at worst, yet demand that they be perfect. They should be well organized, decisive, open, warm, attentive, strong, sensitive, good listeners, efficient, collaborative, and most certainly charismatic. And, if they are less than we expect, we simply confirm our predisposition. Added to this general stereotype are our own personal stereotypes about leaders of a particular sex, age, educational background, dress, and personal style. One does not have to be brilliant to understand why in many unstructured groups, the statement that someone is attempting to be the leader will be interpreted as an accusation, rather than a compliment, which will inevitably be followed by a rash of denials. After all, who in their right mind would want to be a leader, knowing what everyone knows? It appears that we are in an age of counter-authoritarianism. Whenever there is power or control to be distributed in a group, there will be created an almost immediate resistance or blockage.

People are afraid of being manipulated, of not receiving their just due for feeding the personal gain of someone else. We no longer trust the motivation of those with power and almost immediately look for their personal agenda. Leaders who do not accept the assumption of distrust and suspicion present in many small groups from the outset will almost certainly be in trouble before they know it. But the initial pessimism can be overcome, the watchfulness, apparent apathy, low level of risk taking, and limited participation can be turned around. The authority or storming stage can be abated if the leader will:

1. Involve the group from the beginning in defining the nature of the group's goals.
2. Define clearly his or her own domain of power and control.
3. Establish how decisions will be made in the group, which decisions the leader will make only with appropriate input from others, and in what situations the group as a group or individuals will have the power to make decisions.
4. Clarify with those involved specific areas of responsibility and how their roles relate to the roles of others.
5. Develop with the group a means of monitoring the life of the group, both in relation to its ongoing process and in relation to outcomes the group agrees are essential.
6. Make a sincere effort to know the personal needs of individual members and consider them along with the needs of the group as a whole.

Effective leadership is knowing how to utilize power and control in a manner that benefits the group and its members. Control is seldom an issue if individual needs are being met. Thus, we know that people need success, that they thrive when interpersonal relationships are positive and supportive. We also know that individuals need to feel heard, to have access to those with power, and to feel that they are personally valued or respected. A cohesive group is one in which members desire to be together, to return to the group itself. Feelings of involvement, personal influence, and progress toward a goal are all ingredients of cohesiveness. Add to this a little humor or fun, and authority issues will diminish rapidly. Although leaders should not minimize the importance of their roles, all too often they take themselves much too seriously and literally turn off the positive energy that waits behind the initial cautions or even hostile behavior.

READER ACTIVITY

We would like you to consider two groups of which you are a member, one of which you feel is successful and meeting your personal needs as well as those of the group, and another group in which there is tension and resistance and where cohesion is minimal at best. Think of each group in relation to the following questions.

1. Do members of the group feel their ideas are actively solicited?
2. Do individuals feel they have the opportunity to influence the life of the group in areas that directly impact their lives?
3. Has the leader(s) defined his or her areas of power and control or do members feel control to be arbitrary and at the whim or personal discretion of the authority figure?
4. Can you think of times recently when members of the group have had a good laugh together? Can the group laugh at itself, at the leadership? Is humor usually at the expense of someone else?

Positive Structured Interventions

There is an increasing body of research confirming the belief that carefully designed interventions into a working group can alter levels of problem-solving effectiveness, cohesion, risk taking, and productivity (Hall and Williams, 1970; Evensen, 1976; Bednar et al., 1976; Stogdill, 1972). Although much of this work is in the formative stages and there remain more questions than answers, it seems appropriate to explore some of the tentative findings here.

In one series of studies exploring the impact of structure on self-study groups, it was noted that "Empirical evidence supports structure as a robust variable which positively effects interpersonal behavior, group attraction and client improvement" (Evensen, 1976, p. 152). Thus, in groups whose purpose is to explore the nature of group behavior and interpersonal relationships, effectively placed structural interventions by leaders allowed various group goals to be attained more rapidly even though leaders in such groups traditionally are a focus of group attention, anxiety, and frustration. More specifically, it was found that by providing more structure, members could risk more, would be more disclosing of feelings and attitudes and, as a result, the level of measured cohesion in the group would increase. Apparently, the structure gave members legitimate permission to say what was on their minds. This is particularly promising because individuals measured as low risk takers were consistently helped toward greater participation and involvement.

It is not difficult to see the implications of such findings for more formalized task groups. By structuring legitimate activities that focus on the process of the group, many of the blocks that often develop in the interpersonal realm of group life could be anticipated and resolved before they restrict the natural development of the group. Encouraging participation from quiet members who tend to be low risk takers might add significantly to the richness of group problem solving as well as to group morale and cohesion.

Other research (Stogdill, 1972) showed that pressures on productivity in a group inevitably lead to some reduction in the levels of measured group cohesion. This is a simple fact of life in that energies usually directed into the area of production are not directed into activities that would tend to increase cohesion. The trick is to keep levels of productivity, drive, and cohesion in some optimum state. This is a critical role of management, because without such a balance one inevitably pays a heavy price. So, with lower worker drive (motivation) and cohesion there is a tendency toward greater amounts of down time, absenteeism, strikes, and work slowdowns, all of which, over time, can have a dramatic impact on the cost of production.

In a group like a problem-solving committee, we have probably all experienced how low cohesion and poor morale can result in overt and covert resistance and affect the quality of the work as well as the energy that people are willing to contribute. Ways of increasing group productivity and cohesion are addressed more particularly in Chapter Seven on problem solving and in Chapter Six on meetings.

Finally, there is an increasing amount of evidence that groups given specific training in problem solving tend to be more effective. A classic experimental study by Hall (1972) exemplifies this type of research and implies that groups can translate such learnings directly into their work efforts. Thus, large numbers of working groups performed work of greater quality on new tasks of a similar nature after training in a rational approach to problem solving. This simply supports what many group facilitators have learned from experience. Groups that are provided appropriate structure, models for work, experience in problem solving, and guidelines for maintaining their own process tend to perform with less tension and with greater productivity than groups not receiving such support.

READER ACTIVITY

The next time you are with a group that has a problem-solving task to accomplish, try the following:

1. Ask the group to write down on a small piece of paper the four types of behavior they believe might block the group in its effective development. Don't have them sign these papers and don't mention anyone by name.
2. Also have them note on the paper the four types of behavior each has experienced in other work groups that have had a positive influence on the group problem solving.
3. Tally the responses under headings of helpful and nonhelpful behaviors, and present them back to the group. If the group is composed of six members, there will be twenty-four helpful and twenty-four nonhelpful behaviors. Of these, there will be seven or eight in each category that will stand out in common across at least three of four members.
4. Now, have the group spend ten or fifteen minutes establishing some clear ground rules to insure the maintenance of the positive behaviors and the removal of those that tend to block group progress.

This kind of activity will not alter people's personalities or even have a lasting impact on the group unless the ground rules are continually utilized and the blocking and supportive behaviors regularly brought to the attention of the group. The activity, however, will raise participant awareness to the point that immediate negative behaviors will diminish. Clearly this is an approach that goes after symptoms rather than causes. Yet, it may help get the job done, make people feel better, and be a step toward more permanent changes in the life of the group.

Maximizing the Group's Potential

The effective group leader or member tends to have an awareness of the group, its needs, and present level of development. Just a few probing questions, a structure for observing, and the willingness to look carefully at what is happening will be all that is required to contribute to a group's effectiveness. Armed with an understanding of stages of development, group needs, and critical events that may occur in any working group increases the possibility of responding in appropriate and constructive ways that may facilitate or unblock a group and help it be more effective. The most frustrated members of groups are those who fail to understand what is happening and feel victimized by what may be a predictable course of events. Knowledge of developmental trends allows for a proactive rather than defensive response and brings some modicum of control back to those privy to such insight.

Few people have ever experienced a working group with the participative approach outlined earlier in the discussion of the right conditions facilitating people's involvement in groups. Nevertheless, knowledge of these qualities provides a basis for comparison as well as ideas for improvement. Developing such a working climate is rare indeed because ignorance rather than awareness of effective group process is the rule, not the exception. It can be likened to being in a foreign country where everything appears to be familiar but it seems next to impossible to communicate effectively. Not only this, but most of us are used to strong leaders who control rewards, establish the ground rules of a particular task, and provide the necessary push to get the job done. We expect to be directed, motivated, intellectual, impersonal, and rational in our approach to problem solving.

As a result, we tend to see ourselves as separate from the group, often competing with other members for recognition and responding to authority rather than to member peers. Such a climate is not conducive to establishing free and open communication, role flexibility, and a truly nonevaluative atmosphere. It is this kind of atmosphere that helps to predetermine the kind of development possible for a group. We are used to being dependent and, even though we do not like it, will often demand behaviors from those in control that insure its presence. Even when a work group is responsive to democratic principles, members too often become the victims of the majority vote, the conflict-reducing option which, if used indiscriminately, may polarize a group and erase the vital thread of compromise upon which the effective decision-making group must be based.

If a group has never had experience outside the confines of a rigid time schedule, agenda, and parliamentary procedure, it is doubtful that it will ever develop the trust necessary for processing its own behaviors or for the interdependence necessary to see issues as other than politically expedient and strategic. Certainly decisions will be made and groups will function, sometimes in an extraordinarily efficient manner. The price paid, however, will be in terms of participant involvement, interest, cooperation, and member accountability. The group, like a growing child, responds best to patience, freedom within limits, concerns from others, and a climate that encourages spontaneity and authenticity. It is a nonquantifiable mixture that varies from group to group, with intangibles often determining the difference between success and failure. Yet, more and more success can be assured if the leader-facilitator is able to formulate the necessary questions to help him or her understand the group with which he or she is to work. This, added to a familiarity with diagnostic techniques and a few basic approaches to working with the task and emotional problems that inevitably face any working group, is essential. Much more than the use of gimmicks and techniques, success seems geared to how effectively the group is able to respond to its very human needs in a manner that exploits no one and maximizes its own potential.

Not only this, but we are becoming increasingly familiar with the value of positive, structured interventions into the life of a working group that can facilitate its development, improve cohesion, open communication, reduce the threat of participant risk taking while improving the quality of problem solving. The biggest drawback now is convincing those individuals stuck in ineffective patterns of leadership that both they and the group will benefit from the adoption of these new approaches. Not an easy task when we realize that individuals have been rewarded most of their lives for behaving just as they are behaving.

EXERCISE 1 Creating a Theory of Group Development from Experience

Objectives

To explore the developmental aspects that seem to occur in most groups

To help the participants think conceptually and to organize their learnings into a meaningful and systematic analysis of groups and their development

Setting

This activity will be explained in terms of (*a*) an ideal and (*b*) the average or common situation.

Ideal: The study of groups should ideally be done over a period of time that allows individual participants to internalize their insights from experience and practice within a developing conceptual framework. Learning about groups is often facilitated when individuals are placed with members they do not know and over a period of weeks asked to participate together in different tasks. Given time and a variety of experiences together, a certain pattern of development will begin to be evident. This will be especially true if the group is expected to work toward goals that they must decide, to make decisions in light of the decision-making procedures that they create, to generate participant roles as a result of the changing needs of the group, and to establish patterns of communication and leadership that reflect these changing needs. Such leaderless groups are able to develop important understandings about group process if a facilitator is available occasionally to help them look closely at a particular issue or concept that they might not naturally focus on without assistance. Thus, a few hours a week over a period of several months can be most helpful in developing a clear perspective of group development. This, of course, is not to say that intensive time blocks together are not also valuable, only that it is difficult to grasp the significance of all that is happening in a short time period.
Average: Most often it seems that groups do not have the time to gain a conceptual understanding of how groups operate. They are most often brought together for a short (perhaps intensive) period, and it is hoped that they "get something out of it."

This particular exercise is aimed more particularly at groups that have had the opportunity to work and learn together and those that have some understanding of what to look for in terms of the process of working groups. For the purpose of this example, it is assumed that three small groups have worked together over a period of weeks both as a single large group in some activities and as individual work groups of perhaps seven to nine members.

Step 1. Three new groups of from seven to nine participants are created. They are composed of two or three members from each of the old working groups. The facilitator gives the three groups the following task:

> As a group, look over your experiences of the past several months (weeks, days) and develop a theory of group development that may be applied to many new groups as their members work over a period of time to become effective problem solvers. Are there particular stages of development that most new groups seem to move through? Can certain behaviors be expected from the members? You will have sixty to ninety minutes to build your group development theory and present it back to the total group (all three groups in this case) in some graphic manner. For exam-

ple, you may wish to present your ideas using pictures, you may wish to develop a skit, or you may design another method for transferring your ideas to the other participants.

Note: It is very important that the groups have enough time, since the tendency is that they will spend considerable time discussing what has happened to them in their own groups and little time attempting to integrate it into a theoretical framework. Thus, at the end of 45 minutes, the groups should be warned that they have only about 20 or 30 minutes to complete their task and be ready to make their presentation. This usually leads to the same kinds of problems that arise in most decision-making groups under pressure—forced decisions and reduced participation.

Step 2. At the end of the hour or hour and a half, the groups are asked to make their presentation. It is important to keep the presentations to no more than 10 minutes and preferably closer to 5. (They should be made aware of this during their planning because it pressures them to make a more precise presentation.) At the end of this time it is most helpful for the facilitator to integrate many of the ideas presented and to supplement them with theoretical concepts about development with which he or she is familiar and which seem appropriate to the discussion.

Step 3. The three groups that started the day and from which the three theories were generated are reconvened. They are given 30 minutes to process their own work together. The group discusses what they saw occur and how they feel about it. They also may wish to analyze what happened in terms of the theory developed from the total resources of the three groups. They are also asked to compare the new groups with the performance of their old groups.

Step 4. Briefly, the three groups are asked to share with all the groups any learnings they gained from the last discussion. The facilitator may wish to explore how many of the characteristics of long-term group development can be seen in the initial phases of a new group in a problem-solving situation.

Variation

With a group that is interested in group process but has not worked together previously, it is possible to give the same instructions suggesting that they look at groups with which they have worked in the past. It has been found that they too hold keys to useful theory, and they will also generate important data within their work group that can later be analyzed in terms of the theory that is developed.

EXERCISE 2 Maslow's[2] Hierarchy of Needs: A Means of Understanding the Problem of Creating Open Communication and Feedback within a Developing Group

Objectives

To help a group begin to understand the kinds of restraints inhibiting individual members

To focus on the emotional aspects of group process

To provide a theoretical frame of reference for understanding the group process

Rationale

According to Maslow, individuals tend to pass through certain stages of development, and the focus of their needs tends to change. How a person acts in a variety of situations will depend partly on the demands and uniqueness of the present moment and partly on the general developmental level in which the individual is functioning. It is not that we ever are able to satisfy all of our needs; in fact, even as mature adults many are still unsatisfied. The point is that as we mature developmentally, from childhood, through adolescence, and through adulthood, the focus of our needs and our inability to see beyond them changes. Thus, the first year or so of a baby's life is dominated by physiological needs. But, even though a preoccupation with such things as food, water, and sex are brought under rational control, there are times when even the most mature individuals feel the overpowering push of some physiological need. Similarly, the growing child is often overly concerned about the safety of his or her environment. Whether he or she is safe from harm or threat can prove to be a dominant theme in his or her developmental system of needs. For the adult, however, familiarity and experience have provided the feeling of safety except in unusual crisis situations. Actually, it is the higher-order needs such as love and self-esteem that seem most difficult for people to handle. If these needs were met to a satisfactory degree, people would no longer have to expend a great amount of time fulfilling them at a period in their lives when they should be able to accept themselves as they are and concentrate on developing their own potentials. Included with this exercise is the well-known Maslow diagram of hierarchical needs.

[2] A. H. Maslow, *Motivation and Personality,* New York: Harper and Row, 1954.

Theoretically, a person moves upward with a tendency to satisfy one level of need before tackling the next. It is a rare person who feels fulfilled in the areas of love and self-esteem. How few are the people who are able to give and accept love and feel unconditionally accepted. Even fewer have resolved the needs to dominate and control, to achieve and be important.

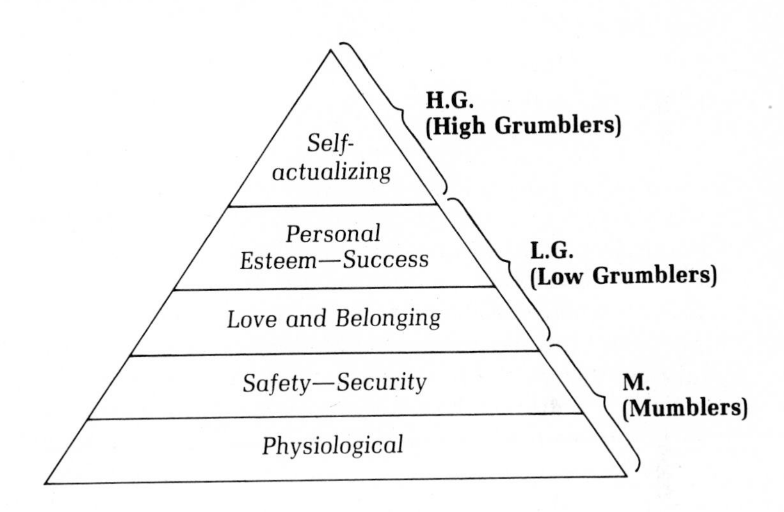

Setting

In any group there are individuals who are controlled to some degree by their own needs in one of these areas. The person who blatantly says, "Why can't we be open here and just say what we really feel?" is either completely unaware of the forces restricting an individual's ability to be open or is sending up a smoke screen to hide his or her own apprehensions. Personal needs act as an important inhibiting factor in the communication and development of any group. This is in itself neither good nor bad, merely an important reality. When individuals within a group push too rapidly for openness, being personal, and giving freely of feedback, it may lead to considerable pain as the group struggles to protect itself from itself. People do not wish to be rejected or to fail in a particular task. Thus, most will play it relatively safe. Sometimes it is important to help a group gain new respect for the various levels at which people operate. If such is the case, then this exercise may provide some insights. This particular activity assumes that the facilitator is in some sort of authority role and wields considerable influence. It also assumes that he or she is willing to be open about his or her own role and willing to alter his or her own behavior if it will benefit the functioning of the group.

Action

The group is broken into trios to discuss specifically the role of the facilitator, his or her strengths and limitations (with examples), and how he or she might improve his or her role in ways that would facilitate the working of the group. After the initial groan, an unusually loud and animated discussion usually takes place with nearly everyone participating and sharing ideas. There is no doubt that they are dissecting the facilitator, and probably enjoying it. It is essential to remind them that they need to be specific on all counts. After 10 or 15 minutes (hopefully that will be enough) the facilitator stands by the board and asks the group to list the limitations that were being discussed with such zest. It is quite possible that no one will volunteer. Those comments that are made will tend to be rather vague and indirect unless the facilitator's rapport with the group is unusually good. Even when asking for good points or ways in which his or her performance could meet the group's goals, the discussion will probably drag, with few individuals participating.

Discussion and Reference to Maslow's Theory

It is important to reaffirm the significance of the discussion and the data presented (hopefully the facilitator will take notes and expand on them). But, it is also clear that the discussion lacked something vital that was present during the discussions in the trios. If possible, the group should develop reasons for the obvious change in spontaneity and involvement. There are many possible reasons: fear of being evaluated by the facilitator or other members, discomfort in making direct criticism (for fear of getting it back), the leader-subordinate relationship that exists, and, perhaps, the fact that it is easier to be candid with two or three rather than the whole group.

At this point it is helpful to give a brief description of Maslow's theory of needs and how it relates to the behaviors that people are willing to exhibit. It seems to be true of most groups that when individuals do not like something about how the group is being conducted, sense becomes the better part of valor, and a disgruntled "mumble" is the most that rises to the surface. Some individuals will actually "grumble" to another person or to the group about what is happening, but it is the rare person who will take the bull by the horns and do something about the situation. This condition is particularly obvious in many classrooms where a teacher may be ineffective or not meeting the needs of the students, and yet nothing is said or done. The same is true with the boss. To say something places the person in jeopardy of losing the boss's good graces and losing his or her own esteem because of reaction in the job (an even lower level need). As a result, people tend to mumble (see M. on diagram) or grumble (L.G.=Low Grumble), but few are willing to pay the potential price and say it as it is (H.G.=High Grumble). High grumblers in a group are usually either those who have worked through their own needs and feel accepted and secure (self-actualizing level) or individuals who are striving for personal recognition (often need for esteem). The two types are easily recognized by the group, and although examples should not be made, it is useful to suggest that the general categorizing of needs and subsequent performance is limited. The theory does, however, raise an important question. How can a group climate be developed that minimizes the threat to personal needs and helps open up a free expression of opinions and ideas? The facilitator may wish to explore this question in a variety of possible ways, or he or she may choose to let the implications go for a future discussion.

EXERCISE 3 Tower Building[3]

Objectives

To aid a group in focusing on the aspects of helping within the context of a particular task

To alert a group to the problems that may develop when working under severe pressures (in this case time and competition)

To deal directly with the inhibiting norms, roles, and leadership practices that can minimize the use of the group's resources

Setting

This exercise is as close to real life as one can get and still remain in the laboratory concept. It involves all the important aspects of group planning and decision making, such as the allocation of human and

[3]The idea for this exercise was originally developed by Clark Abt in his simulation activity, "An Education System Planning Game," originally played in 1965 at Lake Arrowhead, Calif.

material resources, working against time deadlines, altering strategies in the face of unexpected crisis, and seeking the problems inherent in negotiations. It provides an opportunity to view the evolution of conflict and compromise as it exists in many working groups, particularly if they have not developed skills in group problem solving.

From 2½ to 3 hours should be allowed for this exercise. It can be undertaken with as few as twelve participants divided into two groups (two observers and two groups of five). However, the activity generates greater excitement and involvement among the participants if there are four or six groups with from seven to nine members and an observer or two in each group. It should be noted that, given space, materials, and an effective communication system, virtually any even number of groups can participate (on one occasion the authors witnessed twenty groups with two hundred students and teachers enthusiastically involved for 3½ hours). As in many of the other exercises that focus on the dimensions of group process, it may prove interesting and enjoyable for groups unsophisticated in group analysis, but many of the potential learnings will pass them by. Participants with at least an understanding of basic group concepts will take considerably more away from the exercise. The following description is based on four groups of seven members each, with an observer in each group. The room should be large enough for each group to meet with some privacy. In most cases heterogeneous groups seem to be especially effective.

Action

After the large group has been divided into four groups of equal size (some stratification may be necessary if groups are to be truly heterogeneous), the facilitator says:

> You are a group of architects [each group] who have won one of four contracts to build a tower. Although your tower will be built independently of the other three, it is to be judged in competition with the others. There is to be a planning deadline and a construction deadline. The criteria to be used in judging the four towers include height, beauty, strength, and message [symbol or motto, and so forth]. A prize will be distributed to each of the three best towers. Independent judges will appraise the towers, and their decisions will be final.
>
> Your groups now have thirty minutes to develop a diagrammatic plan of the tower you are going to build. You should know that you will have to plan in terms of certain materials that will be made accessible to you only during the building phase. The judges will also be asked to consider the completeness of this initial plan in their overall judging. The materials available for building will include [for each group]:

2 rolls of masking tape (1″ wide)
2 rolls of colored streamers
1 pair of large scissors
1 ruler
1 roll of colored toilet paper
150 sheets of medium newsprint (36″)
4 magic markers (different colors)
12 large sheets of construction paper (different colors)

It may be possible to negotiate with other groups for certain additional materials. Also, other materials from the person of the participants may be used; however, no artificial bases, for example from chairs, wastebaskets, and so forth, may be used. The tower must stand alone on its own support—it may not be attached from the ceiling.

You may begin your planning. Since this is a competitive situation, no additional directions will be given during the planning period.

After 25 minutes of the planning period, the following announcement is made to the four groups.

We are disappointed to have to announce that because of the existing crisis in government funding, we have not been allocated all of the materials we originally anticipated. As a result, there are only enough materials available for building two towers. We are extremely sorry for this inconvenience. Rather than selecting two groups from the plans developed at this stage, we have decided to allow two groups to merge, decide on one plan, and build a tower together. Thus, there will be two towers each constructed by two teams from a plan on which they both agree. You will have an additional thirty minutes planning time to integrate your ideas or to come up with a new plan. Remember, both teams must agree on the final plan [the facilitator selects the teams that are to join forces].

After 30 minutes, the following announcement is made.

You now have thirty minutes to build your tower. It must be completed by _____ at which time the judging will take place. Remember the criteria of height, beauty, strength, and message. The judges will also circulate among the groups to make certain that all regulations are being followed. Good luck.

Observation and Data Collection

There are four major sources of information that will be discussed with the participants.

1. There is information gained by the observer during the initial planning phase. He or she will find it most useful to focus upon a few salient features of the groups such as: (a) communication pat-

tern (who to whom), (*b*) leadership styles exhibited, (*c*) specific behaviors that seem to inhibit or facilitate the group during the planning.

2. The second planning session provides another distinct period of observation. One of the observers should collect the same type of data gathered during the first observation period, except now on the combined groups.

The other observer should turn his or her attention to the developmental aspects of the new combined group. This should be carried through the actual construction period. For example, he or she might ask the following questions as he or she observes the groups:

a. How is the potential leadership conflict either resolved or not resolved?

b. What norms are developing within this group, and are they different from those that existed in the original smaller group you observed?

c. Have individuals with certain roles in the first group taken new ones in the larger group, and, if so, what is their reaction to this state of affairs?

d. How is the decision-making process developed in this second planning session? Is consensus actually reached, or is one plan forced on the total group? What are the responses of the group members? How is resistance exhibited (actively or passively)?

e. Is there a real attempt to involve members from both groups through the allocation of responsibility during the building phase? How?

3. During the actual building phase, both observers should concentrate on the questions in *b* above. Special attention should be given to the quality of participation of the various group members. Is there any difference in the quality of participation among those in leadership (control) roles and those less involved in the planning? Are there signs of withdrawal and resistant behaviors by certain members? Can these behaviors be explained as a result of the historical development of the two groups, that is, the merger?

Do the two groups tend to work together as one, or are the original relationships developed before the merger still the basis for most communication and participation during the building stage?

4. The final source of data is collected from a reaction sheet handed to all participants at the time the tower is finally completed. Information from this instrument will be used by the facilitator in a summary of the activity. This may provide a wide range of data. Below are a number of questions that might be useful for a summary discussion.

1. Please note the letter of the original planning group you joined: A B C D

2. My opinions were valued and solicited in the first planning group. (circle one)

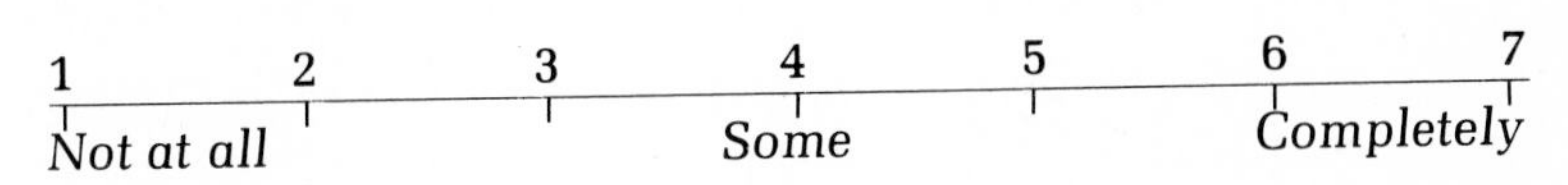

3. My opinions were valued and solicited in the second planning group.

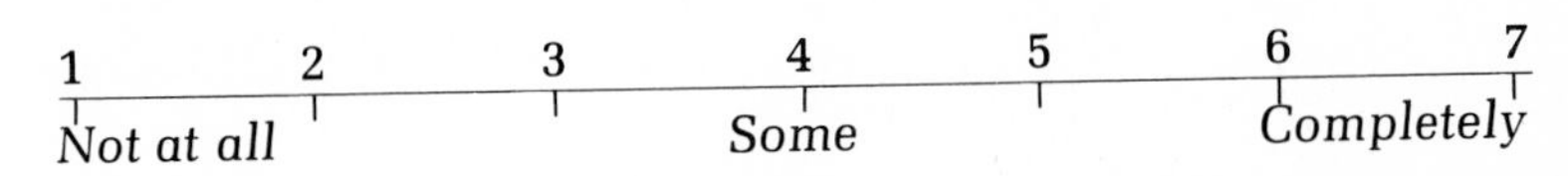

4. If there was a difference in the problem-solving climate that developed in one group, briefly explain why.

5. How satisfied are you with the product of your merged group?

1 2 3 4 5 6 7

Not at all *Somewhat* *Competely*

6. What specific behaviors hurt your group's (merged) efforts to work together and be as successful as it might have been?

7. What behaviors were facilitative as the two merged groups attempted to work together?

Judging and Discussion Period

Immediately after the towers have been completed, a period of about 20 minutes is taken to (*a*) have participants complete the brief reaction sheet, (b) have about a 10-minute break, and (*c*) have the judges (a team of three) decide on the winning tower. The winner and prize are not announced until after the discussion period.

JUDGE'S EVALUATION SHEET

On a one- (low) to-seven (high) scale, rate the towers on the following criteria.

	Height	Beauty	Strength	Message	Total
Team One (A & B)					
Team Two (C & D)					

After the break, each merged team (A & B and C & D) meets together to discuss what happened. It is assumed that during the break the two observers compared notes and have agreed upon a procedure for processing what occurred during the planning session prior to merging as well as the one after the two teams had to join forces.[4] If the group is helped with stimulating questions about what happened as the participants worked together, the discussion should require at least 45 minutes. Special attention should be given to the behaviors or situations that helped or did not help the two groups merge into a working team.

Summary

During this discussion, the facilitator should be organizing the data from the participant reaction sheets. His or her 10- or 15-minute presentation of these data should underline the learnings being discussed in the two groups. He or she should be free to draw both inter- and intragroup hypotheses from the data and then be willing to ask for confirmation or denial of his or her views. This type of presentation and discussion can help show the importance of reaction sheets and process periods after a period of problem solving and group involvement.

Conclusion

A natural end to the exercise is to report the judges' results. This can be done with humor and zest. If possible, the winning team can be given a prize that is easily divided (Cokes, and so forth) while the losers receive a booby prize that can also be divided (large candy bar, and so forth). Occasionally, this award period offers new information for discussion.

[4] If time allows, it is usually very helpful for the facilitator to spend as much as 30 minutes (prior to the exercise) with the observers, giving them the sequence of events and their role in the post-session processing. It is important that they interject their process data as a stimulus for further discussion and underline a point being made in the group. They should not just report data and await a reaction.

EXERCISE 4 The Bag or the Box: The Stress of Ambiguity upon the Development of a New Group

Objectives

To demonstrate the impact of an ambiguous and potentially threatening environment on a developing group

To observe the building of tension-reducing mechanisms within a climate of heightened anxiety

Rationale

Individuals or groups, when placed in unfamiliar surroundings and faced with unpredictable variables, will tend to become disoriented. They will then organize that environment in a manner that is safest for them at that moment. Under these conditions, leadership behavior becomes exaggerated and dependency magnified as the group struggles to bring order and predictability to their situation. It is rare in our experience that we are in a position to stand back and observe a group under such stress—usually we are too much involved to even recognize what is happening. Once having observed it in others, there is a tendency, however, to become more aware of similar conditions developing and of our responses to them.

Setting

This is one of the few exercises that requires special physical conditions. It is included because of its increasing availability and the willingness of facilitators to use it. Of first importance is to secure an observation room with a one-way viewing mirror. This room should be large enough to hold fifteen or twenty people. Also, the participants' room should be large enough to allow movement. (In most cases this would rule out observation rooms used by counselors). A small circle of seven to ten chairs is placed at one end of the room. If possible, the participants should be a group of from seven to ten students (preferably boys and girls) between about nine and twelve years of age.[5] Although it is necessary that the children

[5] It should be noted that children are suggested for this exercise because they are so spontaneous with their responses. Adult participants respond similarly, but less dramatically and less overtly. The young students also provide an interesting change of pace for the usually adult-centered groups and give them a new perspective on group development.

be unacquainted, it is best that there is some heterogeneity in the group and that relationships are, to some degree, still evolving.

Action-Directions

The group of students are met outside the observation room and told simply that they are to enter the room and work as a group for the next 30 minutes. They are requested not to leave the room.

When they enter the room, they find a variety of conditions that will tend to be disorienting. First, there will be no authority figure to give them directions or specific written directions to follow. Second, there will be the one-way mirror and the suspicion that they are being observed. Finally, there will be an object in one corner (at some distance from the circle of chairs). The object will be either a large canvas bag with a person in it (covered) or a blanket with a person under it. The person does not move or make any sound. On other occasions a large box may be used which is made to resemble a cage. In it sits a person (apparently unable to get out) who does not speak or move. A sign on the cage suggests only "Do not move."

Directions to Observers

As in many groups or in ongoing groups faced with an anxiety-producing situation, the developmental process will be influenced and a certain amount of regression is predictable. Those observing should note the specific tension-reducing behaviors used by individuals within the group and by the group as a whole. Questions such as the following may be helpful to the group as it observes the unfolding events:

How is the potential leadership in the group identified?

How do these leaders orient themselves to: (*a*) the lack of structure, (*b*) the bag, and (*c*) the mirror?

How do they orient themselves to the group?

What conflicts seem to arise because of the uncertainty of the situation?

How are these conflicts resolved or at least reduced?

What are the various individual and group roles that tend to develop as the participants attempt to cope with the anxiety-producing situation?

Are particular rules of behavior perceived as acceptable by the group (norms), and do these change during the 30-minute observational period?

Does the group move through definable stages as it attempts to cope with the ambiguity and make sense out of the unfamiliar environment?

Interview

Following the 30 minutes, the children should be given an opportunity to move into the observers' room and watch the observers as they move into the room with the bag. Perhaps with Cokes and in the most relaxed and informal atmosphere, groups of observers and children may discuss what happened. The participants are bound to have many feelings, and it takes skill, patience, and real interest to draw these out. Many of the questions that the observers have been asking themselves can now be broadened to include the perspective of the children. It is important to note, however, that the children may have many questions of their own and it may be helpful to give them a chance to ask them so that the process becomes one of exchanging ideas and less like the typical adult-student interrogation. The "bag" should also respond to questions from the students as a means of breaking the ice during the interview period. At some point the person should be allowed out of his or her bag and allowed to participate, although on several occasions, after a period of questions by the children, the authors have seen the "bag" walk out of the room.

Post-Session Analysis

Following the departure of the children, the observers should (perhaps in clusters of three or so) summarize their learnings in relation to parallel situations in other groups, particularly in terms of the process of group development. The sharing of these ideas based on concrete examples can be most stimulating.

EXERCISE 5 Tinker Toy Exercise

Objectives

To help focus a group upon the importance of nonverbal communication in its developmental process

To familiarize the group with how easily the group climate is influenced by particular behaviors of the participants

Setting

As in many of the other exercises, competition between groups can be used as a means of bringing a sense of reality to a particular task. How individuals behave and what their impact is upon a group will tend to be characteristic of other situations that stimulate, annoy, challenge, or bore them. In this example, it is assumed that there are two groups of randomly selected participants with from seven to ten in each group. (Note: if the group has been working together for some time, an interesting dimension may be added by assigning the individuals who tend to verbally dominate a group—gathered from previous data—into one subgroup and those who are less active in the other.)

Each group is placed around a table (no chairs). The two tables are close enough so that each may see what the other is doing (about 10 to 15 feet apart). There is nothing on the table. They are told to await instructions.

Materials

One can of Tinker Toys is needed for each group.

Action

The groups are told that they will have 45 minutes to develop the best possible product from the materials distributed to them. It is important that the product represent a group effort. The groups may begin as soon as they have the assigned materials, but the members may *not* speak to one another during the exercise.

At this point, the Tinker Toys are spread on the center of the table. The can and all instructions are taken away. The participants must use *all* of the materials in their project. They may also use any materials they may have brought with them, although they may not leave the table to obtain such materials.

Observation of the Task

It is assumed that the observers have some skill and experience in observing groups. If there are three for each group, they should have 15 minutes prior to the beginning of the exercise to decide what observation procedures will give the participants the best picture of what actually happened. The following questions may help the observers in selecting their observational instruments and strategies.

1. How do individuals in the group arrange themselves around the table (randomly, friendship subgroups, and so forth)?

2. At the end of the task, how has this initial ordering been changed? In front of whom is the final product? Why? Has this person the most skill? Is he or she the best organizer? Is he or she the most popular? Does he or she have the most power?
3. To which people in the group was most of the nonverbal communication directed? Were others encouraged to share their ideas in some manner?
4. Did members "jump" on one of the first ideas communicated or did they seek to explore alternatives? Did all the participants really understand what they were building when they started, and did they seem to agree with the idea?
5. How were tasks distributed during the work period? Was there any organization of labor or did individuals just do what came naturally?
6. Were particular roles (facilitating and inhibiting) identified within the group?
7. What did individuals who were obviously not directly involved in the decisions or perceived as resource people do to compensate for feelings of noninvolvement, impotency, or even inadequacy in this particular group?
8. Did particular behaviors by individuals (no names are necessary, but it is helpful if the group can accept direct feedback) create defensive responses in other members? Responses should be specific about the action and reaction parts to this question.
9. How did the group respond to the pressure of time and to the pressure of the group working next to them?
10. What norms developed as the group began to work together? Did these change as the task progressed?
11. What types of leadership were exhibited within the group and what impact did these have upon the developing group climate?

Other observations relating to membership, subgrouping patterns, tension release, reward, and punishment in the group may be explored. It is also important to observe how the two groups use each other as outlets for their own feelings.

Follow-up and Discussion

Depending on the particular objectives of the facilitator, there are a number of possible interventions that may be used following the action phase of the exercise.

Option 1 Still within the nonverbal framework, each group is asked to move to an opposite end of the room and in a relatively open space (ideally, about 15 × 15 feet), form themselves into a "group shape" that should symbolically represent a picture of how the individuals within the group feel about their own participation and the product of the group. They should try out as many shapes as they feel are necessary until they can all agree on one. Again, this activity should be observed in terms of group pressure, decision making, leadership, and so forth. Do the members really have the courage to represent the feelings generated during the activity? Once the shape is formed, the members sit down where they are and discuss the implications of their shape and how it reflects the process of their work together. After this discussion, the observers may wish to bring the group into a circle and share with them some of the data (again, there may be a tendency to overwhelm them with too much information) they feel is most useful. They should help draw the participants into the discussion and to have them, using their own experience from participation, respond to the questions the observers were asking. Observers are most effective when they are used to supplement the insights of the participants themselves.

Option 2 The facilitator announces that the groups will now have an opportunity to present their products to the other group. Each group will be asked to explain briefly what the product was and how the idea was arrived at. Then they will be asked to make a number of observations about their work together. The observers should carefully note who responds for the group, how this was decided, and how accurate the description is in terms of what actually happened. Also, it should be noted which behaviors occur while one group is listening to the other and while participants in the group being described hear a version of what happened according to their representative. Often the product is not what some of the participants thought, nor is the description of what happened the same as their own. After both groups have presented their product to the other group, each group meets to discuss the observers' data and explore what happened during (and after) the activity.

Summary

It is quite possible that the discussion period after either of the two options may take as much as an hour, but it can take much less if there is a more systematic presentation of the data and discussion is minimized. Time should be allowed at the end of the program so that each group can share with the total group specific learnings that seem to have developed out of the particular activity. Special emphasis should be placed on those behaviors characteristic of most groups as they develop during a specific task.

Other Possible Materials

The same basic design can be used with a wide range of materials. If the group is outdoors and near a wooded area, it is possible to build interesting products from only what is available in nature. Or, a brief trip to a dime store or supermarket can result in a seemingly endless supply of materials that could be distributed to the groups (clothes pins, tape, paper, plastic hair curlers, hair pins, rubber bands, crayons, pipe cleaners, and so forth). The main thing is to give the group enough so that decisions must be made as to the allocation of materials, but the materials should be different enough so they do not naturally form a product.

BIBLIOGRAPHY

Bednar, R. L., and C. Battersby. "The Effects of Specific Cognitive Structure on Early Group Development." *Journal of Applied Behavioral Science*, 12 (1976), 513–522.

Bednar, R. L., J. Melnick, and T. Kaul. "*Risk, Responsibility and Structure: A Conceptual Framework for Initiating Counseling and Psychotherapy.*" *Journal of Counseling Psychology*, 21, No. 1 (1974), 31–37.

Bennis, W., and H. Shepherd. "A theory of group development." *Human Relations*, 9 (1956), 418–419.

Bion, W. R. *Experiences in Groups*. New York: Basic Books, 1961.

Cattell, R. B., and D. R. Sanders, and G. F. Stice. "The dimensions of syntality in small groups." *Human Relations*, 6 (1953), 331–336.

Cohen, A. M. and R. D. Smith. *Critical Incidents in Growth Groups: Theory and Techniques*. La Jolla, Calif.: University Associates, 1976.

Cooley, C. H. *Social Organization*. New York: Charles Scribners' Sons, 1909.

Erikson, E. *Childhood and Society*. New York: W. W. Norton, 1950.

Evensen, P. E. *Effects of Specific Cognitive and Behavioral Structures on Early Group Interactions*. Louisville, Ky.: University of Kentucky Press, No. 76–20, 1976, p. 152.

Golenbiewski, R. T. *The Small Group*. Chicago: University of Chicago Press, 1962, pp. 193–200.

Hall, J. and M. S. Williams. "Group Dynamics Training and Improved Decision Making," *Journal of Applied Behavioral Science*, 6 (1970), 39–68.

Hare, A. P. *Handbook of Small Group Research.* 2nd ed. New York: Free Press, 1976.

Homans, G. C. *The Human Group.* New York: Harcourt, Brace, 1950.

Homans, G. C. *The Nature of Social Science.* New York: Harcourt, Brace, 1967.

Lundgran, D. C. "Developmental Trends in the Emergence of Interpersonal Issues in T-Groups." *Small Group Behavior,* 8, No. 2 (1977), 179–200.

Mabry, E. A. "Exploratory Analysis of a Developmental Model for Task-Oriented Small Groups." *Human Communications,* 2, No. 1 (1975), 66–74.

Mann, R. D. *Interpersonal Styles and Group Development.* New York: Wiley, 1967.

Mills, T. *The Sociology of Small Groups.* Englewood Cliffs, N.J.: Prentice-Hall, 1967.

Parsons, T. *The Social System.* Glencoe, Ill.: The Free Press, 1951.

Redl, F. *When We Deal with Children.* New York: The Free Press, 1966.

Schutz, W. *The Interpersonal Underworld.* Palo Alto, Calif.: Science & Behavior Books, 1966.

Stogdill, R. M. "Group Productivity, Drive and Cohesiveness." *Organizational Behavior and Human Performance,* 8 (1972), 26–43.

Thelen, H. *Dynamics of Groups at Work.* Chicago: Phoenix Books, 1954.

Theodorsen, G. A. "Elements in the progressive development of small groups." *Social Forces,* 31 (1953), 311–320.

Tuckman, B. W. "Developmental Sequence in Small Groups." *Psychological Bulletin,* 6396 (1965), 384–399.

Tuckman, B. W. and M. A. C. Jensen. "Stages of Small-Group Development Revisited." *Group and Organizational Studies,* 2, No. 4 (1977), 419–27.

Nine

The Use of Humor in Small Groups

How serious we are. How important everything becomes as we exaggerate our dilemmas, overreact to problems, and make normally complex situations larger than life. Cynics deplore the fact that we have never lived in such a humorless period. Watergate, a purposeless war, the deterioration of once stable social institutions such as the family and the church, tax revolts, inflation, recession, unemployment, fuel shortages, and changing morality all tug at our national psyche, leaving many with a resounding sense of pessimism and futility. For some, the solution seems to be in seeking an elixir—something we can take to minimize the pain, something that goes down easily. Thus, increasing numbers of people allow themselves to be drugged by sedentary, nonparticipative sports, by television and the predictable situation comedy, by soap opera voyeurism, or by drugs themselves. We become unwary victims of a patterned and highly predictable life style in which we expect others to entertain us, where we perceive ourselves as powerless, and where "tying one on" becomes a source of pleasure itself, rather than drawing pleasure from each moment, and humor from life itself.

As people see themselves as less and less in control of society, politics, work, and school, it is only natural that they become less creative, less motivated, less willing to risk, and less open to the very conditions that spawn much that is humorous. What humor there is begins to reflect our own spirit. The clever play on words, the double entendre, even the pun, give way to sarcasm, the put down, "black humor." Instead, we have "The $1.98 Beauty Show" brand of humor at the expense of somebody else, "The Gong Show" and the celebrity roast, where in "good humor" we poke fun or deride another friend. Laughing in relief at someone else's misfortune or failure is commonplace. It somehow becomes easier to get ahead as a result of someone's failure than as a result of one's own outstanding work.

Graduate business school students at a respected eastern institution of higher learning are schooled in appearing humorless and in controlling their emotions so as not to give an advantage to their peers who are exploring the same advancement opportunities. The image to be cultivated is serious and spontaneity is condemned. Strategic positioning for future success pushes away opportunities for maximizing the moment or for dealing with the process that is occurring within a group or organization in order to smooth the way for a more effective future. And with that goes our ability to laugh at what is happening.

READER ACTIVITY

Please take a moment and consider three or four factors in your life that give you concern and a sense of pessimism. Our society is so complex and many conditions impinge upon us over which we have little control that you should easily be able to isolate a few of these. Now the question becomes, how do you relieve the natural stress that develops around these situations? Are there ways in which you turn such stress into humor, draw from tense realities enough laughter that allows you to live with the stress in relative comfort? Do you tend to be a creator of humor, a seeker of it, an observer?

WHAT IS HUMOR?

Humor, like a feather in the wind, will sail gently earthward. Just when we think its life is exhausted, it will float lightly up and away propelled by an expected breath of air, changing everything. Unpredictable humor can ease our embarrassment, calm our anger, and relax our tensions. Most of all, it can free us from pedantic, ritualized, and thoroughly predictable behaviors or events.

In a group, humor can often be a scarce natural resource whose presence can very literally change the life of the group. It can make what is merely tolerable both interesting and even exciting and what is boring bearable, while keeping reality in perspective. Although it may be planned and carefully staged, most often humor is generated spontaneously from unpredictable turns in events that flow from the natural development of the group itself, from the success and crises its members experience together.

Most of us seek order, comfort, predictability, and stability in our lives, and if we succeed we may well drive away what is the essence of most humor, which often springs forth from uncertainty, discomfort, and disorder. All of us have experienced dull, predictable, and boring relationships, committees, and organizations where there is little room for new and exciting ideas, for play or outrageous spontaneity, or for release from the challenge of shared risk taking. Too often we are lulled into familiar, patterned, humorless responses that we help to create by the norms we allow to dominate the groups in which we participate.

In reality, a marvelous source of humor that is readily available to any group member is the group itself. Group members share basic human characteristics, such as the following:

Power and impotence

Anger, joy, and sadness

Security and insecurity

Likes and dislikes

Interest and boredom

Success and failure

These, and more, are the heart of any group. People attempt to maintain their personal sense of integrity, to look "good" rather than "foolish," to walk away wanting to return and knowing they are wanted. The constant struggle within any group of individuals attempting to meet their own needs as well as those of the group creates marvelous dynamics and provides a platform for a never-ending source of humor. Whether the humor is recognized or used as a constructive part of the life of the group, is another question. For now, it is enough to know that within virtually any group, for those knowing how to observe, there is a bottomless reservoir of rich anecdotes, or vignettes, which can provide as much humor as any theater. By actively developing and utilizing the humor that exists in any group in a positive manner, both leaders and participants may have an important influence on attitudes, performance, and membership, not to mention norms, goals, and communication patterns.

Situational humor drawn from shared experience may increase our enjoyment, reduce defenses, increase the willingness to risk, open communication, and, as a result, increase individual feeling of membership. Obviously, humor is no panacea for creating harmony or group success. But as will be shown, not to cultivate the humor that is present will certainly influence how a group develops, the attitude of members, and how the group functions. Humor can become a natural and constructive part of a group's life. Furthermore, if it is true that many groups move into patterns of defensiveness, noncreativity, and even hostility, the utilization of humor in an increasingly supportive atmosphere may be an important ingredient in reversing such a nonconstructive pattern or in reducing the opportunity for its initial development.

Humor is simply the response to discovering something that is laughable. Most of us relate it closely to outcomes such as being entertained or amused. For our purposes, we are attempting to explore what it is in a group that allows humor to occur—humor that is constructive to the development of the group and the individual members. Clearly, we are after much more than ways of tickling a group's psychic funny bone. Optimistically, we wish to increase your awareness, your understanding of humor in groups, and ultimately your ability to capitalize on its use so you will be increasingly willing to risk using it.

Strange as it seems, participants in groups, without being born comics or particularly humorous themselves, have the capability of generating real humor within the group—everything from warm internal smiles to belly laughs. The answer to how lies in the ability to design certain events that allow humor to take shape and emerge out of the ongoing life of the group, its activity and purpose. If a group has developed a certain amount of trust, humor will be easier to exploit because individuals will feel less embarrassed, will find it easier to show feelings, and will be less competitive and therefore able to support a shared, positive experience. Without the development of a certain initial trust, it might be impossible to tap one of the greatest sources of humor available to a group—the members themselves. Their frailties, failures, successes, idiosyncrasies, and personal styles are what makes each of them unique. Laughing with each other rather than at one another can be enormously freeing, even exhilarating. It can turn into positive support and a sort of stroking rather than being used at each other's expense. It appears that humor and trust are reciprocal, each less possible without the other and each building on the other. The same seems to be true of openness in a group, the ability to risk, and even the ability to provide descriptive feedback to members. All seem to be enhanced in a group capable of laughing "with itself."

BARRIERS TO HUMOR

It seems paradoxical, then, that although most people love having fun and enjoy humor, they are often resistant to its use, and some groups are just as humorless as some individuals we know. Actually, the reasons are not too complex. Many groups equate humor with being silly or wasting time. After all, it's suggested that people can't be working if they're busy having fun. In one large corporation with which we are familiar, whenever the boss travels, people appear out of their offices, talk, smile, walk around joking a bit, like ground hogs out of hibernation. They are not aware of their changed behavior, but it is quite evident that his absence provides an implicit permission to relax and to allow natural humor out to play. The people are quite professional and they don't take advantage of the situation. They simply allow themselves to be natural and to enjoy the pleasure derived from well-intentioned humor.

Perhaps the greatest resistance by people to the use of humor is the ingrained sense that school is school and work is work and play is play. Our puritan ethic subscribes to the idea that work should be hard, serious, important, direct, and efficient. Humor and fun suggest lack of focus, failure to have correct priorities, a misuse of time, and a lack of attention to what is really important and to be rewarded.

Not only is humor related to silliness, wasting time, and laziness, but it is often seen as a measure of immaturity. Because work is such an important part of our lives, many individuals will avoid discrediting themselves and being labeled as the "jokester." Historically, the jester was often likened to the fool. Similarly, an individual who thoroughly enjoys humor will be labeled as not being serious or tough-minded, and it is not uncommon to find a career path blocked by stereotyped views such as these. Of course, once again, certain types of humor may well be acceptable, for instance, those with a competitive edge, or used at the expense of others.

THE ROLE OF HUMOR IN GROUPS

Humor is extremely difficult to quantify, a bit like describing the light emitted by a firefly. We see it, know it exists, but how it's created is baffling. So in this initial exploration of humor, we are attempting to develop theory based on the face validity of our own experiences. We know humor plays many roles in any group, but there is little empirical evidence of the exact degree to which humor can influence a group.

For example, we know from personal experience that humor can be used effectively in disorienting individuals, redirecting them away from potentially hostile or aggressive situations, and diffusing a negative event. It may also be used as a means of avoiding much needed conflict or confrontation in a group. Conversely, humor can be used by members to hurt or injure others, or to minimize the value of their contributions. Group leaders diagnostically aware of problems within their own group would ideally intervene into the life of the group in a positive manner to facilitate the group in its own development. Thus, if the group were using humor to avoid needed conflict, the source of humor would have to be identified and in some constructive manner pointed out, and possibly extinguished so the group could be responsible to itself. If, as many believe, humor introduced into a group that is defensive and even hostile will reduce tension and provide a climate that is more positive and constructive, then it only makes sense for leaders to try to draw some levity into such a situation.

In one of the few studies of its kind, Baron (1974) found that of two groups of individuals who were subjected separately to anger-inducing situations, the group that was shown a series of humorous cartoons later exhibited markedly less aggressive, less hostile behavior to the object of anger. Given the opportunity to shock the individual who had angered or provoked them previously, the group tempered by humor apparently diffused some of its anger, whereas those not provided a means of diffusing their anger were significantly more punitive in their responses. There is little doubt that the intervention of humorous material did reduce the anger and distract the focus of tension in an indirect manner. Very simply, by helping one group feel better, the humor was the source of an increase in the group's tolerance for anger or hostility and a parallel reduction in their need to act out on their feelings as strongly as would otherwise occur.

In essence, the study suggests that by not acting positively toward the group (providing humor) certain negative outcomes could be predicted. Thus, group members have real power to redirect the emotions of the group in a positive manner. In this case, to do nothing can be perceived as a negative intervention. Because most of us are leaders at one time or another in groups, it is interesting to note that the leader is often caught between a rock and a hard place. He or she is in the unenviable position of manipulating the group by acting or not acting, assuming the outcomes of his or her behavior are predictable. From our point of view, it is no more manipulative to facilitate a group in an active intervening manner than to omit an item from an agenda that has known consequences for the group. If humor can be used to move a group constructively toward its product and maintenance goals, then the leader has the obligation to do everything possible to help. Because our previous discussion of leadership suggests it is situational and shared among group members, the remainder of this chapter views the group member as an active interventionist, a person capable of using humor to influence the life of a group and in many cases the attitudes of other members. As in everything else, the value of an individual's contribution can be measured in relation to need, intent, honesty toward the group, skill of the intervention, and outcome in relation to how the group was helped and valued the help. Thus, we are going to look at what actively creates humor in a group and the critical interface between the leader and other group members.

SOURCES OF HUMOR

Every time we are drawn into laughter from a play, a film, a song, or a picture, we trace it predictably back to the unpredictable. Catching us off guard is the essence of humor, which is created by presenting us with a paradox, a discrepancy between words and action, a startling event, or a sudden truth or personal insight; all have a common thread. But there are other sources of humor that we will find valuable to consider in our understanding of small-group behavior. Shared success or failure, the absurd or outrageous, something very familiar and personal, or a strong memory may evoke humor. Even the release of being faced with great fear and defeating it can generate both relief and humor.

Of course every situation is different. What we are interested in is the part of any intervention or experience in a group that can be applied in some manner to another rather similar event. It is this ability to generalize that is the foundation of all work in the social sciences. So for each type of humor and for each example, we will attempt to generalize as much as our own limited experience will allow.

READER ACTIVITY

There are many ways to laugh. A belly laugh, a smile, an internal grin that only you know occurs, the giggle of embarrassment, a shriek of delight, or the helpless laughter when everything has failed and all you have left is your sense of humor. Think of the last few times you laughed in a group. Can you remember how it felt and what caused the humor? Was it strictly situational or could you generalize what happened to other situations. In other words, would such humor be predictable? Could it be created intentionally?

Let's take a look at two jokes, both capable of generating humor. The quality and quantity of humor drawn from each will depend first on where you, the reader, are in your life, your previous experience, and personal needs. How humorous you find each joke will depend on the degree you personalize the experience of each of the two stories, whether the style of the joke is appealing, whether you are dealing with any of life's universal questions, whether you anticipate the catch, and whether either provides you with something more than a laugh. A laugh attached to a personal insight will obviously have much more significance.

The first story concerns two American tourists traveling through the English countryside. Stopping at an old monastery, they decide to have lunch. The brother who greets and waits on them brings them each a huge and quite delicious plate of fish and chips. Appreciating the hospitality and the tasty meal, one of the Americans asks the brother whom he might thank for the meal. He nonchalantly points to the kitchen and suggests that the two chefs would be happy to hear the compliment in person. First, there would be the fish friar, and after him there would be the chip monk.

The twist, the catch, the play on words all make the situation believable so you can be slipped the unbelievable. Even though the joke represents a prepackaged effort at humor, it incorporates much of what must be present in a spontaneous situation that might create humor in a small group.

The second story involves a fifty-year-old businessman. After years of working at a break-neck speed without vacations and without enjoying either himself or his family, he began to ask the question, "What's life all about, anyway?" The question represented the beginning of a personal odyssey. He took a leave of absence for a year, read all of the great philosophers, talked to as many great thinkers as he could, took special courses, and became a scholar in his own right. All in the pursuit of the question: "What is the meaning of life?" After nine months, the man had exhausted nearly all of his financial resources in his search, and his leave of absence was coming to an end. Still he had no satisfactory answer. One day, he heard of a famous wise man, a guru, who lived on the top of a mountain in a distant European retreat and was reputed to have found the meaning of life. Worn down, doubtful, he set out to find the famous man and make a final attempt to understand. The journey was a most arduous one, but finally, at sunset one evening, he trudged the last mile to the ancient retreat. Seeing an old man sitting by himself, the searcher urgently asked, "Sir, are you the wise man reputed to have discovered the meaning of life?" "Well, my son, I have certainly given it thought and, yes, some say I have come close to the answer."

His anticipation mounting, the searcher begged, "Please, then tell me. I have looked for so long." Putting a hand on his shoulder, he told the businessman, "My journey, my own personal search tells me that the meaning of life is. . . ." The man was virtually beside himself as the old man's words ". . . . it's a waterfall" came from his lips. The man stood amazed, confused, feeling helpless. "You mean to tell me that I've come all this way, exhausted all my personal resources for you to tell me that the meaning of life is a waterfall? You have the nerve to tell me this?" The wise man suddenly looked terribly pained. He held his head for a moment in thought and, then looking the man in the face, asked in a puzzled tone, "You mean it isn't?"

Who of us has not taken the journey, struggled with the question, felt disillusioned by our life on this earth, prayed for time to think about the bigger questions? Who has not felt the doubt or hoped that there is an answer? Who has not hoped that the work of defining the answer could be done by some guru? And who could not at least smile at the trap into which we had fallen and the renewed insight—that we are the answer. Again, the packaged joke personalized, related to our own experience, to a question of ultimate significance and providing a zinging insight snapped our head around and made us laugh as much at our own foolishness as anything.

The two jokes, although each of a different type and at different levels of seriousness and intensity, had aspects in common and readied us, set us up in a predictable manner so, again, the unpredictable could work on us. With this in mind, we are ready to explore various ways of generating humor in small groups.

The Paradox

Years ago, an American tourist drove his car up a steep European mountainside terraced with level after level of small, stone nuts, each looking the same and equally impoverished. Each sat on a tiny parcel of land looking as if a strong wind would topple it off its precarious perch. The small plot of tilled soil next to each could hardly sustain the occupants and the walk to the valley below and back would take several hours a day. The tourist's thoroughly American middle-class sympathy was slightly assuaged by the smiles and good humor that greeted him at every turn. The road was overrun with men, women, children, donkeys, cows, and goats. Suddenly the road took a sharp turn to the right at the top of a sharply rising precipice. What he saw drew his breath away. Unbelievable beauty! Azure blue sky, emerald green mountainside, all framed by the crystal blue hues of the Mediterranean. Although not wishing to romanticize their poverty, the laugh was on the tourist. Each of these "poor" people had a piece of this jewel, bathed in clean air and sun and a simple life style that seemed to put important things into perspective.

The difference between what might normally be expected and what occurs, or between what we might normally do and what should be done to make a difference can be seen clearly in the following story of paradox. A group of recently divorced men and women had come together in a workshop setting to talk about mutual problems and concerns generated from their recent traumas and to seek commiseration from others who might understand. Such groups are increasingly common and can be tremendously beneficial or, on occasion, terribly dissatisfying depending on the skill of the leader. Used as a platform for blaming the other fallen spouse who is not there to defend him or herself (which is the immediate inclination of almost everyone) can be self-destructive, although momentarily satisfying.

Realizing that blame and self-pity are the food for depression, the leader decided to take the proverbial bull by the horns. She asked the participants to form small clusters and talk about all of the reasons the divorce was the fault of the other spouse. Predictably, the group needed little encouragement. Off they went into the corruption of their own marriages and what their partners had done to the marital bliss they had so desired. After perhaps 20 minutes of catharsis, the leader suggested that they not leave out any bloody details, to make sure the other participants had a clear picture and that it was alright to exaggerate a bit to help make a point. Well, the room was in an absolute uproar of self-justification and blame. As the stories became more exaggerated, laughter began to well up from the various small groups. Cautious levity turned to hilarity as exaggerated truths and one-sided stories of anger, pain, and absurdity filled the air.

Then, with the grace of a ballet dancer, the leader turned the group to look at itself, asking each individual to analyze how much power they had somehow, in all of this exaggerated blaming and storytelling, given away to their partners. The paradox was suddenly all too clear. How little responsibility each was willing to take, how dependent and helpless they sounded, how utterly out of control. The humor drawn from exaggeration surely made many of the participants gasp as much as when the tourist turned not to see the beauty of the Mediterranean but the degree to which his initial sympathy had been narrow and self-serving and distorted. Therefore, the real humor in the workshop did not lie in the initial bursts of energy and laughter, but in the later deep and probing humor of self-realization and insight.

The prepackaged "monk's jokes" have punch lines that grab each listener in a different way depending on their background and readiness. In the workshop for these divorced individuals, the punch line was just as certain, and the humor generated from the exaggerated storytelling and then from the paradox would fit into the personal experience of each individual.

Certainly the leader was not a comic, a joke teller. Rather, she was a skilled facilitator of social interaction who allowed humor to be woven into the fabric of a productive educational experience by those who must live with their own learnings.

Irony

Ironic humor evolves from the discrepancy between what is real and what is perceived. Human beings are capable of unbelievable distortions of reality. That we see what we need to see is obvious to anyone willing to take a hard look at themselves. Obviously, we become masters at self-deception because we somehow are not ready or do not wish to face that which is real. In the extreme, this behavior is psychotic. For most of us it is merely the normal protecting of ourselves from a very difficult world. Taken with a grain of salt and kept in perspective, the discovery of our own self-deceptions can provide us with useful insights about ourselves and our world.

READER ACTIVITY

When was the last time you remember fooling yourself? A time when suddenly you were struck in the face with a new reality—perhaps an idea you had previously denied only to discover it was absolutely true. And there you were—caught with your face hanging out and only a smile to cover it? Sometimes we are so wrong it almost feels good to admit so loudly that "yes, we can be wrong, dead wrong."

Groups as individuals have marvelous ways of deceiving themselves from those things they do not wish to "own." So just as an individual is capable of denying that a problem exists, so too, a group can persuade itself that obvious information has meaning other than that which is most obvious. In the case of an individual, the fact that a person's daughter goes out with other teenagers who take drugs, her grades are declining, and she returns home with her eyes occasionally dilated may mean little since it is quite possible for many parents to read these symptoms and attach totally different meaning to them. For these parents it would be out of the question that their daughter, a product of their home, could possibly be involved in drugs.

The faculty of the elite eastern girl's prep school had somehow denied all the signals that "their girls" were involved in the use of drugs. The symptoms were all there. But somehow, their own fears and fantasies made denial easier than facing the problem. Even though interviews suggested that the problem existed, the outside consultant knew he could not simply tell the faculty that drug use was at a rather severe level in the school. After all, they could easily use the age-old justification, "What could he possibly know about our school—he didn't even go to a boarding school himself." (So much for consultant expertise.) Instead, a group of faculty were gathered together to help develop questions that needed to be asked about the school to help improve the social and academic climates. As good academics, the faculty did not avoid the hard questions. Among them was one framed about drug use. They asked parents, faculty, and students the question, "What percentage of students use alcoholic beverages or drugs (grass or pills) as regularly as every week or two?" They then asked parents and faculty what they thought the students would say and asked the students what they thought the faculty would say. Without quite realizing it, they had set up the possibility for major discrepancies to occur. In the theory of change, one of the major stimuli for change is when our perception of reality differs dramatically from other believable data.

Several weeks later, the entire faculty was gathered together to listen to a report of the data. The data feedback session was turned into a workshop situation exploring each question in depth, with the faculty predicting responses prior to the data being revealed, interpreting the results, and then discussing the resulting information. Later, it was understood that a two-day problem-solving program with both faculty and students would occur to attack major issues cooperatively.

Throughout the feedback session, considerable humor had been generated as predictions were confirmed or denied by the data. Inevitably the greatest laughter would result at points of greatest discrepancy. Finally, the question relating to drug use in the school approached. The reason so much tension arises over such a question is complex. First, parents and faculty take it as a personal failure if their students have gone astray and use drugs. Second, many don't understand the problem since they have not experienced drugs themselves. Thus, they project onto the problem many fears and stereotypes that make drug use the only problem rather than being a major symptom for other problems as well as being a problem in itself. Finally, since there are no simple answers to control or prevention, there is a sense of impotence among adults at dealing with the information if it is perceived as negative. The data were presented on individual sheets of newsprint paper, one at a time.

Faculty perception of students using drugs regularly—32 percent
Parent perception of students using drugs regularly—22 percent

Having digested these two pieces of information, the faculty were asked to write down their prediction of the student response on a small piece of paper and to move into groups of four for discussion. Now the student data were shown:

Student perception of students using drugs regularly—71 percent

Nervous laughter, groans of disbelief. Asked to discuss the implications of the data, many immediately tried to explain away the data. The students exaggerated, they were trying to be smart, they saw others do it and assumed even more took part, and on and on. Finally, the other data were shared.

Faculty perception of what students would say—48 percent
Parent perception of what students would say—33 percent
Student perception of what the faculty would say—36 percent

Asked to summarize the significance of all the data relating to the question and report it back to all the faculty, the various small working groups were seen to be huddled and almost whispering as they shared their points of view. When reporting the data back, virtually every group agreed in a variety of humorous, self-effacing, and generally nondefensive ways that they as a faculty had been shortsighted; that a problem did exist; that they had resembled ostriches more than intelligent faculty, and "in all of their wisdom they needed to be educated" before they could be expected to grasp all the significance of the data. The humor in response to the discrepancies may have been the result of not being able to fool themselves any longer, in the realization of being caught in their own joke and having nowhere else to go but to acceptance and laughter.

Although one assumes the faculty could have been angry—perhaps experiencing a sense of being duped and feeling foolish—our experience in many similar situations is that once reality is grasped, the more prone a group is to perceive itself as rational, the less willing they will be to deny the true significance of the information. In this instance, humor acted as an almost predictable vehicle for admitting imperfection and beginning to deal with the new reality. By working in small, intimate clusters, they could first defuse their own discomfort and possible embarrassment with the data and begin putting it into perspective with humor and candor. They could begin to see they were not alone in their failure to predict successfully. The weight of judgments was lessened and the possibility of guilt diminished, thus freeing the group.

The Unanticipated

The twelve seminary students had been meeting for more than a week readying themselves to move to new posts as the heads of inner-city parishes. Although some had previously worked in poor communities, each saw their new role as a challenge and were somewhat awed by the problems they were anticipating. Part of the training was directed at helping each of the students gain better self-understanding in relation to his own behavior and the impact his own leadership might have on his new parish.

Four days into the program, a climate of work had evolved that was supportive and trusting and allowed no room for the usual game playing that occurs in many meetings of professional peers. On one occasion, Jim was discussing a particular point in his usual forceful and colorful manner. It was his style, as a member of an inner-city parish for many years, to use what might be called the "local vernacular," complete with a wide variety of colorful four-letter words. Every time he spoke, Vic and several other members of the group would grimace. Finally, it became too much for Vic, who shouted, "Do you always have to express everything with such a filthy mouth? Aren't you creative enough to talk as a civilized adult?" For what seemed like an eternity, no one spoke. Then, in the understatement of the year, the group facilitator quietly said, "It appears that some of us have some rather strong feelings about the type of language Jim has been using." For about ten minutes, chaos prevailed as sides were drawn over the use of profanity, especially by representatives of the church.

Seeing that the argument was going nowhere in a hurry and that members were becoming increasingly polarized, the facilitator suggested, "Because we all have strong feelings over this issue, I would like to have each of us think of the most filthy, disgusting statement we have ever actually heard. I would like us all to share these." Several of the students almost fainted away at the thought, others thought it to be a wonderful idea, and still others actually blushed as they half-whispered a patently common four-letter word. Eventually, the group reached a consensus on the word that was most vulgar to them. The facilitator then told the twelve students in a most serious voice, "I would like us to all say this word three times as a group, a bit like a cheer, saying it louder each time." The group was absolutely befuddled and disoriented, not knowing whether to follow the directions, to be serious, or to take it all as a joke. Up to this point, although there had been a few snickers (as when someone had told a racy story in a mixed group), the tone had been quite serious. "On the count of three," the leader said. "One, two"—it was like the moment before a balloon bursts—"three!" As a result of all the pent-up emotion, the group only whispered the word. They looked at each other sheepishly. "Come on fellas, we can do better than that. Now, once again." This time there was real gusto and a bit of pleasure as the word was shouted and most everyone either smiled or laughed. "O.K., one last time, let's really give it a go. Let's hear it." At that point, out came a roar, and the group was beside itself, laughing at the absurdity of it all—discovering together how little it all meant and how foolishly and judgmentally they had acted. It wasn't as though it was necessary or good or bad to swear, but rather that a word of no real importance other than how they themselves interpreted it had made them so uptight and embarrassed.

And the humor, once again, the punch line had been created out of the feelings and emotions of the group itself. Thrown into a totally new and unanticipated situation where there was no prescribed role, being totally disoriented in relation to each other and the previous norms of the group, the individuals gained a new and different perspective. It was from that new vantage point, totally unexpected, that allowed many to see their own narrow biases that surely would pose problems for them in their new parishes, where such words were part of the culture and certainly could not be passed off as filthy. The facilitator knew the humor was there, thought the group would find it, and then simply designed the particular intervention to tap it. Without the predictable humor, the group would have been left in the win-lose battle, leading to further polarization.

READER ACTIVITY

Take a break and ask someone close to you (both in proximity and personally if possible) if he or she can remember a situation or two in which something happened that was so unanticipated, perhaps so paradoxical that humor was the last thing he or she would expect but he or she exploded in laughter. Why was it so funny? Was it emotion? fatigue? surprise? frustration? or was there nothing left to do? Again, can you generalize this? Relate it to other situations. Does it have any implications for you as a group member?

Sudden Awareness

The parents of the junior high school were outraged at what they perceived as a total breakdown in discipline in the school and blamed it mainly on the teachers. The teachers felt that they had been blamed unfairly and been made scapegoats, and that the real problem lay in permissive parents who could care less what happened to their children once they left home. The students felt totally alienated from both the faculty and parents, whom they felt simply didn't understand them. All three groups were feeling unappreciated and didn't trust the others.

A day had been planned to bring together about one hundred parents, children, and faculty to look at problems that could be solved collaboratively and to begin developing workable solutions. But the participants were expected to be defensive and unable to listen. Since beginnings are so crucial, it seemed important for everyone to stop riding their own bandwagons of hurts and anger, to put things in perspective, and to begin working in a cooperative manner. If humor could not be injected into the situation, one could predict a quick departure of the various groups into stereotyped role behavior, with kids acting as parents expected them; parents acting as kids expected; and teachers acting as everyone expected. Humor was needed as an equalizer. But how?

The fact is that most adults are so busy acting adultlike around adolescents that adolescents seldom see their fun, childlike side, And similarly, adolescents are so busy acting like adolescents for themselves and adults that they seldom have the opportunity to be more adult. If both groups could be placed in situations that showed the other group a new side, then the chances for understanding and trust developing would be greatly increased.

Everybody met in the gym feeling totally strange and uncomfortable. Mixed groups of between six and eight were formed with three or four boys and girls, and the rest teachers and parents. The intention was to provide more support for the students than for the teachers or parents because age can be an inhibiting factor. After several warm-up activities designed to get people talking, each individual was provided a piece of drawing paper and access to coloring materials. The task was for each person to draw two pictures representing school—one thing he or she disliked about school when he or she had been a teenager and one thing he or she really liked (other than gym, lunch, and recess). They were to draw these impressions as best they could and share them back with the group. Immediately, the discomfort level in the room rose, but there was no way to withdraw from the task easily. So the task prevailed and teenagers and adults began an equalizing experience in which almost all would fail a bit (the drawing) and all would succeed a bit (sharing experiences). When the individuals reconvened into their small groups to share their drawings, a profound awareness took hold of most groups. They realized that virtually everyone could appreciate the likes and dislikes of each other regardless of age, and that teachers held some of the same fears and apprehensions that resulted in their disliking school as did the kids in the group. Not only that, but they also felt a little embarrassed to share not only their inadequate drawings but their past feelings. The awareness that we are all in this boat together provided a common point of discussion and understanding. The relief at no longer feeling as uncomfortable with each other and having their discomfort mutually channeled into a common task created a release of tension so that most groups were laughing with each other—uproariously over drawings, personal experiences, or the simple satisfaction that they were surviving a terribly uncomfortable experience.

Finally, each group, either from their original discussion or later exploration, was asked to design a skit. These would represent a common problem that existed today in the school that would take the cooperation of students, teachers, and perhaps parents to solve. These were performed for the entire workshop. Common problem themes were identified and a structure was then developed to begin generating potential solutions.

Without the use of a humor-producing format designed to reduce the members' tensions and resistances toward each other, without creating equalizers and a climate in which laughter was legitimate, it is doubtful whether the productive outcomes of this situation could have occurred. The awareness the group gained of common roots, insights, and feelings freed them to listen, to take themselves less seriously, and to be less reactive and more proactive with each other. It was this illumination that freed the latent humor and good feelings in the group. Not only this, but within the situation of the workshop and the climate of humor and good will that prevailed, new perceptions could be established as both student and adult stereotypes came crashing down. As equal partners in the task, working toward a common goal, there was no need to preach or talk about good communications. The positive common experience created by focusing on similarities, not differences, made a profound impact.

Shared Risk Taking

Mutual risk taking is one of the surest ways of increasing cohesion and a sense of membership in a group. An initiation is in fact a ceremony of introduction to membership. In this country it often is associated with hazing or activities incorporating fear, the unknown, or adventure, so that an individual has to earn the privilege of membership. People who have gone through initiation rites as part of a group know that the sharing of the common experience will draw the group together and create a bond that seldom will be experienced under less stressful conditions. Within the initiation rite, whether it be in a fraternity, marine boot camp, or a social club, the common goal and performance anxiety of the group provides a natural and expected release for humor. It is not unusual that we translate such severe examples to other group settings, but, for example, if team building is a group goal, it will be enhanced if some of the characteristics of the initiation rite can be incorporated in the program. Here is a case in point.

A group of counselors, teachers, and administrators came together to learn to be more effective leaders. To learn from each other in the most effective manner, it was believed necessary to help them become a team as rapidly as possible. Thus, on each of five days, the group was faced with a task that required individual ingenuity and courage and that would be easier if assistance could be acquired from the group. On day one, small groups were taken deep into unfamiliar woods and given brief orienteering lessons and the directions to their next meal (whenever they arrived). Although people did not fear being abandoned, the situation provoked enough anxiety that humor became a common tool for coping. Because the humor was generated by the situation and not directed toward any individual, the group established an implicit norm of supportive humor rather than humor at someone's expense. On the second day, each individual was required to climb a sheer cliff, supported by the good will of the group and a rope tied to another member. As each person successfully completed the ordeal gales of laughter would be heard as people laughed about their fears, the overcoming of potential failure, and the good feelings of success. Similarly, a third day saw small groups taken to a river with boards, rope, inner tubes, and a bit of canvas and told that camp was five miles down the river. They were to construct a raft and meet the rest of the group there. Swimmers and nonswimmers alike were drawn together by the challenge, excitement, fear, and fun of the task, with some people playing counselor to the nonswimmers, some acting as architects, and others playing less defined support roles. The laughter and merriment as each raft was launched and stayed afloat (some did not) was like a glue drawing the group together. In all of these experiences, humor not only reduced anxiety but became a sign of the group's support for each other and confidence in overcoming the odds they faced.

Another example involves a group of university students who agreed that each would do something in front of the group that he or she had always wanted to do, but had been too afraid to do in the past. Although the group knew each other and thought doing this would provide a good night's entertainment, they had no understanding of the degree to which it would unite the group in a bond of experience that would never be forgotten. One of the first members stood before the group, asked them not to laugh at his effort to read "Don Quixote," because he had wanted to do it most of his life in front of an audience. He proceeded to render a rather ineffective reading. There wasn't a smile, not the glimmer of a laugh until he was done. When he finished, the group cheered and laughed with him at overcoming his own great fear. The laughter was in the spirit of caring and having made it through the rite. Another person, who had always wanted to be an athlete of fame, had the group cheer his mock heroics, which allowed him to act out a fantasy and put away a dream by admitting his own limitation. Again, the scene produced tremendous laughter and humor, but at what? Not at the scene, but at the never-to-be-reached dream found in all of us that we may yearn for until we die. Here he was putting it to rest.It was the laughter at themselves that spilled over into good will for him. Finally, one woman had each individual in the group promise to say something about her they felt she could improve since she perceived that she had spent her life being "nice" and avoiding conflict or arguments. The group did, and then laughed at how inconsequential it all was in light of other things they felt. But, the initiation rite of it all was real and the experience of shared anxiety and success was also real.

The two examples here were both extreme, one on an emotional level and the other combining the physical and the emotional. At a less intense level, some of the same feelings and team-building benefits can be gained by taking a group of people who work in the same department away for a day, and after anonymously identifying problems that block their effective work together, provide a structure for working several of the issues through to solution. Five small groups working on the same problem that people were previously afraid even to mention can have a releasing effect and provide a sense of trust not experienced before. Of course, the problem solving will be for naught if the group does not make a commitment to follow-through and action. However, if people believe they are being heard and that solutions are possible, the outcome in terms of the life of the group will generate an enormous outpouring of humor and good will as people risk their view of common problems and try to develop mutually satisfying solutions together.

READER ACTIVITY

Have you ever been in a group that is caught in the grip of fear? Most of us have. It seems that there is always someone who, even in the worst possible situation, can find something humorous. At times, the humor just creates a ripple through the group while on another occasion it bursts through the group, removing a heavy burden. What is it that draws this type of humor from deep inside? Is it simply a raw, primal response to uncontrolled anxiety, or are there other ingredients? What is your experience? Can you recount an incident where such humor changed the entire complexion of a situation? How? What did you experience personally?

Generating Personal Truths

The most important person to each of us is ourselves. We do an amazing job of building defenses against an insensitive world. Any time a group leader wishes to gain a group's attention and increase interest and motivation, he or she has simply to develop an activity around the sharing of each person's personal ideas on the subject, or, better yet, the sharing of something personal and meaningful from each individual's life experience. The interest others show in us will increase our ability to show interest in someone else. The fact is that in each of us is a never-ending stream of enjoyment, because out of our personal pain of growing older emerges the secret to our growth, understanding, development, and above all, humor. Consider an example.

A group of thirty professionals in the social sciences and related fields spent ten days attempting to understand themselves and others better. The group had been together about five days and the leader knew about half of the group well, although some he didn't know at all. The room was darkened and the group was asked to lie on the floor with their heads toward the center like the spokes on a wheel (admittedly not a common activity for a social get-together). The members were then asked to shut their eyes and try to remember the things that happened in their lives that they really didn't like but couldn't do anything about. They were especially encouraged to consider events from their families, the way they were treated as kids that they resented, or people who created special pain for them. When they felt like it, they were to describe in a word or a sentence or more the situation in a way that would help each of them understand it.

They didn't move for two hours. After five minutes, the group was one. The laughter was so intense, it was so easy to recognize their own pain in virtually everyone there. An uncle, a grandmother, a sister, grandparents, teachers—the commonality of it all brought a sense of ecstasy, along with the pain of bad memories and the joy of having overcome most of them like everyone else. It was like the combination of delicious memories and sadness when a vacation is over. But, the realization important to us here is that within each of us are a thousand stones that merely have to be turned over to discover huge quantities of humor and shared experiences with other very tough, fragile survivors. Lying on one's back in a darkened room is obviously not the everyday method of unlocking this type of humor. But people are forever ready to talk about themselves, their successes and failures, their joy and sadness in ways that will inevitably allow them to laugh through the sharing of it. As long as people feel protected from ridicule or contempt, they will share, especially in small clusters of people where they don't feel too vulnerable. It is for us as group leaders and members to create events that can spring free some of this humor. The activity is much more than traipsing back over the clutter of time or rummaging through nostalgia for its own sake. It is reconnecting ourselves to important parts of the past and present with others who share common elements of history and who have other unique dimensions from which we can learn.

How well the stage had been set by the leader. How easily each member received permission to share and to enjoy the experiences of others. The situation did not just occur. It was carefully developed with predictable results.

READER ACTIVITY

What would you have thought to say had you been lying there on the floor that night? Was there any pain or sadness in your past that tickles you now or that would make others laugh? Think for a moment of the rituals, the expectations, the defeats and later victories, the demands and personal embarrassments suffered that were later triumphs or at the very least put into a less traumatic perspective. How do they look now? Can you catch the feeling of both the pain and the pleasure?

Similarly, asking a group of business people in a cluster of three or four to share a significant learning from a major success that might help someone or a mistake that influenced the course of their careers that they wouldn't want others to experience, would both enrich the listeners and help them get to know the other individual in a different way than would normally be possible. It is the personal connection made through one's own experiences and the new insights gained through others that will set humor into motion. However, by conforming always to the nonpersonal norms of the corporate mentality, business people risk losing their senses of humor. The well of humor will dry up, leaving little left but "one-ups-man" jokes, sarcasm, put downs, and macho locker room humor.

USING HUMOR IN GROUPS

Our effort in this chapter has been to explore the concept of humor in small groups, how to understand it and utilize it in a manner that is constructive to the life of the group itself. Although not meant to be prescriptive, there do appear to be some rather basic generalities that can be made that will prove helpful as individuals consider their role in the group, either in a support or leader role.

First, although much humor will flow naturally and developmentally from a group as it evolves, it has been shown that humor can actually be built systematically into the life of the group. By being sensitive to the diagnostic needs of the group, individuals in positions of responsibility can take active, interventionist roles that take advantage of the reality that within almost any situation there is an underlying platform of humor that can be tapped.

Second, people have an enormous capacity to have fun, but are also prone to misery and negative attitudes. Because groups are complex and not always easy to manage, it is not difficult for the miserable side to take over and gain the upper hand. By looking closely at the process of the group, one can always find opportunities to cultivate humor. By taking advantage of:

paradoxes within the group
discrepancies
the unpredictable
the unanticipated
universal truths
the absurd
the familiar and the memorable

it is always possible to capitalize on a never-ending source of humor.

Finally, it appears that a healthy group will be a humorous group. Groups able to laugh at their failures will be able to take risks together, will be prone to communicate openly and without fear, will be sensitive to the membership needs of the participants, and will be open to change. Humor appears to play an integral part in these and other aspects of a well-functioning group.

EXERCISE 1 Understanding Humor

Objectives

To involve group members in the development of humorous situations

To develop an understanding of various types of humor

To explore the positive and negative aspects of various attempts to inject humor in a group situation

Rationale

Most of us take humor for granted. Because most humor is spontaneous, we tend to be reactive, seldom if ever considering the underlying causes in a particular situation that helped to set the stage for what developed. Was it an accident springing from an unpredictable situation? Was it carefully choreographed? Was success or failure in the style of a key individual? If we wish to use humor to our advantage in group settings, we must take careful aim at such questions and begin to determine what types of humor can intentionally be developed among a group of individuals through the implementation of planned activities.

Action

Ideally, there would be twenty to thirty-five people present in a large group, although the design could be easily adapted for larger or smaller numbers. The facilitator should divide the large group into random clusters of from three to five. Then he or she instructs the clusters to:

> Develop one or possibly two activities that will involve every member of the large group in a humorous experience. Each activity should be self-contained and can last anywhere from one to five minutes. You will have fifteen minutes to design your activity and prepare yourselves. Other than what I have said, there are no rules, so be as creative as possible within the bounds of some social propriety.

At the end of fifteen minutes, the facilitator draws from a hat the number of one group at a time. The small cluster then involves the rest of the large group (they can participate in their own activity if possible) in their activity. At the end of the allotted five minutes and before the next cluster has been drawn from the hat, the facilitator should lead a large group discussion stimulated by such questions as:

1. What were the planned sources of humor designed into the activity?
2. If you felt the situation that developed actually evoked humor, describe why. Was it in the event or in the behavior of the facilitator, or both?
3. If the situation or activity failed to generate a humorous response, why do you believe it failed?
4. What could have been done in retrospect to increase the effectiveness of the design to gain the humorous outcome that was intended?

Follow-up Discussion

After each of the clusters has had the opportunity to present, a brief summary session of perhaps ten minutes should occur. The purpose of this large-group discussion is to draw from those present a series of concise statements that reflect the principles of developing humor in groups that might be generalized from one group to another. It might prove useful to let individuals meet in informal clusters of three or four for perhaps five minutes to help formulate their statements and then present them to the large group. Our experience is that a rather insightful and sophisticated list results that covers many of the points raised in the chapter. For this reason, the activity is best utilized before the chapter is presented. The combination of design and the development of theory and its application seem to make this an appropriate introductory activity.

EXERCISE 2 Moderating Tension-Producing Issues with Humor

Objectives

To provide practice in designing activities for dealing with stressful issues without polarizing a group or antagonizing the participants

Rationale

There are always issues among groups that by their very nature are going to create tension and stress. It is not uncommon for us to avoid such issues until a crisis occurs and dealing with them becomes a necessity. One reason avoidance is not unusual is that individuals simply are not familiar with methods for minimizing the conflict and maximizing the positive attributes of a situation. Humor is one means of reducing stress and allowing individuals to maintain perspective.

Action

The facilitator should survey the participants and discover a number of issues of social and personal significance to the group members. These could include topics such as ERA, fidelity, sexism, the problems with a two-party system, grading, organizational racism, or many others. From a group of say twenty-four, the facilitator divides the group into four groups of six. Two of the groups should be instructed to select one high-stress issue and the other two groups should take a second issue. At this time, the facilitator assigns each group to work independently to design a process for exploring the issue they have selected

1. in a manner that raises points of view but not defenses.
2. in a manner that is direct and open.
3. so that humor is allowed to develop and is utilized as a means of maintaining member perspective.

After 15 minutes of planning, each group is allowed 15 minutes to involve the other eighteen (or more if desired) members of the large group in a structured process for looking at their particular issue. Thus, if groups A and B each are designing around the issue of organizational racism, group A would present their activity followed immediately by group B's presentation around the same issue and involving the total group. At this point, there should be a 15-minute open discussion that compares the designs of the two groups and ways that each helped or hindered the promotion of an open climate for discussion and learning. The discussion might focus on

1. identifying the role of humor in each design and how it did or did not create some perspective.
2. comparing and contrasting the two designs in relation to strengths and benefits, again focusing particular attention on planned or unplanned humor.
3. focusing on ways of improving each design given the goals of opening communication and reducing antagonism.

After this set of presentations and the period of discussion, the final two groups (which have a different issue) are given 5 minutes to caucus and revise their designs, if necessary, based on what they have now learned. The process is then repeated for the final two groups, with similar comparisons being made.

This activity focuses on the participants "doing" so that they develop the belief that humor can be facilitated by their planning and structured designing.

BIBLIOGRAPHY

Baron, R. A. "The Aggression-inhibiting influence of nonhostile humor." *The Journal of Experimental Social Psychology*, 10 (1974), 23–33.

Chapman, A. and H. Foot, eds. *Humor and Laughter: Theory, Research and Applications*. New York: Wiley, 1976.

Freud, Sigmund. (James Strachey, ed. and translator). *Jokes and Their Relation to the Unconscious*. New York: Norton, 1963.

Goldstein, J. and P. McGhee, eds. *The Psychology of Humor*. New York: Academic Press, 1972.

Hassett, J. and J. Houlihan. "Different jokes for different folks." *Psychology Today*, (January 1979), 64–71.

Author Index

Subject Index

Student Response Form

We would like to find out what your reactions are to the second edition of GROUPS: THEORY AND EXPERIENCE. Your evaluation of the book will help us respond to the interests and needs of the readers of future editions. Please complete the form and mail it to College Marketing, Houghton Mifflin Company, One Beacon Street, Boston, MA 02107.

1. We would like to know how you rate our textbook in each of the following areas:

	Excellent	Good	Adequate	Poor
a. Selection of topics	____	____	____	____
b. Detail of coverage	____	____	____	____
c. Order of topics	____	____	____	____
d. Writing style/readability	____	____	____	____
e. Reader Activities	____	____	____	____
f. Explanation of concepts	____	____	____	____
g. End-of-chapter exercises	____	____	____	____

2. Please cite specific examples that illustrate any of the above ratings.

____

____

____

3. Describe the strongest feature(s) of the book.

____

____

____

4. Describe the weakest feature(s) of the book.

____

____

____

5. What other topics should be included in this text?

____

____

____

6. What recommendations can you make for improving this book?

CDEFGHIJ-H-82